Lecture Notes in Computer Science　8146

Commenced Publication in 1973
Founding and Former Series Editors:
Gerhard Goos, Juris Hartmanis, and Jan van Leeuwen

Advanced Research in Computing and Software Science

Subline of Lectures Notes in Computer Science

Berthold Vöcking (Ed.)

Algorithmic Game Theory

6th International Symposium, SAGT 2013
Aachen, Germany, October 21-23, 2013
Proceedings

 Springer

Volume Editor

Berthold Vöcking
RWTH Aachen University
Department of Computer Science
Aachen, Germany
E-mail: voecking@cs.rwth-aachen.de

ISSN 0302-9743　　　　　　　　　　　e-ISSN 1611-3349
ISBN 978-3-642-41391-9　　　　　　　e-ISBN 978-3-642-41392-6
DOI 10.1007/978-3-642-41392-6
Springer Heidelberg New York Dordrecht London

Library of Congress Control Number: 2013949603

CR Subject Classification (1998): I.6, H.5.3, J.1, K.6.0, H.3.5, J.4, K.4.4, G.1.2, F.2.2

LNCS Sublibrary: SL 3 – Information Systems and Applications, incl. Internet/Web and HCI

Typesetting: Camera-ready by author, data conversion by Scientific Publishing Services, Chennai, India

Printed on acid-free paper

Springer is part of Springer Science+Business Media (www.springer.com)

Preface

This volume contains the proceedings of the Sixth International Symposium on Algorithmic Game Theory (SAGT) held in Aachen, Germany, in October 2013.

The program of SAGT 2013 consists of three invited lectures and 25 presentations of refereed submissions. The Program Committee selected 25 out of 65 submissions after a careful reviewing process.

The accepted submissions were invited to these proceedings. They cover various important aspects of algorithmic game theory such as solution concepts in game theory, efficiency of equilibria and the price of anarchy, computational aspects of equilibria and game theoretical measures, repeated games and convergence of dynamics, evolution and learning in games, coordination and collective action, network games and graph-theoretic aspects of social networks, voting and social choice, as well as algorithmic mechanism design.

We would like to thank all authors who submitted their research work and all Program Committee members and external reviewers for their effort in selecting the program for SAGT 2013.

August 2013   Berthold Vöcking

Organization

Program Committee

Krzysztof R. Apt	CWI Amsterdam, The Netherlands
Moshe Babaioff	Microsoft Research Silicon Valley, USA
Felix Brandt	TU Munich, Germany
Ioannis Caragiannis	University of Patras, Greece
Shuchi Chawla	University of Wisconsin, USA
Giorgos Christodoulou	University of Liverpool, UK
Xiaotie Deng	University of Liverpool, UK
Edith Elkind	NTU Singapore
Amos Fiat	Tel Aviv University, Israel
Dimitris Fotakis	NTU Athens, Greece
Tobias Harks	Maastricht University, The Netherlands
Jochen Koenemann	University of Waterloo, Canada
Annamaria Kovacs	Frankfurt University, Germany
Dov Monderer	Technion Haifa, Israel
Luca Moscardelli	University of Pescara, Italy
Noam Nisan	Hebrew University of Jerusalem, Israel and Microsoft Research Silicon Valley, USA
Michael Schapira	Hebrew University of Jerusalem, Israel
Alexander Skopalik	University of Paderborn, Germany
Berthold Vöcking	RWTH Aachen University, Germany - Chair

Steering Committee

Elias Koutsoupias	University of Athens, Greece
Marios Mavronicolas	University of Cyprus, Cyprus
Dov Monderer	Technion Haifa, Israel
Burkhard Monien	University of Paderborn, Germany
Christos Papadimitriou	UC Berkeley, USA
Giuseppe Persiano	University of Salerno, Italy
Paul Spirakis	University of Patras, Greece - Chair
Berthold Vöcking	RWTH Aachen University, Germany

Organizing Team

Johannes Dams	RWTH Aachen University, Germany
Oliver Göbel	RWTH Aachen University, Germany
Martin Hoefer	Saarland University and MPI, Germany - Publicity Chair
Rebecca Reiffenhäuser	RWTH Aachen University, Germany

Erika Schlebusch   RWTH Aachen University, Germany
Berthold Vöcking   RWTH Aachen University, Germany - Chair

Additional Reviewers

Ardenboim, Alon
Babichenko, Yakov
Badanidiyuru, Ashwinkumar
Bilò, Vittorio
Brill, Markus
Bu, Tian-Ming
Candogan, Ozan
Chen, Zhou
Cord-Landwehr, Andreas
Cornaz, Denis
De Keijzer, Bart
Devanur, Nikhil
Dombb, Yair
Drees, Maximilian
Faliszewski, Piotr
Fanelli, Angelo
Farczadi, Linda
Feldman, Moran
Feldmann, Andreas Emil
Friggstad, Zachary
Fu, Zhiguo
Geist, Christian
Georgiou, Konstantinos
Hamann, Heiko
Kanellopoulos, Panagiotis
Karanikolas, Nikos
Kardel, Keyvan
Karlin, Anna
Kling, Peter
Kontogiannis, Spyros

Koutsoupias, Elias
Kyropoulou, Maria
Lenzner, Pascal
Lu, Pinyan
Lucier, Brendan
Meir, Reshef
Monaco, Gianpiero
Nikolova, Evdokia
Obraztsova, Svetlana
Oren, Sigal
Paes Leme, Renato
Panagopoulou, Panagiota
Schäfer, Guido
Seedig, Hans Georg
Sgouritsa, Alkmini
Shah, Nisarg
Simon, Sunil Easaw
Slivkins, Alex
Tang, Bo
Thakur, Subhasis
Thompson, David
van Den Brink, Rene
Ventre, Carmine
Xiao, Tao
Yang, Chun-Lei
Zhang, Chihao
Zhang, Jinshang
Zhang, Qiang
Zhang, Xiaodong
Zohar, Aviv

Table of Contents

Alternative Solution Concepts

Social Networks

Mechanism Design

Abstracts

The Complexity of Fully Proportional Representation for Single-Crossing Electorates

Piotr Skowron[1], Lan Yu[2], Piotr Faliszewski[3], and Edith Elkind[2]

[1] University of Warsaw, Poland
[2] Nanyang Technological University, Singapore
[3] AGH University of Science and Technology, Poland

Abstract. We study the complexity of winner determination in single-crossing elections under two classic fully proportional representation rules—Chamberlin–Courant's rule and Monroe's rule. Winner determination for these rules is known to be NP-hard for unrestricted preferences. We show that for single-crossing preferences this problem admits a polynomial-time algorithm for Chamberlin–Courant's rule, but remains NP-hard for Monroe's rule. Our algorithm for Chamberlin–Courant's rule can be modified to work for elections with *bounded single-crossing width*. To circumvent the hardness result for Monroe's rule, we consider single-crossing elections that satisfy an additional constraint, namely, ones where each candidate is ranked first by at least one voter (such elections are called narcissistic). For single-crossing narcissistic elections, we provide an efficient algorithm for the egalitarian version of Monroe's rule.

1 Introduction

Parliamentary elections, i.e., procedures for selecting a fixed-size set of candidates that, in some sense, best represent the voters, received a lot of attention in the literature. Some well-known approaches include first-past-the-post system (FPTP), where the voters are divided into districts and in each district a plurality election is held to find this district's representative; party-list systems, where the voters vote for parties and later the parties distribute the seats among their members; SNTV (single nontransferable vote) and Bloc rules, where the voters cast t-approval ballots and the rule picks k candidates with the highest approval scores (here k is the target parliament size, and $t = 1$ for SNTV and $t = k$ for Bloc); and a variant of STV (single transferable vote). In this paper, we focus on two voting rules that, for each voter, explicitly define the candidate that will represent her in the parliament (such rules are said to provide *fully proportional representation*), namely, Chamberlin–Courant's rule [5] and Monroe's rule [11]. Besides parliamentary elections, winner determination algorithms for these rules can also be used for other applications, such as resource allocation [11,16] and recommender systems [9].

Let us consider an election where we seek a k-member parliament chosen out of m candidates by n voters. Both Chamberlin–Courant's rule and Monroe's rule work by finding a function Φ that assigns to each voter v the candidate that is to represent v in the

B. Vöcking (Ed.): SAGT 2013, LNCS 8146, pp. 1–12, 2013.
© Springer-Verlag Berlin Heidelberg 2013

parliament. This function is required to assign at most k candidates altogether[1]. Further, under Monroe's rule each candidate is either assigned to about $\frac{n}{k}$ voters or to none. The latter restriction does not apply to Chamberlin–Courant's rule, where each selected candidate may represent an arbitrary number of voters, and, as a consequence, the parliament elected in this manner may have to use weighted voting in its proceedings. Finally, each voter should be represented by a candidate that this voter ranks as highly as possible.

To specify the last requirement formally, we assume that there is a global *dissatisfaction function* α, $\alpha \colon \mathbb{N} \to \mathbb{N}$, such that $\alpha(i)$ is a voter's dissatisfaction from being represented by a candidate that she views as i-th best. (A typical example is Borda dissatisfaction function α_B given by $\alpha_B(i) = i - 1$.) In the utilitarian variants of Chamberlin–Courant's and Monroe's rules we seek assignments that minimize the sum of voters' dissatisfactions; in the egalitarian variants (introduced recently by Betzler et al. [3]) we seek assignments that minimize the dissatisfaction of the worst-off voter.

Chamberlin–Courant's and Monroe's rules have a number of attractive properties, which distinguish them from other multiwinner rules. Indeed, they elect parliaments that (at least in some sense) proportionally represent the voters, ensure that candidates who are not individually popular cannot make it to the parliament even if they come from very popular parties, and take minority candidates into account. Unfortunately, these rules do have one flaw that makes them impractical: It is NP-hard to compute their winners [13,9,3]. Nonetheless, these rules are so attractive that there is a growing body of research on computing their winners exactly (e.g., through integer linear programming formulations [12], by means of fixed-parameter tractability analysis [3], or by considering restricted preference domains [3,18]) and approximately [9,16,15]. We continue this line of research by considering the complexity of finding Chamberlin–Courant and Monroe winners for the case where voters' preferences are single-crossing. Our results complement those of Betzler et al. [3] for single-peaked electorates.

Recall that voters are said to have single-crossing preferences if it is possible to order them so that for every pair of candidates a, b the voters who prefer a to b form a consecutive block on one side of the order and the voters who prefer b to a form a consecutive block on the other side. For example, it is quite natural to assume that the voters are aligned on the standard political left-right axis. Given two candidates a and b, where a is viewed as more left-wing and b is viewed as more right-wing, the left-leaning voters would prefer a to b and the right-leaning voters would prefer b to a. While real-life elections are typically too noisy to have this property, it is plausible that they are often close to single-crossing, and it is important to understand the complexity of the idealized model before proceeding to study nearly single-crossing profiles.

Our main results are as follows: for single-crossing elections winner determination under Chamberlin–Courant's rule is in P (for every dissatisfaction function, and both for the utilitarian and for the egalitarian version of this rule), but under Monroe's rule it is NP-hard. Our hardness result for Monroe's rule applies to the utilitarian setting with Borda dissatisfaction function. Our algorithm for Chamberlin–Courant's rule extends

---

[1] Under Monroe's rule we are required to pick exactly k winners. Some authors also impose this requirement in the case of Chamberlin–Courant's rule, but allowing for smaller parliaments appears to be more consistent with the spirit of this rule and is standard in its computational analysis (see, e.g., [9,3,16,15,18]).

to elections that have bounded *single-crossing width* (see [6,7]). Our proof proceeds by showing that for single-crossing elections Chamberlin–Courant's rule admits an optimal assignment that has the *contiguous blocks property*: the set of voters assigned to an elected representative forms a contiguous block in the voters' order witnessing that the election is single-crossing. This property can be interpreted as saying that each selected candidate represents a group of voters who are fairly similar to each other, and we believe it to be desirable in the context of proportional representation.

The NP-hardness result for Monroe's rule motivates us to search for further domain restrictions that may make this problem tractable. To this end, we focus on the egalitarian version of Monroe's rule and, following the example of Cornaz et al. [6], consider elections that, in addition to being single-crossing, are *narcissistic*, i.e., have the property that every candidate is ranked first by at least one voter. In parliamentary elections, narcissistic profiles are very natural: we expect all candidates to vote for themselves. We provide a polynomial-time algorithm for the egalitarian version of Monroe's rule for all elections that belong to this class. Our algorithm is based on the observation that for single-crossing narcissistic elections under the egalitarian version of Monroe's rule there is always an optimal assignment that satisfies the contiguous blocks property.

In a sense, our result for single-crossing narcissistic elections is not new: it can be shown that such elections are single-peaked (this result is implicit in the work of Barberà and Moreno [1]), and Betzler et al. [3] provide a polynomial-time algorithm for the egalitarian version of Monroe's rule for single-peaked electorates. However, our algorithm has two significant advantages over the one of Betzler et al.: First, it has considerably better worst-case running time, and second, it produces assignments that have the contiguous blocks property.

We omit some of the proofs due to space constraints; the full version of the paper is available as a technical report [17].

2 Preliminaries

For every positive integer s, we let $[s]$ denote the set $\{1, \ldots, s\}$. An *election* is a pair $E = (C, V)$ where $C = \{c_1, \ldots, c_m\}$ is a set of candidates and $V = (v_1, \ldots, v_n)$ is an ordered list of voters. Each voter $v \in V$ has a *preference order* $\succ_v$, i.e., a linear order over C that ranks all the candidates from the most desirable one to the least desirable one. For each voter $v \in V$ and each candidate $c \in C$, we denote by $\text{pos}_v(c)$ the position of c in v's preference order (the top candidate has position 1 and the last candidate has position $\|C\|$). We refer to the list V as the *preference profile*. A list U is said to be a *sublist* of a list V (denoted by $U \subseteq V$) if U can be obtained from V by deleting voters. An election (C', V') is said to be a *subelection* of an election (C, V) if C' is a subset of C and V' is obtained by taking a sublist of V and restricting the preferences of the voters in that sublist to C'. Given a subset of candidates A, we denote by $A_{\rightarrow}$ a fixed ordering of candidates in A and by $A_{\leftarrow}$ the reverse of this ordering. Given two sets $A, B \subset C$, we write $\cdots \succ A \succ B \succ \ldots$ to denote a vote where all candidates in A are ranked above all candidates in B.

Chamberlin–Courant's and Monroe's Rules. Both Chamberlin–Courant's rule and Monroe's rule rely on the notion of a *dissatisfaction function*. This function specifies, for each $i \in [m]$, a voter' dissatisfaction from being represented by candidate she ranks in position i. Formally, for an m-candidate election, a *dissatisfaction function* is a non-decreasing function $\alpha\colon [m] \to \mathbb{N}$ with $\alpha(1) = 0$. We will typically be interested in families of dissatisfaction functions, $(\alpha^m)_{m=1}^{\infty}$, with one function for each possible number of candidates. In particular, we will be interested in Borda dissatisfaction function $\alpha_B^m(i) = \alpha_B(i) = i - 1$. We assume that our dissatisfaction functions are computable in polynomial time with respect to m.

Let k be a positive integer. A *k-CC-assignment function* for an election $E = (C, V)$ is a mapping $\Phi\colon V \to C$ such that $\|\Phi(V)\| \leq k$. A *k-Monroe-assignment function* for E is a k-CC-assignment function that additionally satisfies the following constraints: $\|\Phi(V)\| = k$, and for each $c \in C$ either $\|\Phi^{-1}(c)\| = 0$ or $\lfloor \frac{n}{k} \rfloor \leq \|\Phi^{-1}(c)\| \leq \lceil \frac{n}{k} \rceil$. That is, both assignment functions select (up to) k candidates, and a k-Monroe-assignment function additionally ensures that each selected candidate is assigned to roughly the same number of voters. For a given assignment function Φ, we say that voter $v \in V$ is *represented* (in the parliament) by candidate $\Phi(v)$. There are several ways to measure the quality of an assignment function Φ with respect to a dissatisfaction function α; we use the following two:

1. $\ell_1(\Phi) = \sum_{i=1,\ldots,n} \alpha(\mathrm{pos}_{v_i}(\Phi(v_i)))$;
2. $\ell_{\infty}(\Phi) = \max_{i=1,\ldots,n} \alpha(\mathrm{pos}_{v_i}(\Phi(v_i)))$.

Intuitively, $\ell_1(\Phi)$ takes the utilitarian view of measuring the sum of voters' dissatisfactions, whereas ℓ_{∞} takes the egalitarian view of looking at the worst-off voter only.

We are now ready to define the voting rules that will be the subject of this paper.

Definition 1. *For every family of dissatisfaction functions* $\alpha = (\alpha^m)_{m=1}^{\infty}$, *every* $R \in \{CC, Monroe\}$, *and every* $\ell \in \{\ell_1, \ell_{\infty}\}$, *an* α-ℓ-R voting rule *is a mapping that takes an election* $E = (C, V)$ *and a positive integer* k *with* $k \leq \|C\|$ *as its input, and returns a* k-R-assignment function Φ *for* E *that minimizes* $\ell(\Phi)$ *(if there are several optimal assignments, the rule is free to return any of them).*

Chamberlin and Courant [5] and Monroe [11] proposed the utilitarian variants of their rules and focused on Borda dissatisfaction function (though Monroe also considered so-called k-approval dissatisfaction functions). Egalitarian variants of both rules have been recently introduced by Betzler et al. [3].

Single-Crossing Profiles. The notion of single-crossing preferences dates back to the work of Mirrlees [10]; we also point the reader to the work of Saporiti and Tohmé [14] for some settings where single-crossing preferences are studied. Formally, this preference domain is defined as follows.

Definition 2. *An election* $E = (C, V)$, *where* $C = \{c_1, \ldots, c_m\}$ *is a set of candidates and* $V = (v_1, \ldots, v_n)$ *is an ordered list of voters, is* single-crossing *(with respect to the given order of voters) if for every pair of candidates* a, b *such that* $a \succ_{v_1} b$, *there exists a* $t \in [n]$ *such that* $\{i \in [n] \mid a \succ_{v_i} b\} = [t]$.

Definition 2 refers to the ordering of the voters provided by V. Alternatively, one could simply require existence of an ordering of the voters that satisfies the single-crossing property. The advantage of our approach is that it simplifies notation, yet does not affect the complexity of the problems that we study: one can compute an order of the voters that makes an election single-crossing, or decide that such an order does not exist, in polynomial time [8,4].

3   Chamberlin–Courant's Rule

We start our discussion by considering the complexity of winner determination under Chamberlin–Courant's rule, for the case of single-crossing profiles. A key observation in our analysis is that for single-crossing profiles there always exists an optimal k-CC-assignment function where the voters matched to a given candidate form contiguous blocks within the voters' order. In what follows, we will say that assignments of this form have the *contiguous blocks property*. We believe that this property is desirable from the social choice perspective: it means that voters who are represented by the same candidate are quite similar, which makes it easier for the candidate to act in a way that reflects the preferences of the group he represents. Later, we will see that the contiguous blocks property also has useful algorithmic implications.

Lemma 1. *Let $E = (C, V)$ be a single-crossing election, where $C = \{c_1, \ldots, c_m\}$, $V = (v_1, \ldots, v_n)$, and v_1 has preference order $c_1 \succ \cdots \succ c_m$. Then for every $k \in [m]$, every dissatisfaction function α for m candidates, and every $\ell \in \{\ell_1, \ell_\infty\}$, there is an optimal k-CC-assignment Φ for α-ℓ-CC such that for each candidate $c_i \in C$, if $\Phi^{-1}(c_i) \neq \emptyset$ then there are two integers, t_i and t'_i, $t_i \leq t'_i$, such that $\Phi^{-1}(c_i) = \{v_{t_i}, v_{t_i+1}, \ldots, v_{t'_i}\}$. Moreover, for each $i < j$ such that $\Phi^{-1}(c_i) \neq \emptyset$ and $\Phi^{-1}(c_j) \neq \emptyset$ it holds that $t'_i < t_j$.*

Proof. Fix a single-crossing election $E = (C, V)$ with $C = \{c_1, \ldots, c_m\}$ and $V = (v_1, \ldots, v_n)$, and let Φ be an optimal k-CC-assignment function for E under α-ℓ-CC. We assume without loss of generality that for each voter v_i in V, the candidate $\Phi(v_i)$ is v_i's most preferred candidate is $\Phi(V)$. Let c_j be v_1's least preferred candidate in $\Phi(V)$. Now consider some voter v_i such that $\Phi(v_i) = c_j$. We have $\Phi(v_{i'}) = c_j$ for every voter $v_{i'}$ such that $i' > i$. Indeed, suppose for the sake of contradiction that $\Phi(v_{i'}) = c_k$ for $k \neq j$. By our choice of c_j we have $c_k \succ_1 c_j$. On the other hand, we have $c_j \succ_i c_k$ and $c_k \succ_{i'} c_j$, a contradiction with E being a single-crossing election. Hence, the voters that are matched to c_j by Φ form a consecutive block at the end of the preference profile.

To see that for each $c \in \Phi(V)$ it holds that voters in $\Phi^{-1}(c)$ form a consecutive block, it suffices to delete c_j and the voters that are matched to c_j from the profile, decrease k by one, and repeat the same argument.   □

Lemma 1 suggests a dynamic programming algorithm for Chamberlin–Courant's rule.

Theorem 1. *For every family α of polynomial-time computable dissatisfaction functions and for $\ell \in \{\ell_1, \ell_\infty\}$, there is a polynomial-time algorithm that given a single-crossing election E and a positive integer k finds an optimal k-CC-assignment for E under α-ℓ-CC.*

Following the ideas of Cornaz et al. [6,7], we can extend our algorithm to profiles with so-called *bounded single-crossing width*.

Definition 3. *A set D, $D \subseteq C$, is a* clone set *in an election $E = (C, V)$ if each voter in V ranks the candidates from D consecutively (but not necessarily in the same order).*

Definition 4. *An election $E = (C, V)$ has* single-crossing width *at most w if there exists a partition of C into sets $D_1, \ldots, D_t$ such that (a) for each $i \in [t]$ the set D_i is a clone set in E and $\|D_i\| \leq w$, and (b) if we contract each D_i in each vote to a single candidate d_i, then the resulting preference profile is single-crossing.*

Profiles with small single-crossing width may arise, e.g., in parliamentary elections where the candidates are divided into (small) parties and the voters have single-crossing preferences over the parties, but not necessarily over the candidates. Using the same techniques as Cornaz et al., we obtain the following result.

Proposition 1. *For every family α of polynomial-time computable dissatisfaction functions and for every $\ell \in \{\ell_1, \ell_\infty\}$, there is an algorithm that given an election $E = (C, V)$ with $C = \{c_1, \ldots, c_m\}$, $V = (v_1, \ldots, v_n)$ whose single-crossing width is bounded by w, a partition of C into clone sets that witnesses this width bound, and a positive integer k, finds an optimal k-CC-assignment for E under α-ℓ-CC, and runs in time $\mathrm{poly}(m, n, k, 2^w)$.*

Naturally, for this result to be useful, we need an efficient algorithm that computes single-crossing width of a profile and an appropriate division into clone sets. Fortunately, such an algorithm is provided by Cornaz et al. [7]. (Interestingly, a very similar problem of finding a division into clones that results in a single-crossing election with as many candidates as possible is NP-hard [8]). As a consequence, we have the following corollary.

Corollary 1. *For every family α of polynomial-time computable dissatisfaction functions and for every $\ell \in \{\ell_1, \ell_\infty\}$, the problem of winner determination for α-ℓ-CC is fixed-parameter tractable with respect to the single-crossing width of the input profile.*

4   Monroe's Rule

The results of Betzler et al. [3] suggest that winner determination under Monroe's rule tends to be harder than winner determination under Chamberlin–Courant's rule. In this section, we show that this is also the case for single-crossing profiles: we prove that for the utilitarian variant of Monroe's rule with Borda dissatisfaction function (perhaps the most natural variant of Monroe's rule) computing winners is NP-hard, even for single-crossing elections. We then complement this hardness result by showing that for the egalitarian version of Monroe's rule winner determination is easy if we additionally assume that the preferences are narcissistic.

4.1   Hardness for General Single-Crossing Profiles

This section is devoted to proving that winner determination under Monroe's rule is NP-hard. The main idea of the proof is to reduce the problem of winner determination for unrestricted profiles to the case of single-crossing profiles.

Table 1. The profile used in the proof of Theorem 2. For each voter list V_i, $1 \leq i \leq 5$, and for each voter v in V_i we list the (sets of) candidates in the order of v's preference (we omit the "$\succ$" symbol for readability). Whenever we list a set of candidates as a part of an order, we assume that the candidates in this set are ordered in some fixed, easily-computable way (for candidates in H we fix this order to be $h_1 \succ \cdots \succ h_{m-k}$). Further, when in a line describing a preference order of an entire collection of voters $V_r = (v_1, \ldots, v_s)$ (specifically, for us r is either 2, 4, or 5, and s is the number of voters in this list of voters) we include a profile $V' = (v'_1, \ldots, v'_s)$ (in our case V' is either an R-profile or an Adj-profile), then we mean that for each voter v_i, $i \in [s]$, in V_r, this part of this voter's preference order is the preference order of v'_i in V'.

$$V_1 : H \ R^{-1}(F_1 \ \ldots \ F_m \ E \ E_m \ \ldots \ E_1) \, c_1 \ \ldots \ c_m \ D_1 \ \ldots \ D_m \ G_1 \ldots G_m \ G$$
$$V_2 : H \ R(R^{-1}(F_1 \ \ldots \ F_m \ E \ E_m \ \ldots \ E_1)) \, c_1 \ \ldots \ c_m \ D_1 \ \ldots \ D_m \ G_1 \ldots G_m \ G$$

$$v_3^1 : H \ F_1 \cdots F_m \ E \ E_m \ \ldots \ E_2 \, c_1 \ E_1 \, c_2 \ \ldots \ c_m \ D_1 \ \ldots \ D_m \ G_1 \ \ldots \ G_m \ G$$
$$v_3^2 : H \ F_1 \cdots F_m \ E \ D_1 \ E_m \ \ldots \ E_3 \, c_2 \ E_2 \, c_3 \ \ldots \ c_m \, c_1 \ E_1 \ D_2 \ \ldots \ D_m \ G_1 \ \ldots \ G_m \ G$$
$$\vdots$$
$$v_3^m : H \ F_1 \cdots F_m \ E \ D_1 \ \ldots \ D_{m-1} \, c_m \ E_m \, c_{m-1} \ \ldots \ c_1 \ E_{m-1} \ \ldots E_1 \ D_m \ G_1 \ \ldots \ G_m \ G$$

$$V_4 : H \ D_1 \ \ldots \ D_m \ \mathrm{Adj}(F_1, c_m, G_1) \ \ldots \ \mathrm{Adj}(F_m, c_1, G_m) \ E \ E_m \ \ldots \ E_1 \ G$$
$$V_5 : H \ R(D_1 \ \ldots \ D_m \ G_1 \ \ldots \ G_m \ G) \, c_m \ \ldots \ c_1 \ F_1 \ \ldots \ F_m \ E \ E_m \ \ldots \ E_1$$

Theorem 2. *Finding a set of winners under $\alpha_B\text{-}\ell_1$-Monroe voting rule is NP-hard, even for single-crossing preferences.*

The proof of this theorem is somewhat involved. We first need the following two technical lemmas.

Lemma 2. *Let $E = (C, V)$ be an election, where $C = \{c_1, \ldots, c_m\}$ and $V = (v_1, \ldots, v_n)$. Let A and B be two disjoint sets of candidates such that $\|A\| = \|B\| = mn$. For each $c_i \in C$, there is a single-crossing election $\mathrm{Adj}_V(A, c_i, B)$ with candidate set $A \cup B \cup \{c_i\}$ and voter list $V' = (v'_1, \ldots, v'_n)$ such that $pos_{v'_j}(c_i) = mn + pos_{v_j}(c_i)$ for each $j \in [n]$, and the profile $(v'_0, v'_1, \ldots, v'_n, v'_{n+1})$, where v'_0 has preference order $a_1 \succ \cdots \succ a_{\|A\|} \succ c_i \succ b_1 \succ \cdots \succ b_{\|B\|}$ and v'_{n+1} has preference order $b_1 \succ \cdots \succ b_{\|B\|} \succ c_i \succ a_1 \succ \cdots \succ a_{\|A\|}$, is also single-crossing.*

Lemma 3. *For every pair of positive integers k, n such that k divides n, and every set $C = \{c_1, \ldots, c_m\}$ of candidates, there is a single-crossing profile $R(C)$ with $(\frac{n}{k} + 1)m$ voters such that each candidate $c_i \in C$ is ranked first by exactly $(\frac{n}{k} + 1)$ voters.*

We also need some additional notation. If $C_\rightarrow$ is an order of candidates in C, then by $R(C_\rightarrow)$ we denote the election that we would construct in Lemma 3 if the first voter's preference order was $C_\rightarrow$ (i.e., if we took c_1 to be the top candidate according to $C_\rightarrow$, c_2 to be the second one, and so on). By $R^{-1}(C_\rightarrow)$ we denote an order $C'_\rightarrow$ of the candidates in C such that $R(C'_\rightarrow)$ produces an election where the last voter has preference order $C_\rightarrow$. We are now ready to give our proof of Theorem 2.

Proof of Theorem 2. Let I be an instance of the problem of finding k winners under $\alpha_B\text{-}\ell_1$-Monroe rule and let (C, V) be the election considered in I. Set $n = \|V\|$ and

$m = \|C\|$. We assume that n is divisible by k and that $n > k$ (computing α_B-ℓ_1-Monroe winners is still NP-hard under these assumptions [3,16]). We will show how to construct in polynomial time an instance I_{sc} of the problem of finding winners under α_B-ℓ_1-Monroe where the election is single-crossing so that it is easy to extract the set of winners for I from the set of winners for I_{sc}.

We construct I_{sc} in the following way. First, we define the candidate set C_{sc} to be the union of the following disjoint sets (we provide names of the candidates only where relevant and abbreviate $\sum_{i=1}^{m}$ to $\sum_i$):

1. $H = \{h_1, \ldots, h_{m-k}\}$, where $\|H\| = m - k$;
2. $F_1, \ldots F_m$, where $\|F_i\| = mn$ for each $i \in [m]$;
3. $E_1, \ldots E_m$, where $\|E_i\| = 2m^2n + m + (m - i)(2mn + 1)\frac{n}{k}$ for each $i \in [m]$;
4. E, where $\|E\| = m^2n + m$;
5. $D_1, \ldots, D_m$, where $\|D_i\| = \|E_i\|$ for each $i \in [m]$;
6. $G_1, \ldots, G_m$, where $\|G_i\| = \|F_i\| = mn$ for each $i \in [m]$;
7. G, where $\|G\| = (\sum_i \|F_i\| + \|E\|)$;
8. $C' = C = \{c_1, \ldots, c_m\}$.

The ordered list V_{sc} of voters consists of the following five sublists (we only give names to those voters to whom we will refer directly later; whenever sufficient, we only give the number of voters in a given list):

1. V_1, $\|V_1\| = \|H\|\frac{n}{k} = (m - k)\frac{n}{k}$;
2. V_2, $\|V_2\| = (\sum_i \|F_i\| + \sum_i \|E_i\| + \|E\|)(\frac{n}{k} + 1)$;
3. $V_3 = (v_3^1, \ldots, v_3^m)$, $\|V_3\| = m$;
4. V_4, $\|V_4\| = n$;
5. V_5, $\|V_5\| = (\sum_i \|D_i\| + \sum_i \|G_i\| + \|G\|)(\frac{n}{k} + 1)$.

We give the preferences of the voters in Table 1. In the thus-defined profile our goal is to find a parliament of size $k_{sc} = \|C_{sc}\| - (m - k)$. Consequently, each selected candidate should be assigned to $\frac{n}{k} + 1$ voters.

We claim that each optimal solution for I_{sc} satisfies the following conditions; we let Φ_{sc} denote one such optimal solution.

(i) Each candidate $c \in F_1 \cup \cdots \cup F_m \cup E \cup E_m \cup \cdots \cup E_1$ is a winner and is assigned to those voters from $V_1 + V_2$ that rank c in position $\|H\| + 1 = m - k + 1$ (note that only one of these candidates can be assigned to (some of the) voters in V_1).

(ii) Each candidate $c \in D_1 \cup \cdots \cup D_m \cup G_1 \cup \cdots \cup G_m \cup G$ is a winner and is assigned to those voters from V_5 that rank c in position $\|H\| + 1 = m - k + 1$.

(iii) Each candidate $h_i \in H$ is a winner and is assigned to $\frac{n}{k} + 1$ voters from $V_1 + V_2 + V_3$ (exactly $\|H\|$ voters from V_3 have some candidate from H assigned to them); each such voter ranks h_i in position i.

(iv) Exactly k candidates from C' are winners. Each of them is assigned to $\frac{n}{k}$ voters in V_4 and to one voter in V_3 that ranks him highest.

(v) The k winners from C' (let us call them $w_1, \ldots, w_k$) are also α_B-ℓ_1-Monroe winners in I and each of them is assigned in I_{sc} to the voters corresponding to those from the I-solution.

Let us now show that indeed the optimal solution is of this form. First, we make the following observations:

(a) By a simple counting argument, at least k of the candidates from C' must be included in the optimal solution.
(b) For each candidate h_i in H, if h_i is part of the optimal solution then h_i is ranked in the i-th position in the preference order of the voters to which h_i is assigned (candidates from H are always ranked first, in the order $h_1 \succ \cdots \succ h_m$).
(c) For each candidate $c \in C_{sc} \setminus (C' \cup H)$, if c is included in the optimal solution then each voter to which c is assigned ranks c in position $m - k + 1$ or worse (this is because every voter's top $m - k$ positions are taken by the candidates from H).
(d) Each voter in $V_1 + V_2 + V_5$ ranks each candidate in C' in position worse than $p_1 = \|H\| + \sum_i \|E_i\| + \sum_i \|F_i\| + \|E\| > \|H\| + \sum_i \|E_i\| + 2m^2 n + m$.
(e) Each voter in V_4 ranks each candidate in C' in position better than $p_2 = \|H\| + \sum_i \|D_i\| + \sum_i \|F_i\| + \sum_i \|G_i\| + m = \|H\| + \sum_i \|E_i\| + 2m^2 n + m$, but worse than $p_3 = \|H\| + \sum_i \|E_i\|$.
(f) $p_1 > p_2$.
(g) For each candidate $c \in C'$, there is exactly one voter in V_3 that ranks c in a position no worse than $p_4 = \|H\| + \sum_i \|F_i\| + \sum_{i<m} \|E_i\| + \|E\| + 1 = \|H\| + \sum_{i<m} \|E_i\| + \|E\| + m^2 n + 1 < \|H\| + \sum_i \|E_i\| = p_3$; all other voters in V_3 rank c in a position worse than $p_5 = \|H\| + \sum_i \|E_i\| + \sum_i \|F_i\| + \|E\| = \|H\| + \sum_i \|E_i\| + 2m^2 n + m = p_2$.

Let Φ be an optimal assignment function among those that use exactly k candidates from C'. We claim that Φ satisfies conditions (i)–(iv). This is so, because assigning voters from V_4 to candidates other than those in C' will result in a strictly worse assignment (the assignment would get worse for the candidates in C' because of points (d), (e), (f) and (g), and it would not improve for the other candidates because of points (b) and (c)). Similarly, each of the k selected candidates from C' should be assigned to exactly one voter from V_3—the one that ranks this candidate highest. Once we assign the k winners from C' to the voters in V_4 and to k voters in V_3, the optimal way to complete the assignment is to do so as described in conditions (i)–(iv).

Let Φ be an optimal assignment function for I_{sc} that uses exactly k candidates from C' and that satisfies conditions (i)–(iv). We now prove that it also satisfies condition (v). Consider a candidate $c_i \in C'$ that is included in the set of winners under Φ. Let $V_4^{c_i}$ be the sublist of the voters from V_4 that are assigned to c_i under Φ (naturally, $\|V_4^{c_i}\| = \frac{n}{k}$). Let V^{c_i} be the sublist of V containing the voters corresponding to those in $V_4^{c_i}$ (again, $\|V^{c_i}\| = \frac{n}{k}$). Let $s(V_4^{c_i})$ be the dissatisfaction of the voters in $V_4^{c_i}$ under Φ and let $s(V^{c_i})$ denote the dissatisfaction the voters in V^{c_i} would have if they were assigned to c_i (in I). The total dissatisfaction of the voters assigned to c_i under Φ is:

$$(\|H\| + \sum_j \|E_j\| + \sum_j \|F_j\| + \|E\| - \|E_i\|) + s(V_4^{c_i}) =$$
$$(\|H\| + \sum_j \|E_j\| + \sum_j \|F_j\| + \|E\| - 2m^2 n - m - (m - i)(2mn + 1)\tfrac{n}{k})$$
$$+ \tfrac{n}{k}(\|H\| + \sum_j \|D_j\| + (m - i)(2mn + 1) + mn) + s(V^{c_i})) =$$
$$(\tfrac{n}{k} + 1)(\|H\| + \sum_j \|E_j\| + mn) + s(V^{c_i}),$$

which shows that the dissatisfaction of the voters in I_{sc} that are assigned to c_i under Φ differs from the dissatisfaction of the respective voters in I, had they been assigned to c_i, only by a value that depends on n, m, and k (but not on i). Thus condition (v) holds.

It remains to show that an optimal assignment function for I_{sc} uses exactly k candidates from C'. We omit this part of the proof (see the full version of the paper [17]).

We conclude that an optimal assignment function Φ_{sc} assigns voters to exactly k candidates from C' and that the dissatisfaction of the voters in I_{sc} under Φ_{sc} is equal to the optimal dissatisfaction of the voters in I plus an easily computable value that depends on m, n, and k only. This completes the proof.   □

Betzler et al. [3] have shown a similar hardness result for single-peaked elections; however, their construction uses an artificial dissatisfaction function rather than Borda. The complexity of winner determination under α_B-ℓ_1-Monroe for single-peaked elections is still an open question. As our result answers this question in the case of single-crossing elections, it is tempting to ask if our proof approach could be used for single-peaked elections. Unfortunately, this does not seem to be the case. The difficulty lies in jointly implementing voters $V_3 + V_4$ (and, in particular, positioning the candidates $c_1, \ldots, c_m$).

4.2 ℓ_∞-Monroe for Single-Crossing Narcissistic Profiles

Given our hardness result for Monroe's rule, it is natural to ask if we can further restrict the problem of computing Monroe winners to obtain tractability. To this end, we focus on the egalitarian version of Monroe'e rule (the results of Betzler et al. [3] suggest that it is likely to be more tractable than the utilitarian version of this rule), and consider an additional domain restriction, namely, *narcissistic* preferences.

An election is said to be *narcissistic* if every candidate is ranked first by at least one voter. Intuitively, such elections arise when candidates are allowed to vote for themselves. This notion was introduced by Bartholdi and Trick [2], and was used in the context of fully proportional representation by Cornaz et al. [6]. It turns out that it is useful in our setting, too: we will show that the egalitarian version of Monroe's rule admits an efficient winner determination algorithm under single-crossing narcissistic preferences.

Lemma 4. *Let $E = (C, V)$ be a single-crossing narcissistic election with the candidate set $C = \{c_1, \ldots, c_m\}$ and voter collection $V = (v_1, \ldots, v_n)$, where v_1 has preference order $c_1 \succ \cdots \succ c_m$. For every $k \in [m]$ and every dissatisfaction function α for m candidates, there is an optimal k-Monroe assignment Φ for E under α-ℓ_∞-Monroe such that for each candidate $c_i \in C$, if $\Phi^{-1}(c_i) \neq \emptyset$ then there are two integers, t_i and t_i', $t_i \leq t_i'$, such that $\Phi^{-1}(c_i) = \{v_{t_i}, v_{t_i+1}, \ldots, v_{t_i'}\}$. Moreover, for each $i < j$ such that $\Phi^{-1}(c_i) \neq \emptyset$ and $\Phi^{-1}(c_j) \neq \emptyset$ it holds that $t_i' < t_j$.*

Based on Lemma 4, it is easy to construct a dynamic programming algorithm for ℓ_∞-Monroe.

Theorem 3. *For every family α of polynomial-time computable dissatisfaction functions, there is a polynomial-time algorithm that given a single-crossing narcissistic election E and a positive integer k finds an optimal k-Monroe assignment for E under α-ℓ_∞-Monroe.*

Unfortunately, the same approach does not work for the utilitarian variant of the Monroe's rule: the full version of this paper [17] contains an example showing that Lemma 4 no longer holds for the utilitarian setting (with Borda dissatisfaction function).

In a way, Theorem 3 is not new: It can be shown that narcissistic elections are necessarily single-peaked (this is implicit in the work of Barberà and Moreno [1]), and for single-peaked elections Betzler et al [3] provide a polynomial-time algorithm for ℓ_∞-Monroe (Proposition 5 in [3]). Thus, if we only care about polynomial-time computability, Theorem 3 does not appear to be useful. However, there are two reasons to prefer the algorithm described in Theorem 3. First, our algorithm is considerably faster: the running time of Betzler et al.'s algorithm is $O(n^3 m^3 k^3)$, while for our algorithm it is $O(nm^2 k)$. Second, our algorithm produces an assignment that has the contiguous blocks property. In contrast, in the full version of our paper [17]. we show that this is not necessarily the case for the algorithm of Betzler et al.

5   Conclusions

We have investiagted the complexity of winner determination under Chamberlin–Courant's and Monroe's rules, for the case of single-crossing profiles. We have presented a polynomial-time algorithm for Chamberlin–Courant's rule for single-crossing elections (and for elections that are close to being single-crossing in the sense of having bounded single-crossing width), and an NP-hardness proof for Monroe's rule for the same setting. Our results further strengthen the intuition that Monroe's rule is algorithmically harder than Chamberlin–Courant's rule. Similar conclusions follow from the work of Betzler et al. [3] and Skowron et al. [16]

Inspired by our negative result for Monroe's rule, we have sought further natural restrictions on voters' preferences. To this end, we considered single-crossing narcissistic profiles and developed an efficient algorithm for the egalitarian version of Monroe's rule under this preference restriction. However, our approach does not extend to general single-crossing elections, or to the utilitarian version of Monroe's rule (see the full version of our paper [17]).

Perhaps the most obvious direction for future research that is suggested by our work is understanding the computational complexity of the utilitarian version of Monroe's rule for single-crossing narcissistic elections and of egalitarian version of Monroe's rule for single-crossing elections. While we have shown that approaches based on the contiguous blocks property are bound to fail, other approaches may be more successful. Going in another direction, perhaps it is possible to obtain efficient algorithms for our restricted domains when using dissatisfaction functions other than Borda.

Acknowledgments. We would like to thank the reviewers for their helpful comments. Lan Yu and Edith Elkind were supported by National Research Foundation (Singapore) under grant 2009-08. Piotr Skowron and Piotr Faliszewski were supported in part by Polands National Science Center grant UMO-2012/06/M/ST1/00358. Piotr Faliszewski was also supported in part by grant DEC-2011/03/B/ST6/01393 and by the AGH University grant 11.11.230.015.

References

1. Barberà, S., Moreno, B.: Top monotonicity: A common root for single peakedness, single crossing and the median voter result. Games and Economic Behavior 73(2), 345–359 (2011)
2. Bartholdi III, J., Trick, M.: Stable matching with preferences derived from a psychological model. Operations Research Letters 5(4), 165–169 (1986)
3. Betzler, N., Slinko, A., Uhlmann, J.: On the computation of fully proportional representation. Journal of Artificial Intelligence Research 47, 475–519 (2013)
4. Bredereck, R., Chen, J., Woeginger, G.: A characterization of the single-crossing domain. Social Choice and Welfare (to appear, 2012)
5. Chamberlin, B., Courant, P.: Representative deliberations and representative decisions: Proportional representation and the borda rule. American Political Science Review 77(3), 718–733 (1983)
6. Cornaz, D., Galand, L., Spanjaard, O.: Bounded single-peaked width and proportional representation. In: Proceedings of the 20th European Conference on Artificial Intelligence, pp. 270–275 (2012)
7. Cornaz, D., Galand, L., Spanjaard, O.: Kemeny elections with bounded single-peaked or single-crossing width. In: Proceedings of the 23rd International Joint Conference on Artificial Intelligence, pp. 76–82 (2013)
8. Elkind, E., Faliszewski, P., Slinko, A.: Clone structures in voters' preferences. In: Proceedings of the 13th ACM Conference on Electronic Commerce, pp. 496–513 (2012)
9. Lu, T., Boutilier, C.: Budgeted social choice: From consensus to personalized decision making. In: Proceedings of the 22nd International Joint Conference on Artificial Intelligence, pp. 280–286 (2011)
10. Mirrlees, J.: An exploration in the theory of optimal income taxation. Review of Economic Studies 38, 175–208 (1971)
11. Monroe, B.: Fully proportional representation. American Political Science Review 89(4), 925–940 (1995)
12. Potthoff, R., Brams, S.: Proportional representation: Broadening the options. Journal of Theoretical Politics 10(2), 147–178 (1998)
13. Procaccia, A., Rosenschein, J., Zohar, A.: On the complexity of achieving proportional representation. Social Choice and Welfare 30(3), 353–362 (2008)
14. Saporiti, A., Tohmé, F.: Single-crossing, strategic voting and the median choice rule. Social Choice and Welfare 26(2), 363–383 (2006)
15. Skowron, P., Faliszewski, P., Slinko, A.: Achieving fully proportional representation is easy in practice. In: Proceedings of the 12th International Conference on Autonomous Agents and Multiagent Systems, pp. 399–406 (2013)
16. Skowron, P., Faliszewski, P., Slinko, A.: Fully proportional representation as resource allocation: Approximability results. In: Proceedings of the 23rd International Joint Conference on Artificial Intelligence, pp. 353–359 (2013)
17. Skowron, P., Yu, L., Faliszewski, P., Elkind, E.: The complexity of fully proportional representation for single-crossing electorates. Technical Report arXiv:1307.1252 [cs.GT], arXiv.org (2013)
18. Yu, L., Chan, H., Elkind, E.: Multiwinner elections under preferences that are single-peaked on a tree. In: Proceedings of the 23rd International Joint Conference on Artificial Intelligence, pp. 425–431 (2013)

New Results on Equilibria in Strategic Candidacy

Jérome Lang[1], Nicolas Maudet[2], and Maria Polukarov[3]

[1] LAMSADE, Université Paris-Dauphine, Paris, France
`lang@lamsade.dauphine.fr`
[2] LIP6, Université Pierre et Marie Curie, Paris, France
`nicolas.maudet@lip6.fr`
[3] University of Southampton, United Kingdom
`mp3@ecs.soton.ac.uk`

Abstract. We consider a voting setting where candidates have preferences about the outcome of the election and are free to join or leave the election. The corresponding candidacy game, where candidates choose strategically to participate or not, has been studied in very few papers, mainly by Dutta et al. [5,6], who showed that no non-dictatorial voting procedure satisfying unanimity is candidacy-strategyproof, or equivalently, is such that the joint action where all candidates enter the election is always a pure strategy Nash equilibrium. They also showed that for voting trees, there are candidacy games with no pure strategy equilibria. However, no results were known about other voting rules. Here we prove several such results. Some are positive (a pure strategy Nash equilibrium is guaranteed for Copeland and the uncovered set, whichever is the number of candidates, and for all Condorcet-consistent rules, for 4 candidates). Some are negative, namely for plurality and maximin.

1 Introduction

The two main criteria for the evaluation of voting rules are their ability to resist various sorts of strategic behaviour and to adapt to changes in the environment. Many (if not most) papers in computational social choice deal with (at least) one of these issues. Typically, strategic behaviour is shown by the voters reporting insincere votes (*manipulation*); by a third party, usually the chair, acting on the set of voters or candidates (*control*), or on the votes (*bribery* and *lobbying*), or on the voting rule (*e.g., agenda control*)[1]; finally, it can arise among the candidates themselves, who may also have preferences about the outcome of the election. However, the latter case has received little attention in (computational) social choice comparing to the former two. One form thereof involves choosing optimal political platforms, but probably the simplest form comes from the very ability of candidates to *decide whether to run for the election or not*, which is the issue we address in this paper. The following table summarises this rough classification of strategic behaviour in voting, according to the identity of strategising agent(s) and also to another relevant dimension, namely what the strategic actions bear on—voters, votes or candidates (we omit the agenda to keep the table small).

---

[1] There are also some forms of strategic behaviour that are specific to multiwinner elections, such as gerrymandering (by the chair) or vote pairing (by the voters).

B. Vöcking (Ed.): SAGT 2013, LNCS 8146, pp. 13–25, 2013.
© Springer-Verlag Berlin Heidelberg 2013

actions → agents ↓	*voters*	*votes*	*candidates*
voters	strategic participation	manipulation	-
third party / chair	voter control	bribery, lobbying	candidate control, cloning
candidates	-	-	*strategic candidacy*

Strategic candidacy does happen frequently in real-life elections, both in large-scale political elections and in small-scale, low-stake elections (*e.g.*, electing a chair in a research group). Throughout the paper we consider a finite set of *potential candidates*, which we simply call *candidates* when this is not ambiguous, and we make the following assumptions:

1. each candidate may choose to run or not for the election;
2. each candidate has a preference ranking over candidates;
3. each candidate ranks himself on top of his ranking;
4. the candidates' preferences are common knowledge among them;
5. the outcome of the election as a function of the set of candidates who choose to run is common knowledge among the candidates.

With the exception of 3, these assumptions were also made in the original model of Dutta et al. [5] which we discuss below. Assumption 2 amounts to saying that a candidate is interested only in the winner of the election[2] and has no indifferences or incomparabilities. Assumption 3 (considered as optional in [5]) is a natural domain restriction in most contexts. Assumptions 4 and 5 are common game-theoretic assumptions: note that we do not have to assume that the candidates know precisely how voters will vote, nor even the number of voters—they just have to know the choice function mapping every subset of candidates to a winner.

Existing work on strategic candidacy is rather scarce. It starts with [5] and [6], that formulate the strategic candidacy game and prove the following results (among others): (i) no non-dictatorial voting procedure satisfying unanimity is candidacy-strategyproof—or equivalently, is such that the joint action where all candidates enter the election is always a pure strategy Nash equilibrium; (ii) for the specific case of voting trees, there are candidacy games with no pure strategy Nash equilibria. These results are discussed further (together with simpler proofs) [7], and extended to voting correspondences [9,15] and to probabilistic voting rules [14].

Many questions remain unsolved. In particular, studying the solution concepts (such as Nash equilibria or strong equilibria) of a candidacy game would help predict the set of actual candidates and hence, the outcome of the vote, and therefore help design better elections. However, little is known about this: we only know that for any reasonable voting rule, there are some candidacy games for which the set of all candidates is not a Nash equilibrium, and that for voting trees, there exist a candidacy game with no pure strategy Nash equilibrium.

---

[2] In some contexts, candidates may have more refined preferences that bear for instance on the number of votes they get, how their score compares to that of other candidates etc. We do not consider these here.

In this paper, we go further in this direction and prove some positive as well as some negative results. We first consider the case of 4 candidates and show that a pure strategy Nash equilibrium always exists for Condorcet-consistent rules. Then we show that for Copeland and the uncovered set there is always an equilibrium in pure strategies, whichever is the number of candidates (although *strong* equilibria are not guaranteed to exist). On the negative side, we show that for plurality, for at least 4 candidates, and for maximin for at least 5 candidates, there are candidacy games without Nash equilibria.

Although it seems that strategic candidacy has not been considered yet in computational social choice, it is related to some questions that have received some attention in this community. First, the existence of strong equilibria is related to a stronger variant of candidate control (see the last paragraph of the conclusion). Other somewhat less related works that also consider a dynamic set of candidates are candidate cloning [8], possible winners with new candidates [3], and the unavailable candidate model [12].

The paper unfolds as follows. In Section 2 we define the strategic candidacy games and give a few preliminary results. In Section 3 we focus on the case of 4 candidates, whereas the case of 5 or more candidates is considered in Section 4. Finally, in Section 5 we discuss further issues, including the relation to candidate control.

2 Model and Preliminaries

In this section, we formally define the model of strategic candidacy and show that it induces a normal form game. We then present two simple results on the existence of Nash equilibria and strong equilibria in this setting.

2.1 Voting Rules

For completeness, we first define the common voting rules that we study in this paper.

There is a set of n voters electing from a set of m candidates. A single vote is a strict ordering of the candidates. A voting rule takes all the votes as input, and produces an outcome—a candidate, called *the winner* of the election. Although voting rules are usually defined for a fixed number of candidates, here we naturally extend the definition to an arbitrary number of candidates. All voting rules we consider in this work are *resolute*: we first define their irresolute version and assume that ties are broken up according to a fixed priority relation over candidates. Since voting rules are applied to varying sets of candidates, we assume that the tie-breaking rule is defined for the whole set of potential candidates, and projected to smaller sets of candidates; in other terms, if x has priority over y when all potential candidates run, this will still be the case for any set of candidates that contains x and y.

The *plurality* winner is the candidate that is ranked first by the largest number of voters. The *Borda* winner is the candidate who gets the highest Borda score: for each voter, a candidate c receives $q - 1$ points (where q is the number of candidates that are actually running) if it is ranked first by that voter, $q - 2$ if it is ranked second, and so on; the Borda score $B(c)$ of c is the total number of points he receives from all the voters.

Let $N(c, x)$ be the number of votes that rank c higher than x. The majority graph associated with a set of votes is the graph whose vertices are the candidates and containing an edge from x to y whenever $N(x, y) > \frac{n}{2}$ (when this holds we say that x

"beats" y). A candidate c is a *Condorcet winner* if x beats y for all $y \neq x$. A voting rule is *Condorcet-consistent* if it always elects a Condorcet winner when one exists.

The *maximin* rule chooses the candidate c for whom $\min_{x \in X \setminus \{c\}} N(c, x)$ is maximal. The Copeland[0] (resp., Copeland[1]) rule elects the candidate c maximising the number of candidates x such that $N(c, x) > \frac{n}{2}$ (resp., $N(c, x) \geq \frac{n}{2}$). The *uncovered set* (UC) rule selects the winner from the "uncovered set of candidates": a candidate c belongs to the uncovered set if and only if, for any other candidate x, if x beats c then c beats some y that beats x.

2.2 Strategic Candidacy

There is a set $X = \{x_1, x_2, \ldots x_m\}$ of m potential candidates, and a set $V = \{1, 2, \ldots n\}$ of n voters. We assume that these sets of voters and candidates are disjoint. As is classical in social choice theory, each voter $i \in V$ has a *preference* relation P_i, over the different candidates—i.e., a strict order ranking the candidates. The combination $P = (P_1, P_2, \ldots, P_n)$ of all the voters' preferences defines their preference *profile*.

Furthermore, each candidate also has a strict preference ordering over the candidates. We naturally assume that the candidates' preferences are *self-supported*—that is, the candidates rank themselves at the top of their ordering. Let $P^X = (P_c^X)_{c \in X}$ denote the candidates' preference profile. Following P^X, the potential candidates may decide to enter an election or withdraw their candidacy. Thus, the voters will only express their preferences over a subset $Y \subseteq X$ of the candidates that will have chosen to participate in the election, and we denote by $P^{\downarrow Y}$ the restriction of P to Y. We assume that the voters are *sincere*.

Given a profile P of the voters' preferences, a *voting rule* r defines a (single) winner among the *actual* candidates—i.e., given a subset $Y \subseteq X$ of candidates, it assigns to a (restricted) profile $P^{\downarrow Y}$ a member of Y. Each such voting rule r induces a natural *game form*, where the set of players is given by the set of potential candidates X, and the strategy set available to each player is $\{0, 1\}$ with 1 corresponding to entering the election and 0 standing for withdrawal of candidacy. A *state* s of the game is a vector of strategies $(s_c)_{c \in X}$, where $s_c \in \{0, 1\}$. For convenience, we use s_{-z} to denote $(s_c)_{c \in X \setminus \{z\}}$—i.e., s reduced by the single entry of player z. Similarly, for a state s we use s_Z to denote the strategy choices of a coalition $Z \subseteq X$ and s_{-Z} for the complement, and we write $s = (s_Z, s_{-Z})$.

The outcome of a state s is $r\left(P^{\downarrow Y}\right)$ where $c \in Y$ if and only if $s_c = 1$.[3] Coupled with a profile P^X of the candidates' preferences, this defines a normal form game $\Gamma = \langle X, P, r, P^X \rangle$ with m players. Here, player c prefers outcome $\Gamma(s)$ over outcome $\Gamma(s')$ if ordering P_c^X ranks $\Gamma(s)$ higher than $\Gamma(s')$.

2.3 Game-Theoretic Concepts

Having defined a normal form game, we can now apply standard game-theoretic solution concepts. Let $\Gamma = \langle X, P, r, P^X \rangle$ be a candidacy game, and let s be a state in Γ.

---

[3] When clear from the context, we use vector s to also denote the set of candidates Y that corresponds to state s; e.g., if $X = \{x_1, x_2, x_3\}$, we write $\{x_1, x_3\}$ and $(1, 0, 1)$ interchangeably.

We say that a coalition $Z \subseteq X$ has an *improving move* in s if there is s'_Z such that $\Gamma(s_{-Z}, s'_Z)$ is preferable over $\Gamma(s)$ by every player $z \in Z$. In particular, the improving move is *unilateral* if $|Z| = 1$. A *(pure strategy) Nash equilibrium (NE)* [13] is a state that has no unilateral improving moves. More generally, a state is a k-NE if no coalition with $|Z| \le k$ has an improving move. A *strong equilibrium (SE)* ([1]) is a state that has no improving moves.

Example 1. Consider the game $\langle \{a, b, c, d\}, P, r, P^X \rangle$, where r is the Borda rule, and P and P^X are as follows[4]:

	P						P^X			
1	1	1	1	1	1	1	a	b	c	d
b	c	c	a	d	b	a	a	b	c	d
d	d	d	c	a	c	b	d	a	b	a
a	a	b	b	c	d	c	b	d	a	c
c	b	a	d	b	a	d	c	c	d	b

The state $(1,1,1,1)$ is not an NE: $abcd \mapsto c$, but $abc \mapsto a$, and d prefers a to c, so for d, leaving is an improving move. Now, $(1,1,1,0)$ is an NE, as nobody has an improving move neither by joining (d prefers a over c), nor by leaving (obviously not a; if b or c leaves then the winner is still a). It can be checked that this is also an SE.

2.4 Preliminary Results

Regardless of the number of voters and the voting rule, a straightforward observation is that a candidacy game with *three* candidates is guaranteed to possess an NE.[5] This, however, is not true for SE.[6] For *any* number of candidates, the following result holds.

Proposition 1. *Let $\Gamma = \langle X, P, r, P^X \rangle$ be a candidacy game where r is Condorcet-consistent. If P has a Condorcet winner c then for any $Y \subseteq X$,*

$$Y \text{ is a SE} \Leftrightarrow Y \text{ is an NE} \Leftrightarrow c \in Y.$$

Proof. Assume c is a Condorcet winner for P and let $Y \subseteq X$ such that $c \in Y$. Because r is Condorcet-consistent, and because c is a Condorcet winner for $P^{\downarrow Y}$, we have $r\left(P^{\downarrow Y}\right) = c$. Assume $Z = Z^+ \cup Z^-$ is a deviating coalition from Y, with Z^+ the candidates who join and Z^- the candidates who leave the election. Clearly, $c \notin Z$, as $c \in Y$ and c has no interest to leave. Therefore, c is still a Condorcet winner in $P^{\downarrow(Y \backslash Z^-) \cup Z^+}$,

---

[4] In our examples, we assume a lexicographic tie-breaking. We also use the simplified notation $Y \mapsto x$ to denote that rule r applied to the subset of candidates $Y \subseteq X$ is x, and we omit curly brackets. The first row in P indicates the number of voters casting the different ballots.

[5] This can be easily seen: Let $X = \{a, b, c\}$ and suppose w.l.o.g. that $abc \mapsto a$. If $\{a, b, c\}$ is not an NE, then either (1) $ab \mapsto b$ and c prefers b to a, or (2) $ac \mapsto c$ and b prefers c to a. Since b and c play symmetric roles, w.l.o.g., assume (1). Then $\{a, b\}$ is an NE.

[6] Here is a counterexample (for which we thank an anonymous reviewer of the previous version of the paper). The selection rule is $abc \mapsto b$; $ab \mapsto a$; $ac \mapsto c$; $bc \mapsto c$; it can be easily implemented by the scoring rule with scoring vector $(5, 4, 0)$ with 5 voters. Preferences of candidates are: $a : a \succ b \succ c$; $b : b \succ c \succ a$; $c : c \succ a \succ b$. The group deviations are: in $\{a, b, c\}$, c leaves; in $\{a, b\}$, b leaves and c joins; in $\{a, c\}$, b joins; in $\{b, c\}$, a joins; in $\{a\}$, c joins; in $\{b\}$, c joins; in $\{c\}$, a and b join.

which by the Condorcet-consistency of r implies that $r\left(P^{\downarrow(Y\setminus Z^-)\cup Z^+}\right) = c$, which contradicts the assumption that Z wants to deviate. We thus conclude that Y is an SE, and *a fortiori* an NE. Finally, let $Y \subseteq X$ such that $c \notin Y$. Then, Y is not an NE (and *a fortiori* not an SE), because c has an interest to join the election.   □

Now, if P has no Condorcet winner, the analysis becomes more complicated. We provide results for this more general case in the following sections. Interestingly, as we demonstrate, some Condorcet-consistent rules (e.g., Copeland and UC) do always possess a Nash equilibrium in this case, while some other (e.g., maximin) do not.

3   Four Candidates

With only 4 potential candidates, we exhibit a sharp contrast between Condorcet consistent rules, which all possess an NE, and scoring rules.

3.1   Scoring Rules

To study scoring rules, we make use of a very powerful result by Saari [16]. It states that for almost all scoring rules, any conceivable choice function can result from a voting profile. This means that our question boils down to checking whether a *choice function*, together with some coherent candidates' preferences, can be found such that no NE exists with 4 candidates. We solve this question by encoding the problem as an Integer Linear Program (ILP), the details of which can be found in Appendix. It turns out that such choice functions do exist: it then follows from Saari's result that counterexamples can be obtained for "most" scoring rules. We exhibit a profile for plurality.

Proposition 2. *For plurality and $m = 4$, there may be no NE.*

Proof. We exhibit a counterexample with 13 voters, whose preferences are contained in the left part of the table below. The top line indicates the number of voters with each particular profile. The right part of the table represents the preferences of the candidates.

3	1	1	1	1	1	1	2	2	a	b	c	d
d	d	d	a	a	a	b	b	c	a	b	c	d
c	b	a	b	c	d	c	a	b	b	a	d	a
a	c	b	c	b	b	d	c	d	c	c	a	b
b	a	c	d	d	c	a	d	a	d	d	b	c

□

Similar constructions of profiles can thus be obtained for other scoring rules. However, Borda comes out as a very peculiar case [16] among scoring rules[7]. This is also verified for the case of strategic candidacy.

---

[7] For a more detailed statement of this result, we point the reader to the work of Saari, in particular [17]. For the case of 4 candidates, families of scoring rules such that, when the scoring vector for 3 candidates is of the form $\langle w_1, w_2, 0\rangle$, the vector for 4 candidates is of the form $\langle 3w_1, w_1 + 2w_2, 2w_2, 0\rangle$ (for instance, $\langle\langle 3, 1, 0, 0\rangle, \langle 1, 0, 0\rangle\langle 1, 0\rangle\rangle$) are an exception in the sense that not all choice functions are implementable with them.

Proposition 3. *For Borda and $m = 4$, there is always an NE.*

We could check this by relying on the fact that Borda rule is represented by a weighted majority graph, and by adding the corresponding constraints into the ILP. The infeasibility of the resulting set of constraints shows that no instances without NE can be constructed. However, it takes only coalitions of pairs of agents to ruin this stability.

Proposition 4. *For Borda and $m = 4$, there may be no 2-NE.*

Proof. Consider the following game:

1 1 1 1 1	a b c d
b c d a b	a b c d
d d a b c	c a a b
c a c c d	d c d a
a b b d a	b d b c

Here, only $s_1 = (0, 1, 1, 1)$ and $s_2 = (1, 1, 0, 1)$ are NE, with $bcd \mapsto b$, and $abd \mapsto d$. But from s_1 the coalition $\{a, c\}$ has an improving move to s_2 as they both prefer d over b. Now take s_2: if b leaves and c joins, they reach $(1, 0, 1, 1)$, with $acd \mapsto c$ and both prefer c over d. $\qquad\square$

3.2 Condorcet-Consistent Rules

We now turn our attention to Condorcet-consistent rules. It turns out that for *all of them*, the existence of an NE can be guaranteed.

Proposition 5. *For $m = 4$, if r is Condorcet-consistent, there always exists an NE.*

Proof. We start with a remark: although we do not assume that r is based on the majority graph, we nevertheless prove our result by considering all possible cases for the majority graphs (we get back to this point at the end of the proof). There are four graphs to consider (all others are obtained from these ones by symmetry).

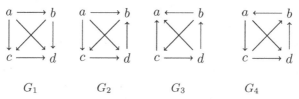

$$G_1 \qquad\qquad G_2 \qquad\qquad G_3 \qquad\qquad G_4$$

For G_1 and G_2, any subset of X containing the Condorcet winner is an NE (see Proposition 1). For G_3, we note that a is a Condorcet loser. That is, $N(a, x) < N(x, a)$ for all $x \in \{b, c, d\}$. Note that in this case, there is no Condorcet winner in the reduced profile $P^{\downarrow\{b,c,d\}}$ as this would imply the existence of a Condorcet winner in P (case G_1 or G_2). W.l.o.g., assume that b beats c, c beats d, and d beats b. W.l.o.g. again, assume that $bcd \mapsto b$. Then, $\{b, c\}$ is an NE. Indeed, in any set of just two candidates, none has an incentive to leave. Now, a or d have no incentive to join as this would not change the winner: in the former case, observe that b is the (unique) Condorcet winner in $P^{\downarrow\{a,b,c\}}$, and the latter follows by our assumption. There is always an NE for G_3.

The proof for G_4 is more complex and proceeds case by case. Since r is Condorcet-consistent, we have $acd \mapsto a$, $bcd \mapsto c$, $ab \mapsto b$, $ac \mapsto a$, $ad \mapsto a$, $bc \mapsto c$, $bd \mapsto d$ and $cd \mapsto c$. The sets of candidates for which r is undetermined are $abcd$, abc and abd.

We observe the following easy facts: (i) if $abcd \mapsto a$ then acd is an NE, (ii) if $abcd \mapsto c$ then bcd is an NE, (iii) if $abc \mapsto a$ then ac is an NE, (iv) if $abd \mapsto a$ then ad is an NE, (v) if $abc \mapsto c$ then bc is an NE. The only remaining cases are:

1. $abcd \mapsto b$, $abc \mapsto b$, $abd \mapsto b$.
2. $abcd \mapsto b$, $abc \mapsto b$, $abd \mapsto d$.
3. $abcd \mapsto d$, $abc \mapsto b$, $abd \mapsto b$.
4. $abcd \mapsto d$, $abc \mapsto b$, $abd \mapsto d$.

In cases 1 and 3, ab is an NE. In case 2, if a prefers b to c then abc is an NE, and if a prefers c to b, then bcd is an NE. In case 4, if a prefers c to d, then bcd is an NE; if b prefers a to d, then ad is an NE; finally, if a prefers d to c and b prefers d to a, then $abcd$ is an NE. To conclude, observe that the proof never uses the fact that two profiles having the same majority graph have the same winner.[8] $\square$

The picture for 4 candidates shows a sharp contrast. On the one hand, the existence of choice functions shows that "almost all scoring rules" [16] may fail to have an NE. On the other hand, Condorcet-consistency alone suffices to guarantee the existence of an NE. (However, this criterion is not sufficient to guarantee stronger notion of stability: e.g., for Copeland, we could exhibit examples without any 2-NE.)

4   More Candidates

The first question which comes to mind is whether examples showing the absence of NE transfer to larger sets of candidates. They indeed do, under an extremely mild assumption. We say that a voting rule is *insensitive to bottom-ranked candidates* (IBC) if given any profile P over $X = \{x_1, \ldots, x_m\}$, if P' is the profile over $X \cup \{x_{m+1}\}$ obtained by adding x_{m+1} at the bottom of every vote of P, then $r(P') = r(P)$. This property is extremely weak (much weaker than Pareto) and is satisfied by almost all voting rules studied in the literature (a noticeable exception being the veto rule).

Lemma 1. *For any voting rule r satisfying IBC, if there exists $\Gamma = \langle X, P, r, P^X \rangle$ with no NE, then there exists $\Gamma' = \langle X', P', r, P^Y \rangle$ with no NE, where $|X'| = |X| + 1$.*

Proof. Take Γ with no NE, with $X = \{x_1, \ldots, x_m\}$. Let $X' = X \cup \{x_{m+1}\}$, P' the profile obtained from P by adding x_{m+1} at the bottom of every vote, and $P^{X'}$ be the candidate profile obtained by adding x_{m+1} at the bottom of every ranking of a candidate x_i, $i \leq m$, and whatever ranking for x_{m+1}. Let $Y \subseteq X$. Because Y is not an NE for Γ, some candidate $x_i \in X$ has an interest to leave or to join, therefore Y is not an NE either for Γ'. Now, consider $Y' = Y \cup \{x_{m+1}\}$. If $x_i \in X$ has an interest to leave (resp., join) Y, then because r satisfies IBC, the winner in $Y' \setminus \{x_i\}$ (resp., $Y' \cup \{x_i\}$) is the same as in $Y \setminus \{x_i\}$ (resp., $Y \cup \{x_i\}$), therefore $x_i \in X$ has an interest to leave (resp., join) Y', therefore Y' is not an NE. $\square$

---

[8] For instance, we may have two profiles P, P' both corresponding to G_4, such that $r(P) = a$ and $r(P') = b$; the proof perfectly works in such a case.

Corollary 1. *For plurality and $m \geq 4$, there may be no NE.*

We now turn our attention to Condorcet-consistent rules, which all admit NE with 4 candidates. However, 5 candidates suffice to show that NE are not guaranteed any longer.

Proposition 6. *For maximin with $m = 5$, there may be no NE.*

Proof. The counterexample is given by the following pairwise comparison matrix, where the entry corresponding to row x and column y is equal to $N(x, y) - N(y, x)$. From Debord's theorem [4] we know that there exists a profile with such a comparison matrix. The candidates' preference profile is given on the right hand side. The tie-breaking priority is lexicographic.

	a	b	c	d	e		$a\ b\ c\ d\ e$
a	0	-3	3	-1	1		$a\ b\ c\ d\ e$
b	3	0	-3	3	1		$c\ e\ d\ a\ b$
c	-3	3	0	-1	-1		$b\ c\ a\ c\ a$
d	1	-3	1	0	-5		$e\ a\ e\ b\ d$
e	-1	-1	1	5	0		$d\ d\ b\ e\ c$

The proof goes by exhibiting all cases. For each subset we indicate the deviation (the winner being shown using bold font): $ad \rightarrow a\mathbf{b}d \rightarrow ab\mathbf{c}d \rightarrow abc\mathbf{d}e \rightarrow \mathbf{b}cde \rightarrow \mathbf{c}de \rightarrow$ $a\mathbf{c}de \rightarrow ab\mathbf{c}de$; $ab \rightarrow ab\mathbf{c} \rightarrow ab\mathbf{c}e \rightarrow \mathbf{b}ce \rightarrow \mathbf{c}e \rightarrow a\mathbf{c}e \rightarrow ab\mathbf{c}e$; $a\mathbf{b}de \rightarrow \mathbf{b}ce$; $ae \rightarrow a\mathbf{b}e \rightarrow ab\mathbf{c}e$; $ac \rightarrow a\mathbf{c}d \rightarrow ab\mathbf{c}d$; $bd \rightarrow \mathbf{b}cd \rightarrow ab\mathbf{c}d$; $b\mathbf{d}e \rightarrow \mathbf{b}cde$; $bc \rightarrow ab\mathbf{c}$; $\mathbf{b}e \rightarrow \mathbf{b}ce$; $cd \rightarrow \mathbf{c}de$; $ce \rightarrow a\mathbf{c}e$; $de \rightarrow a\mathbf{d}e$. □

Corollary 2. *For maximin and $m \geq 5$, there may be no NE.*

This negative result does not extend to all Condorcet-consistent rules. In particular, next we show the existence of NE for Copeland and the uncovered set (UC), under deterministic tie-breaking, for any number of candidates.

Proposition 7. *For Copeland^0, with any number of candidates, there is always an NE.*

Proof. Let P be a profile and $\rightarrow_P$ its associated majority graph. Let $C(x, P)$ be the number of candidates $y \neq x$ such that $x \rightarrow_P y$. Let $COP^0(P)$ be the set of the Copeland^0 cowinners for P, i.e., the set of candidates maximising $C(\cdot, P)$, and $Cop^0(P) = c$ the Copeland^0 winner—the highest-priority candidate in $COP^0(P)$. Consider $Dom(c) = \{c\} \cup \{y | c \rightarrow_P y\}$. Note that $C\left(c, P^{\downarrow Dom(c)}\right) = |Dom(c)| - 1 = q = C(c, P)$. Also, since any $y \in Dom(c)$ is beaten by c, we have $C(y, P^{\downarrow Dom(c)}) \leq q - 1$.

We claim that $Dom(c)$ is an NE. Note that c is a Condorcet winner in the restriction of P to $Dom(c)$, and *a fortiori*, in the restriction of P to any subset of $Dom(c)$. Hence, c is the Copeland^0 winner in $Dom(c)$ and any of its subsets, and no candidate in $Dom(c)$ has an incentive to leave.

Now, assume there is a candidate $z \in X \setminus Dom(c)$ such that $Cop^0\left(P^{\downarrow Dom(c) \cup \{z\}}\right) \neq c$. Note that $c \not\rightarrow_P z$ as z does not belong to

$Dom(c)$; so, $C(c, P^{\downarrow Dom(c)\cup\{z\}}) = q$. For any $y \in Dom(c)$ we have $C(y, P^{\downarrow Dom(c)\cup\{z\}}) \leq (q-1) + 1 = q = C(c, P^{\downarrow Dom(c)\cup\{z\}})$. If $C(y, P^{\downarrow Dom(c)\cup\{z\}}) < C(c, P^{\downarrow Dom(c)\cup\{z\}})$, then y is not the Copeland^0 winner in $P^{\downarrow Dom(c)\cup\{z\}}$. If $C(y, P^{\downarrow Dom(c)\cup\{z\}}) = C(c, P^{\downarrow Dom(c)\cup\{z\}})$, then $C(y, P) \geq C(c, P)$. That is, either $c \notin COP^0(P)$, a contradiction, or both y, c are in $COP^0(P)$. In that case, the tie-breaking priority ensures that $Cop^0(P^{\downarrow Dom(c)\cup\{z\}}) \neq y$.

Hence, $Cop^0(P^{\downarrow Dom(c)\cup\{z\}}) = z$. By $Cop^0(P) = c$ we have $C(z, P^{\downarrow Dom(c)\cup\{z\}}) \leq C(z, P) \leq C(c, P) \leq q$; therefore, $C(z, P^{\downarrow Dom(c)\cup\{z\}}) = q$, and the tie-breaking priority favours z over c. But then, $C(z, P) = C(c, P)$, i.e., both c and z are in $COP^0(P)$, and the tie-breaking priority ensures that $Cop^0(P^{\downarrow Dom(c)\cup\{z\}}) \neq z$, a contradiction. Therefore, the Copeland^0 winner in $P^{\downarrow Dom(c)\cup\{z\}}$ must be c, which implies that z has no incentive to join $Dom(c)$. □

Note that if the number of voters is odd, we do not have to care about head-to head ties. In this case, all Copeland$^\alpha$ rules, where each agent in a head-to-head election gets $0 \geq \alpha \geq 1$ points in the case of a tie (Copeland^0 being a special cases), are equivalent, and the result above holds. However, if the number of voters is even, this is not necessarily the case. Thus, in particular, for Copeland$^{0.5}$ (more often referred to as Copeland), $Dom(c)$ is generally no more an NE, and we do now know whether the existence of an NE is guaranteed or not.

Proposition 8. *For UC, with any number of candidates, and an odd number of voters, there is always an NE.*

Proof. Let c be the UC winner in P, i.e., the highest-priority candidate in $UC(P)$. Consider (again) $Dom(c) = \{c\} \cup \{y | c \rightarrow_P y\}$. We claim that $Dom(c)$ is an NE.

Since c is a Condorcet winner in the restriction of P to $Dom(c)$, and *a fortiori*, in the restriction of P to any subset of $Dom(c)$, it is the UC winner in $Dom(c)$ and in any of its subsets, and no candidate in $Dom(c)$ wants to leave.

Now, let $z \in X \setminus Dom(c)$. Since $z \notin Dom(c)$, we have $c \not\rightarrow_P z$ and hence, $z \rightarrow_P c$, as n is odd. Since $x \in UC(P)$, there must be $y \in Dom(c)$ such that $y \rightarrow_P z$. This implies that $x \in UC(P^{\downarrow Dom(c)\cup\{z\}})$, which, due to tie-breaking priority, yields that c is the UC winner in $P^{\downarrow Dom(c)\cup\{z\}}$. Thus, z has no incentive to join $Dom(c)$. □

Note that the proofs of Propositions 7 and 8 also show that for Copeland^0 and UC, there always exists an NE *in which the winning candidate is the winner in the full profile* (with all candidates present)[9].

5   Conclusions

In this work, we further explored the landscape of strategic candidacy in elections and obtained several positive results (for Condorcet-consistent rules with 4 candidates; for two versions of Copeland, as well as for the uncovered set, with any number of candidates) and several negative results (for plurality and maximin). Many cases remain open, especially Borda with more than 4 candidates.

---

[9] We thank Edith Elkind for this remark.

Another line for further research is the study of the set of states that can be reached by some improvement path (e.g., best or better response dynamics) starting, say, from the set or all potential candidates. In some cases, even when the existence of NE is guaranteed (e.g., for Copeland), we could already come up with examples showing that no equilibrium point is reachable by a sequence of better responses. But other types of dynamics can also be considered.

Finally, there is an interesting connection between strategic candidacy and control by deleting or adding candidates [2,11], as well as *multimode* control [10] where the chair is allowed both to delete *and* to add some candidates. Strategic candidacy relates to a slightly more demanding notion of control, which we can call *consenting control*, in which candidates have to agree to be added or removed. For instance, s is an SE if there is no consenting destructive control by removing+adding candidates against the current winner $r(X_s)$. Not only this notion is of independent interest, but also, complexity results for control may allow to derive complexity results for the problem of deciding the existence of NE or SE in a strategic candidacy game.

Acknowledgements. We would like to thank Michel Le Breton, Vincent Merlin, Edith Elkind and the anonymous reviewers for helpful discussions.

References

1. Aumann, R.: Acceptable points in general cooperative n-person games. In: Contributions to the Theory of Games IV. Annals of Mathematics Study, vol. 40, pp. 287–324 (1959)
2. Bartholdi, J., Tovey, C., Trick, M.: How hard is it to control an election? Social Choice and Welfare 16(8-9), 27–40 (1992)
3. Chevaleyre, Y., Lang, J., Maudet, N., Monnot, J., Xia, L.: New candidates welcome! possible winners with respect to the addition of new candidates. Mathematical Social Sciences 64(1), 74–88 (2012)
4. Debord, B.: Caractérisation des matrices de préférences nettes et méthodes d'agrégation associées. Mathematiques et Sciences Humaines 97, 5–17 (1987)
5. Dutta, B., Breton, M.L., Jackson, M.O.: Strategic candidacy and voting procedures. Econometrica 69, 1013–1037 (2001)
6. Dutta, B., Breton, M.L., Jackson, M.O.: Voting by successive elimination and strategic candidacy in committees. Journal of Economic Theory 103, 190–218 (2002)
7. Ehlers, L., Weymark, J.A.: Candidate stability and nonbinary social choice. Economic Theory 22(2), 233–243 (2003)
8. Elkind, E., Faliszewski, P., Slinko, A.M.: Cloning in elections: Finding the possible winners. J. Artif. Intell. Res (JAIR) 42, 529–573 (2011)
9. Eraslan, H., McLennan, A.: Strategic candidacy for multivalued voting procedures. Journal of Economic Theory 117(1), 29–54 (2004)
10. Faliszewski, P., Hemaspaandra, E., Hemaspaandra, L.: Multimode control attacks on elections. JAIR 40, 305–351 (2011)
11. Hemaspaandra, E., Hemaspaandra, L., Rothe, J.: Anyone but him: The complexity of precluding an alternative. Artificial Intelligence 171(5-6), 255–285 (2007)
12. Lu, T., Boutilier, C.: The unavailable candidate model: a decision-theoretic view of social choice. In: ACM Conference on Electronic Commerce 2010, pp. 263–274 (2010)
13. Nash, J.: Non-cooperative games. Annals of Mathematics 54(2), 286–295 (1951)

14. Rodriguez-Alvarez, C.: Candidate stability and probabilistic voting procedures. Economic Theory 27(3), 657–677 (2006)
15. Rodriguez-Alvarez, C.: Candidate stability and voting correspondences. Social Choice and Welfare 27(3), 545–570 (2006)
16. Saari, D.: A dictionary of voting paradoxes. Journal of Economic Theory 48 (1989)
17. Saari, D.: Election results and a partial ordering for positional ordering. In: Schofield, N.J. (ed.) Collective Decision-Making: Social Choice and Political Economy, pp. 93–110. Kluwer (1996)

Appendix: ILP Formulation

Let S be the set of $(2^{|X|})$ states, and $A(s)$ be the set of agents candidating in state s.

Choice functions without NE. We introduce a binary variable w_{si}, indicating that agent i wins in state s. We add constraints enforcing that a winner in each state s is unique:

$$\forall i \in X, \forall s \in S : \quad w_{s,i} \in \{0,1\} \tag{1}$$

$$\forall s \in S : \textstyle\sum_{i \in X} w_{s,i} = 1 \tag{2}$$

$$\forall s \in S, \forall i \in X \notin A(s) : \quad w_{s,i} = 0 \tag{3}$$

Now we denote by $D(s)$ the set of *possible deviations* from state s (states where a single agent's candidacy differs from s). We denote by $a(s,t)$ an agent potentially deviating from s to t. Binary variables $d_{s,t}$ indicate a deviation from s to t. In each state, there must be at least one deviation, otherwise this state is an NE:

$$\forall s \in S, \forall t \in S : \quad d_{s,t} \in \{0,1\} \tag{4}$$

$$\forall s \in S : \textstyle\sum_{t \in D(s)} d_{s,t} \geq 1 \tag{5}$$

Now, we introduce constraints related to the preferences of the candidates. For this purpose, we introduce a binary variable $p_{i,j,k}$, indicating that agent i prefers candidate j over candidate k. If there is indeed a deviation from s to t, the deviating agent must prefer the winner of the new state over the winner of the previous state:

$$\forall s \in S, \forall t \in D(s), \forall i \in X, \forall j \in X : w_{s,i} + w_{t,j} + d_{s,t} - p_{a(s,t),j,i} \leq 2 \tag{6}$$

Finally we ensure that the preferences are irreflexive and transitive[10], and respect the constraint of being self-supported:

$$\forall i \in X, \forall j \in X : \quad p_{i,j,j} = 0 \tag{7}$$

$$\forall a \in X, \forall i \in X \forall j \in X, \forall k \in X : p_{a,i,j} + p_{a,j,k} - p_{a,i,k} \leq 1 \tag{8}$$

$$\forall i \in X, \forall j \in X : \quad p_{i,i,j} = 1 \tag{9}$$

---

[10] Notice that this ILP does not necessarily contain complete preferences: the program only needs to check those preference relations that correspond to possible deviations. Any linear extension of these (partial) preferences gives an instance with complete preferences.

Constraints for Borda. We introduce a new integer variable $N_{i,j}$ to represent the number of voters preferring i over j in the weighted tournament. We first make sure that the values of $N_{i,j}$ are coherent throughout the weighted tournament:

$$\forall i \in X, \forall j \in X, \forall k \in X, \forall l \in X : N_{i,j} + N_{j,i} = N_{k,l} + N_{l,k} \tag{10}$$

In each state, when agent i wins, we must make sure that his total amount of points is the highest among all the agents in this state (note that i can simply tie with those agents that i is prioritised over by the tie-breaking; we omit this for the sake of readability):

$$\forall s \in S, \forall i \in A(s), \forall j \in A(s) \setminus \{i\} :$$
$$(1 - w_{s,i}) \times M + \sum_{j \in A(s) \setminus \{i\}} N_{i,j} > \sum_{j \in A(s) \setminus \{k\}} N_{k,j} \tag{11}$$

Here M is an arbitrary large value, used to relax the constraint when $w_{s,i}$ is 0.

Plurality Voting with Truth-Biased Agents[*]

Svetlana Obraztsova[1], Evangelos Markakis[2], and David R.M. Thompson[3]

[1] National Technical University of Athens, Athens, Greece
[2] Athens University of Economics and Business, Athens, Greece
[3] University of British Columbia, Vancouver, Canada

Abstract. We study a game-theoretic model for Plurality, one of the most well-studied and widely-used voting rules. It is well known that the most standard game-theoretic approaches can be problematic in the sense that they lead to a multitude of Nash equilibria, many of which are counter-intuitive. Instead, we focus on a model recently proposed to avoid such issues [2,6,11]. The main idea of the model is that voters have incentives to be truthful when their vote is not pivotal, i.e., when they cannot change the outcome by a unilateral deviation. This modification is quite powerful and recent simulations reveal that equilibria which survive this refinement tend to have nice properties.

We undertake a theoretical study of pure Nash and strong Nash equilibria of this model under Plurality. For pure Nash equilibria we provide characterizations based on understanding some crucial properties about the structure of equilibrium profiles. These properties demonstrate how the model leads to filtering out undesirable equilibria. We also prove that deciding the existence of an equilibrium with a certain candidate as a winner is NP-hard. We then move on to strong Nash equilibria, where we obtain analogous characterizations. Finally, we also observe some relations between strong Nash equilibria and Condorcet winners, which demonstrate that this notion forms an even better refinement of stable profiles.

1 Introduction

We study Plurality-based voting systems from a game-theoretic point of view. Voting mechanisms constitute a popular tool for preference aggregation and decision making in various contexts involving entities with possibly diverse preferences. Ideally in such protocols, one would like to ensure that the voters do not have an incentive to mis-report their preferences in order to get some favorite candidate elected. However, the famous Gibbard-Satterthwaite theorem [4,9] states that under mild assumptions, this is impossible. Strategic behavior is therefore inherent in most voting rules.

In the presence of strategic voting, a large volume of research has emerged that focuses on various aspects of manipulation. This includes manipulation by coalitions, but also equilibrium analysis, where voters are viewed as rational agents participating in a game. In this work we follow the latter approach, which was initiated by [3] and has led to several game-theoretic models for capturing voting behavior. One problem

---

[*] Work supported by the project ALGONOW of the research funding program Thales (co-financed by the European Social Fund-ESF and Greek national funds), and also by a grant of the Russian government (FCP).

B. Vöcking (Ed.): SAGT 2013, LNCS 8146, pp. 26–37, 2013.

however that emerges in some of the models is the fact that they yield a multitude of Nash equilibria and hence they lack predictive power. A typical example of this well known fact is that every candidate (even one who is ranked last by all voters) is a winner in some equilibrium: if everybody votes for him, then no voter can change the outcome by a unilateral deviation.

As a result, the literature has largely concentrated on proposing more realistic models that avoid such issues. Among these, one promising idea has been formalized in a series of recent papers, and in particular in [2,6,11].

The main idea in these works is to model the voters so that they prefer to be truthful when their vote is not pivotal, i.e., when they cannot change the outcome by a unilateral deviation. Such voters are referred to as *truth-biased* voters in [6]. This twist, which is the focus of our work, turns out to be quite powerful. For the Plurality rule, this was empirically evaluated in [11]. Their experiments suggest that the model achieves a significant refinement of equilibrium profiles, i.e., models with truth-bias may have more predictive power than models without. However, there has thus far been no theoretical study on the properties of Nash equilibria with truth-bias. Further, the interaction between truth-bias and other equilibrium refinements such as the concept of strong Nash equilibrium has not yet been investigated.

Contribution. We undertake a theoretical analysis of the model with truth-biased agents under the Plurality rule. We focus on the set of pure Nash and strong Nash equilibria. In Section 3, we obtain a characterization for the existence of a pure equilibrium with a given candidate as a winner. Our characterization is based on understanding some crucial properties regarding the performance of "runner up" candidates. These properties also demonstrate how this model achieves a refinement of equilibrium profiles. Our results can be seen as a complement to the corresponding experimental findings of [11]. On the negative side, we derive an NP-hardness result for determining if an equilibrium exists with a given candidate as a winner, implying that any characterization has to rely on conditions that are not easily checkable. In Section 4, we move on to strong Nash equilibria. We obtain characterizations for the same type of questions as for the case of pure equilibria. Interestingly in this case we can check existence in polynomial time. We also observe some relations between strong Nash equilibria and Condorcet winners, which imply that there can be only one possible winner in all strong Nash equilibria of a game, i.e., this notion forms an even better refinement of stable profiles.

Related Work. Analysis of Nash equilibria in voting is challenging since natural "basic" models for voting games have the problems mentioned in the Introduction, i.e., multiplicity of equilibria and no predictive power over outcomes. Thus, the literature on equilibrium analysis of voting can be viewed as a search for models that get away from these limitations. One approach is to introduce uncertainty, e.g., about how many voters support each candidate (as in [8]). Another approach involves changing the temporal structure of the game. Xia and Conitzer [12], and also [1] consider the case where agents vote publicly and one-at-a-time. Yet another line of work considers the case where voters are allowed to change their votes dynamically [6,5].

A more direct approach is to assume that voters have a slight preference for a particular action, so that in situations where a voter cannot influence the outcome, he will strictly prefer this favored action. Desmedt and Elkind [1] study a model where voters may prefer to abstain if they are not pivotal. Another line of research considers what happens when every voter slightly prefers to vote honestly (i.e, for his most preferred candidate in the case of plurality), when he is not pivotal.

This last approach, introducing a small reward for truthfulness, is the one that we follow. The works of [6] and [2] are two examples of using this approach with plurality. The former studied convergence of iterative best response procedures, whereas the latter demonstrated that pure equilibria do not always exist. More recently, Thompson *et al.* [11] conducted a large-scale computational experiment, testing how frequently pure Nash equilibria existed under this model and studied properties of equilibrium outcomes. Since our work strives to answer similar questions analytically rather than experimentally, some of their findings are particularly relevant, such as the fact that most games in their simulations had at least one pure Nash equilibrium, and that equilibrium outcomes tended to be good, e.g., Condorcet winners often won.

Finally, another way of getting more predictive power is to have a stronger solution concept, e.g., strong equilibrium. Messner and Polborn [7] investigated the possibility of strong equilibria and were able to characterize when such equilibria exist, for the special case of three-candidate plurality elections. Sertel and Sanver [10] also studied strong equilibria under Plurality, and were able to show that strong equilibria outcomes are characterized by a generalized form of Condorcet winners.

2   Definitions and Notation

We consider a set of m candidates $C = \{c_1, \ldots, c_m\}$ and a set of n voters $V = \{1, \ldots, n\}$. Each voter i has a *preference order* (i.e., a ranking) over C, which we denote by a_i. For notational convenience in comparing candidates, we will sometimes use $\succ_i$ instead of a_i. When $c_k \succ_i c_j$ for some $c_k, c_j \in C$, we say that voter i prefers c_k to c_j.

At an election, each voter submits a preference order b_i, which does not necessarily coincide with a_i. We refer to b_i as the vote or ballot of voter i. If we denote by $\mathcal{L}(C)$ the space of all linear orderings over C, then the vector of submitted ballots $\mathbf{b} = (b_1, \ldots, b_n) \in \mathcal{L}(C)^n$ is called a *preference profile*. An *election* is then determined by a pair $(C, \mathbf{b})$. At a profile $\mathbf{b}$, voter i has voted truthfully if $b_i = a_i$. Any other vote from i will be referred to as a non-truthful vote. Similarly the profile $\mathbf{a} = (a_1, \ldots, a_n)$ is the *truthful preference profile*, whereas any other profile is a non-truthful one.

A Basic Game-Theoretic Model. The obvious approach is to view the voters as players whose strategy space is $\mathcal{L}(C)$. It is convenient to associate with each voter i and preference order a_i, a utility function $u_i : C \to \mathbb{R}$. This means that if, e.g., candidate c_j is elected, then voter i derives a utility of $u_i(c_j)$. The specific numerical values of the utility functions are not important as long as the functions are consistent with the truthful vote of each voter. That is, we require $u_i(c_k) \neq u_i(c_j)$ for every $i \in V, c_j, c_k \in C$, and also if $u_i(c_k) > u_i(c_j)$, then $c_k \succ_i c_j$ and vice versa.

Consider now a voting rule $f : \mathcal{L}(C) \to C$ (we consider single-winner elections). The most natural way to define a voting game is to consider that each voter i derives a utility of $u_i(f(\mathbf{b}))$, when $\mathbf{b}$ is the submitted profile. Thus the payoff function of each player when his real preference is a_i will be:

$$p_i(a_i, f(\mathbf{b})) = u_i(c_j), \text{ if } c_j = f(\mathbf{b})$$

We refer to this as the basic model. Given a profile $\mathbf{b}$, we say that a vote $b'_i \in \mathcal{L}(C)$ is a *profitable deviation* of voter i from $\mathbf{b}$, if $p_i(a_i, f(b'_i, \mathbf{b}_{-i})) > p_i(a_i, f(\mathbf{b}))$. A profile $\mathbf{b}$ is then a Nash equilibrium if for every $i \in V$, no profitable deviation exists from $\mathbf{b}$.

There are various problematic issues regarding the equilibria of the basic model, as identified in the Introduction. As a result, it has not received much attention to date.

The Model We Study. Instead, we focus on a slight but powerful modification that was introduced recently. The main idea is that since strategizing always incurs some cost (e.g. cost in time, or effort, for finding how to deviate optimally), voters have a slight preference for voting truthfully when they cannot unilaterally affect the outcome of the election. The twist that was used in order to capture this is that there is always a small extra gain in the payoff function by voting truthfully. This extra gain is small enough so that voters may still prefer to be non-truthful in cases where they can affect the outcome, see e.g. [2,6,11]. Formally, let $\epsilon < \min_{i \in [n], j, k \in [m]} |u_i(c_j) - u_i(c_k)|$. If $\mathbf{a}$ is the real profile and $\mathbf{b}$ is the submitted one, then the payoff function of voter i is given by:

$$p_i(a_i, f(\mathbf{b})) = \begin{cases} u_i(c_j), & \text{if } c_j = f(\mathbf{b}) \wedge a_i \neq b_i, \\ u_i(c_j) + \epsilon, & \text{if } c_j = f(\mathbf{b}) \wedge a_i = b_i. \end{cases} \quad (1)$$

We denote such a game instance by $G(C, \mathbf{a})$. One can now see that with these new payoff functions, voters have an incentive to tell the truth if they cannot change the outcome. As a result, this model eliminates some undesirable Nash equilibria of the basic model, e.g., the bad equilibrium where all voters would vote for a candidate who is ranked last by everybody is no longer a Nash equilibrium.

An even further refinement is achieved with the concept of *strong Nash equilibrium*. Given a profile $\mathbf{b}$, a deviation by a coalition $S \subseteq V$ is given by a vector $\mathbf{b}'_S \in \mathcal{L}(C)^{|S|}$, where $b'_i \neq b_i$ for at least one member of S. Such a deviation is profitable for S if all its members are strictly better off. Hence a strong Nash equilibrium is a profile where there is no coalition with a profitable deviation. This notion is of interest in voting theory since voters may often choose to form coalitions to manipulate the election.

Plurality Voting. Throughout this work, the rule f is taken to be the Plurality rule, along with lexicographic tie-breaking. This is one of the most basic and well-studied voting rules, where the winner is the person with the maximum number of votes that ranked him as a first choice. In case of ties, we assume without loss of generality that tie-breaking is resolved by the linear order $c_1 \succ c_2 \succ ... \succ c_m$.

Given a voter $p \in V$, and his vote b_p under a profile $\mathbf{b}$, we denote by $top(b_p)$ the top choice of the vote. We will use repeatedly the following quantities:

Definition 1. *Given a preference profile* $\mathbf{b}$, *we define*

1. $N_i(\mathbf{b}) = \{p \in V : top(b_p) = c_i\}$, *the set of voters who voted for* c_i *in* $\mathbf{b}$,
2. $N_S(\mathbf{b}) = \{p \in V : top(b_p) \in S\}$
3. $sc(c_i, \mathbf{b}) = |N_i(\mathbf{b})|$, *the number of supporters of* c_i *in* $\mathbf{b}$,
4. $n_i = sc(c_i, \mathbf{a}) = |N_i(\mathbf{a})|$, *the number of supporters of* c_i *at the truthful profile* $\mathbf{a}$.

3 Analysis of Pure Nash Equilibria

Due to lack of space, all proofs along with illustrative examples regarding the results of Section 3 and Section 4 are deferred to the full version of this work.

Before we embark on the study of pure Nash equilibria under truth-biased agents, we note that unlike the basic model, the refinement can cause some games to not admit an equilibrium, as already identified in [11]. Despite this fact however, the extensive simulations in [11] have shown that most of the games they produced had at least one equilibrium. In fact they have also observed in their simulations that the estimated probability of a uniformly at random chosen instance having an equilibrium goes to 1, as the number of voters increases. Hence we do not consider this issue a major concern.

We introduce below some auxiliary notation that we use in this Section.

Definition 2. *Given a preference profile* $\mathbf{b}$, *we define*

- $\mathcal{W}(\mathbf{b}) = \{c_i \in C : sc(c_i, \mathbf{b}) = \max_{j \in C} sc(c_j, \mathbf{b})\}$. *This is the set of candidates who attained the maximum score in* $\mathbf{b}$. *We refer to* $\mathcal{W}(\mathbf{b})$ *as the* winning set *and if* $|\mathcal{W}(\mathbf{b})| > 1$, *the winner is determined from the tie-breaking rule.*
- $\mathcal{H}(\mathbf{b}) = \{c_i \in C : sc(c_i, \mathbf{b}) = \max_{j \in C} sc(c_j, \mathbf{b}) - 1\}$. *We refer to this set of candidates as the* chasing set.

3.1   When Is the Truthful Profile a Nash Equilibrium?

We start our analysis by identifying necessary and sufficient conditions that make truthful voting a Nash equilibrium. Clearly the stability of the truthful profile $\mathbf{a}$, can be threatened either by members of $\mathcal{W}(\mathbf{a})$, other than the winner, or by the members of $\mathcal{H}(\mathbf{a})$ if it is non-empty. This leads to the cases described below.

Theorem 1. *Consider a game* $G(C, \mathbf{a})$, *and let* $c_i = f(\mathbf{a})$ *be the winner under* $\mathbf{a}$. *Then* $\mathbf{a}$ *is a Nash equilibrium if and only if none of the following conditions hold.*

(1) $|\mathcal{W}(\mathbf{a})| > 1$ *and there exists a candidate* $c_j \in \mathcal{W}(\mathbf{a})$ *and a voter* p *such that* $c_j \succ_p c_i$ *and* $c_j \neq top(a_p)$;
(2) $|\mathcal{H}(\mathbf{a})| \geq 1$ *and there exists a candidate* $c_j \in \mathcal{H}(\mathbf{a})$ *and a voter* p *such that* $c_j \succ_p c_i$, $c_j \neq top(a_p)$, *and* $c_j \succ c_i$ *in the tie-breaking linear order.*

Remark 1. Theorem 1 still holds if we use any other deterministic tie-breaking rule instead of the lexicographic one. Condition (2) would now require that tie-breaking favors c_j in the winning set $\mathcal{W}(b_p, \mathbf{a}_{-p})$, where $top(b_p) = c_j$.

3.2   Properties of Non-truthful Nash Equilibria

As we will see, there can be many instances that have non-truthful profiles as Nash equilibria. However, the model and in particular the definition of our payoff function imposes some restrictions on how people can lie at equilibrium profiles. The purpose of this subsection is to identify some important properties that hold for such equilibrium profiles. None of the properties we establish here hold for the basic model, hence the properties demonstrate a clear distinction between the basic model and our model and help us understand the elimination of undesirable equilibria that takes place.

We describe below a running example that will provide some intuition throughout this subsection.

Example 1. In Figure 1, consider the truthful profile **a**, shown in Subfigure 1(a). This is not a Nash equilibrium, since voter 5 can vote for c_2 and make c_2 a winner, a more preferred outcome for voter 5. However, in Subfigure 1(b), we can see that if voter 5 changes his vote to rank c_2 first, the resulting profile is an equilibrium. Finally in Subfigure 1(c), we see that if candidate c_2 collects even more votes, by having voter 6 also support him, then the resulting profile is not an equilibrium any more. The reason is that voter 5 or 6 would be better off by ϵ if they stick to their truthful vote, since c_2 will get elected anyway (i.e., this suggests that at an equilibrium, candidate c_2 should not need more than just the necessary number of votes to get elected).

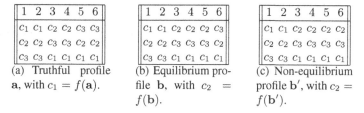

(a) Truthful profile **a**, with $c_1 = f(\mathbf{a})$.   (b) Equilibrium profile **b**, with $c_2 = f(\mathbf{b})$.   (c) Non-equilibrium profile **b'**, with $c_2 = f(\mathbf{b'})$.

Fig. 1. An example with a non-truthful Nash equilibrium

The first property we identify is intuitively very clear and close to what we would expect in real-life scenarios; at an equilibrium profile, people who lie are among the set of people who voted the elected candidate as their first choice. There should be no point in lying otherwise and be in equilibrium (this is unlike the basic model where people may lie in favor of some candidate who does not get elected and still be in equilibrium).

Lemma 1. *Suppose that* **b** $\neq$ **a** *is a non-truthful profile, which is a Nash equilibrium. Let* $c_j = f(\mathbf{b})$. *Then all non-truthful votes in* **b** *have* c_j *as a top candidate.*

The next property, which we will use repeatedly in the sequel, is an important observation about the structure of non-truthful equilibrium profiles. It further highlights the differences with the basic model and it is also very useful for the characterizations we obtain in Subsection 3.3. To state the property, we need first the following definition:

Definition 3. *Given a profile* **b** *with* $c_j = f(\mathbf{b})$, *a candidate* $c_k \neq c_j$ *is called a threshold candidate with respect to* **b** *when the following condition holds:*

(1) if $c_k \succ c_j$, *then* $sc(c_k, \mathbf{b}) = sc(c_j, \mathbf{b}) - 1$,
(2) otherwise $sc(c_k, \mathbf{b}) = sc(c_j, \mathbf{b})$;

Hence a threshold candidate is someone who could win the election if he had one additional vote. As we show below, a feature of all non-truthful equilibria is that there must exist at least one threshold candidate. The intuition for this is that since in our model voters who are not pivotal prefer to vote truthfully, then any equilibrium that arises from manipulation should provide just enough votes to the winner so as to beat the required threshold (as provided by the threshold candidate) and not any more. Hence there cannot be a non-truthful equilibrium where the winner wins by a large margin from the rest of the candidates. This is evident in Example 1, in Subfigures 1(b) and 1(c). Clearly there can be truthful equilibria where the winner wins by a large margin.

Lemma 2. *Consider a game $G(C, \mathbf{a})$, and suppose that $\mathbf{b} \neq \mathbf{a}$ is a Nash equilibrium. Then there always exists at least one threshold candidate c_k with respect to $\mathbf{b}$. Additionally, it holds that $N_k(\mathbf{a}) = N_k(\mathbf{b})$, thus $sc(c_k, \mathbf{b}) = n_k$.*

Remark 2. It is not always the case that the winner of $\mathbf{a}$ is a threshold candidate in an equilibrium $\mathbf{b}$.

3.3   Characterization Results

This subsection contains necessary and sufficient conditions for the existence of equilibria with a specified candidate as a winner. This yields a full characterization of games that admit some Nash equilibrium.

We start first with the toy case of elections with two candidates, which turn out to always have a unique Nash equilibrium.

Theorem 2. *In any game $G(C, \mathbf{a})$ with 2 candidates, truthful voting is a dominant strategy, and $\mathbf{a}$ is the unique Nash equilibrium.*

We consider now elections with at least 3 candidates. We deal first with equilibria where the specified winner is the truthful winner.

Theorem 3. *Consider a game $G(C, \mathbf{a})$, and let $c_i = f(\mathbf{a})$. If there is a Nash equilibrium with c_i as the winner, then $\mathbf{a}$ is the unique such equilibrium and its existence is determined by the necessary and sufficient conditions of Theorem 1.*

From now on, and for the rest of this subsection, fix a preference profile $\mathbf{a}$, with $c_i = f(\mathbf{a})$, and fix also a candidate $c_j \neq c_i$. We want to understand when can there exist a non-truthful equilibrium $\mathbf{b}$ with c_j being the winner. This question cannot have a simple answer since as we show below, it is an NP-complete problem.

Theorem 4. *Consider a game $G(C, \mathbf{a})$, with $c_i = f(\mathbf{a})$ and let $c_j \neq c_i$. Given a score s, deciding if there exists an equilibrium $\mathbf{b}$, with $c_j = f(\mathbf{b})$ and $sc(c_j, \mathbf{b}) = s$, is NP-complete.*

We note here that in our reduction both the number of candidates and the number of voters is non-constant. For a constant number of voters, the problem becomes polynomial time solvable (since one can check all possible configurations of votes in favor of a given candidate c_j). The complexity of the problem when the number of candidates is constant is still unknown.

Despite the NP-hardness, one can still try to obtain characterization results, so as to gain more insights into the difficulty of the problem. To do this, we will utilize the lemmas and intuitions from Subsection 3.2. We first have to understand what values for s can yield an equilibrium $\mathbf{b}$ with $sc(c_j, \mathbf{b}) = s$. One thing we can immediately deduce for example is that s has to belong to the interval $[n_j, n_i + 1]$ (obviously $s \geq n_j$, and the upper bound is by having in worst case c_i as a threshold candidate). We also have to determine which voters decide to non-truthfully support c_j at equilibrium, instead of their top candidate. In light of Lemma 1, we know that there should be exactly $s - n_j$ such voters. Finally, in light of Lemma 2, we need to determine the set of threshold candidates in such an equilibrium (note that these are candidates whose supporters vote truthfully for them at the equilibrium, by Lemma 2).

Given this discussion, we will focus first on determining which candidates could be eligible for being threshold candidates at an equilibrium. Building on Definition 3, we define below the notion of an *s-eligible threshold set*, for a given candidate c_j and a winning score of s.

Definition 4. *Given a game $G(C, \mathbf{a})$, fix a score $s \in [n]$ and a candidate $c_j \in C$. A non-empty set $T \subset C$ is an s-eligible threshold set with respect to c_j, if it can be decomposed into two subsets $T = T^1 \cup T^2$, such that the following conditions hold:*

(i) *For every $c_k \in T^1$, it holds that $n_k = s$ and $c_j \succ c_k$.*
(ii) *For every $c_k \in T^2$ it holds that $n_k = s - 1$ and $c_k \succ c_j$.*
(iii) *For every voter $p \in V$, we have $c_j \succ_p c_k$, $\forall c_k \in T \setminus \{top(a_p)\}$.*

To obtain some intuition about this definition, conditions (i) and (ii) simply correspond to the set of possible threshold candidates at some equilibrium, as in Definition 3. Note that we define these candidates with respect to the real profile $\mathbf{a}$, and look at their score $n_k = sc(c_k, \mathbf{a})$. This is not an issue, because in any equilibrium $\mathbf{b}$, where c_k is a threshold candidate, we know by Lemma 2 that $sc(c_k, \mathbf{a}) = sc(c_k, \mathbf{b})$. Finally, condition (iii) simply ensures stability: in order for T to be a potential set of threshold candidates, every voter should prefer the winner c_j to any candidate from T (except if a member of T is his top choice). Otherwise, some voter would have an incentive to vote for such a candidate from T and we would not have an equilibrium.

In the analysis below, we will often need to argue about candidates from the set $M_{\geq s} = \{c_k \in C | n_k \geq s\}$. This set arises naturally in the analysis, since in any equilibrium where the winning score is s, there must be non-truthful voters whose real preferences were candidates from $M_{\geq s}$.

To continue, we consider two cases for the realization of threshold candidates.

Equilibria for Threshold Sets with $T^1 = \emptyset$. Given c_j and the possible score s, let T be an s-eligible threshold set, with $T^1 = \emptyset$, i.e., $T := T^2$. We will characterize when can there be an equilibrium, such that T is precisely the set of all threshold candidates. For this we establish first some properties that have to hold at equilibrium.

Lemma 3. *Given c_j and s, let T be an s-eligible threshold set w.r.t. c_j, with $T^1 = \emptyset$. Suppose that $\mathbf{b} \neq \mathbf{a}$ is a Nash equilibrium such that $c_j = f(\mathbf{b})$, and T is the set of all threshold candidates in $\mathbf{b}$. Then*

(a) *The only candidate who has s points in $\mathbf{b}$ is c_j;*
(b) *A candidate that has $s - 1$ points in $\mathbf{b}$, either belongs to T^2 or is beaten by c_j under tie-breaking (and hence beaten also by all candidates in T^2).*

We now provide some upper bounds on the scores of the members of $M_{\geq s}$. Define the set $M^1 = \{c_\ell \in M_{\geq s} : \exists c_k \in T^2 \text{ with } c_k \succ c_\ell\}$. Let also $M^2 = M_{\geq s} \setminus M^1$.

Lemma 4. *Under the same assumptions, as in Lemma 3,*

(a) $sc(c_\ell, \mathbf{b}) \leq s - 2$, $\forall c_\ell \in M^1$,
(b) $sc(c_\ell, \mathbf{b}) \leq s - 3$, $\forall c_\ell \in M^2$.

In order to complete the characterization, we also have to argue about candidates who have $s - 1$ or $s - 2$ points in the truthful profile, as they may affect stability too. To this end, in analogy to the sets M^1 and M^2, we define the sets: $U^1 = \{c_\ell \in C : n_\ell = s - 1, \exists c_k \in T^2 \text{ with } c_k \succ c_\ell\}$, $U^2 = \{c_\ell \in C : n_\ell = s - 1, c_\ell \succ c_k \forall c_k \in T^2\}$. Finally we will also need the set $U^3 = \{c_\ell \in C : n_\ell = s - 1, c_j \succ c_\ell, \text{ or } n_\ell = s - 2, c_\ell \succ c_k \forall c_k \in T^2\}$.

The theorem below provides an iff condition for determining existence of an equilibrium with c_j as a winner. The conditions boil down to finding the correct "bookkeeping" for the $s - n_j$ non-truthful supporters of c_j, i.e., determining a lower bound on how much other candidates have to lose from their real supporters.

Theorem 5. *Given c_j and s, let T be an s-eligible threshold set w.r.t. c_j, such that $T^1 = \emptyset$. There exists a non-truthful Nash equilibrium $\mathbf{b}$ with $c_j = f(\mathbf{b})$, $sc(c_j, \mathbf{b}) = s$, and such that T is the set of all threshold candidates in $\mathbf{b}$, if and only if there exists a pair of sets (D, R) with $D \subseteq V \setminus N_T(\mathbf{a})$, $|D| = s - n_j$, $R \subseteq U^3$, such that:*

(i) for every $c_\ell \in M^1$, $|D \cap N_\ell(\mathbf{a})| \geq n_\ell - s + 2$;
(ii) for every $c_\ell \in M^2$, $|D \cap N_\ell(\mathbf{a})| \geq n_\ell - s + 3$;
(iii) for every $c_\ell \in U^1$, $|D \cap N_\ell(\mathbf{a})| \geq 1$;
(iv) for every $c_\ell \in U^2$, $|D \cap N_\ell(\mathbf{a})| \geq 2$;
(v) for every $(p, c_k) \in D \times R$, it holds that $c_j \succ_p c_k$;
(vi) for every $c_\ell \in U^3 \setminus R$, $|D \cap N_\ell(\mathbf{a})| \geq 1$.

Equilibria for Threshold Sets with $T^1 \neq \emptyset$. We now provide analogous results for the case that we are given an s-eligible threshold set T, with $T^1 \neq \emptyset$. In analogy to M^1, we define the set $K^1 = \{c_\ell \in M_{\geq s} : \exists c_k \in T^1 \text{ with } c_k \succ c_\ell\}$. Let also $K^2 = M_{\geq s} \setminus K^1$.

Lemma 5. *Given c_j and s, let T be an s-eligible threshold set w.r.t. c_j, with $T^1 \neq \emptyset$. Suppose that $\mathbf{b} \neq \mathbf{a}$ is a Nash equilibrium such that $c_j = f(\mathbf{b})$ and T is the set of all threshold candidates in $\mathbf{b}$. Then the set of candidates who have s points in $\mathbf{b}$ is $\{c_j\} \cup T^1$.*

Lemma 6. *Under the same assumptions as in Lemma 5,*

(a) $sc(c_\ell, \mathbf{b}) \leq s - 1$ $\forall c_\ell \in K^1$,
(b) $sc(c_\ell, \mathbf{b}) \leq s - 2$ $\forall c_\ell \in K^2$.

As in the previous case, in analogy to the sets U^1, U^2, U^3, here we need the sets: $W^1 = \{c_\ell \in C : n_\ell = s, \forall c_k \in T^1 c_\ell \succ c_k \text{ and } c_j \succ c_\ell\}$, $W^2 = \{c_\ell \in C : n_\ell = s, c_\ell \succ c_j\}$.

Theorem 6. *Consider a game $G(C, \mathbf{a})$ with $c_i = f(\mathbf{a})$, and a candidate $c_j \neq c_i$. Let T be an s-eligible threshold set with respect to c_j, with $T^1 \neq \emptyset$. There exists a Nash equilibrium $\mathbf{b} \neq \mathbf{a}$ such that $c_j = f(\mathbf{b})$, $sc(c_j, \mathbf{b}) = s$, and such that T is the set of all threshold candidates in $\mathbf{b}$, if and only if there exists a pair of sets (D, R) with $D \subseteq V \setminus N_T(\mathbf{a})$, $|D| = s - n_j$, $R \subseteq U^3$, such that:*

(i) for every $c_\ell \in K^1$, $|D \cap N_\ell(\mathbf{a})| \geq n_\ell - s + 1$;
(ii) for every $c_\ell \in K^2$, $|D \cap N_\ell(\mathbf{a})| \geq n_\ell - s + 2$;
(iii) for every $c_\ell \in W^2$, $|D \cap N_\ell(\mathbf{a})| \geq 1$;
(iv) for every $(p, c_k) \in D \times R$, it holds that $c_j \succ_p c_k$;
(v) for every $c_\ell \in W^1 \setminus R$, $|D \cap N_\ell(\mathbf{a})| \geq 1$.

Implications: Sufficient Conditions for Checking Existence. Eventually, we are interested in deciding whether there exists an equilibrium, where c_j is the winner with score s (independent of who are the threshold candidates). In the full version of this work[1], we show how to use the characterizations of Theorems 5 and 6 to derive a simple sufficient condition that is also polynomial time checkable. The condition essentially boils down to checking if the difference $s - n_j$ is within "reasonable" bounds. In fact, we can also establish non-existence for a large range outside these bounds. As a consequence, despite the NP-hardness result of Theorem 4, it is only for a relatively small range of s that we cannot have a polynomial time algorithm for checking existence.

4   Strong Nash Equilibria

For the basic game-theoretic model, a restricted version of the concept of strong equilibrium has been studied in [7], where characterizations were obtained for the case of 3 candidates. In our model, we obtain a complete characterization for an arbitrary number of candidates and voters. We also identify some connections with Condorcet winners. Our results demonstrate that strong Nash equilibria have an even more restricted structure than pure Nash equilibria and manage to further refine the set of stable outcomes (whenever they exist).

4.1   Truthful Strong Nash Equilibrium

We start by characterizing the profiles where $\mathbf{a}$ is a strong Nash equilibrium. In the following, we denote by $N_{i \succ j}$ the set $\{p \in V | c_i \succ_p c_j\}$ (i.e., the definition is with respect to the truthful profile $\mathbf{a}$). Even though at first sight, one may think that we should look at an exponential number of coalitional deviations to check if $\mathbf{a}$ is a strong Nash equilibrium, it turns out that we need to check only a small number of conditions, and therefore it can be done quite efficiently.

Theorem 7. *Consider a game $G(C, \mathbf{a})$ with $c_i = f(\mathbf{a})$. Then $\mathbf{a}$ is a strong Nash equilibrium if and only if the following condition holds: for any candidate c_j with $c_j \succ c_i$ we have $|N_{j \succ i} \setminus N_j(\mathbf{a})| < n_i - n_j$ and for any candidate c_j with $c_i \succ c_j$ we have $|N_{j \succ i} \setminus N_j(\mathbf{a})| \leq n_i - n_j$.*

4.2   Characterization Results and Relations to Condorcet Winners

We start this subsection with a characterization of existence of strong Nash equilibria. To characterize the existence of strong equilibria with a certain candidate c_j as a winner,

---

[1] Available at the authors' websites.

one needs to distinguish the various special cases that may arise regarding coalitional deviations. As a result, the characterization comes in two parts. We present in the theorem below the characterization in the first out of the two cases, namely when c_j beats the truthful winner $c_i = f(\mathbf{a})$ in tie-breaking. Hence, suppose that $c_j \succ c_i$. We eventually need to argue about the following set in our analysis:

$$T = \{c_\ell | c_j \succ c_\ell \succ c_i \text{ and } |N_{\ell \succ j}| \geq n_i\}$$

Theorem 8. *Consider a game $G(C, \mathbf{a})$, with $c_i = f(\mathbf{a})$, and suppose $c_j \succ c_i$. There is no strong Nash equilibrium $\mathbf{b} \neq \mathbf{a}$ with $c_j = f(\mathbf{b})$ if and only if at least one of the following conditions holds.*

 (i) $n < 2n_i$.
 (ii) *There exists a voter in $V \setminus N_i(\mathbf{a})$ who prefers c_i to c_j.*
 (iii) *There exists a candidate c_ℓ such that $|N_{\ell \succ j}| \geq n_i$ and $c_\ell \succ c_j$.*
 (iv) *There exists a candidate c_ℓ such that $|N_{\ell \succ j}| \geq n_i + 1$ and $c_i \succ c_\ell$.*
 (v) *There exists a candidate c_ℓ such that $|N_{\ell \succ j}| \geq n_i + 1$ and $c_j \succ c_\ell \succ c_i$.*
 (vi) $|N_{j \succ i} \cap (\bigcap_{\ell \in T} N_{j \succ \ell})| < n_i$.

An analogous theorem deals with the other case, which we omit due to lack of space. Note that despite the large number of conditions to check in this characterization, they are all verifiable in polynomial time. Hence we have the following corollary.

Corollary 1. *Given a game $G(C, \mathbf{a})$, we can decide in polynomial time if a strong Nash equilibrium exists with a certain candidate as a winner.*

We end this section with the following observation, which shows some interesting connections between strong Nash equilibria and Condorcet winners. This fact can be derived as a special case of the models studied in [10].

Theorem 9. *([10]) If there exists a strong Nash equilibrium $\mathbf{b}$ with c_j as the winner, then c_j is a Condorcet winner.*

Remark 3. The opposite direction of Theorem 9 is not true.

Finally the next corollary shows that we cannot have too many different strong Nash equilibria. Hence the notion of strong Nash equilibrium provides a quite powerful refinement on the set of stable profiles.

Corollary 2. *Given a game $G(C, \mathbf{a})$, the winner in all strong Nash equilibria is the same. Also, if the truthful profile $\mathbf{a}$ is a strong Nash equilibrium then it is the unique strong Nash equilibrium for this game.*

5   Conclusions

We have provided a theoretical analysis for Plurality voting under the truth-biased game-theoretic model of [2,6,11]. Our results complement the empirical work of Thompson, *et al.*, in that they both support truth-bias as an effective method of equilibrium selection. In particular, we have exhibited that certain undesirable equilibria are now filtered out. Finally, we also characterized the set of strong Nash equilibria. Together, truth-bias and strong Nash make for a very strong equilibrium refinement as illustrated in Section 4. One should also be aware however, that the cost we have to pay for such a strong refinement is that there are instances where no equilibrium exists.

There are plenty of avenues for future research. A challenging question is to find other refinements where existence is always guaranteed. Another natural direction is to further exploit the idea of rewarding truthfulness, extending it to other voting rules. Finally, one more interesting idea is to further enrich our model of truth-bias, so that when voters vote non-truthfully, their utility can depend on how far their vote is from their truthful preference. Any notion of distance could be used here.

References

1. Desmedt, Y., Elkind, E.: Equilibria of plurality voting with abstentions. In: Proceedings of the 11th ACM Conference on Electronic Commerce (EC), Cambridge, Massachusetts, pp. 347–356 (June 2010)
2. Dutta, B., Laslier, J.F.: Costless honesty in voting. Presentation in 10th International Meeting of the Society for Social Choice and Welfare, Moscow (2010)
3. Farquharson, R.: Theory of Voting. Yale University Press (1969)
4. Gibbard, A.: Manipulation of voting schemes. Econometrica 41(4), 587–602 (1973)
5. Lev, O., Rosenschein, J.S.: Convergence of iterative voting. In: Proceedings of the 11th International Coference on Autonomous Agents and Multiagent Systems (AAMAS), Valencia, Spain, vol. 2, pp. 611–618 (June 2012)
6. Meir, R., Polukarov, M., Rosenschein, J.S., Jennings, N.R.: Convergence to equilibria of plurality voting. In: Proceedings of the 24th National Conference on Artificial Intelligence (AAAI), Atlanta, pp. 823–828 (July 2010)
7. Messner, M., Polborn, M.K.: Single transferable vote resists strategic voting. International Journal of Game Theory 8, 341–354 (1991)
8. Myerson, R.B., Weber, R.J.: A theory of voting equilibria. The American Political Science Review 87(1), 102–114 (1993)
9. Satterthwaite, M.A.: Strategy-proofness and Arrow's conditions: Existence and correspondence theorems for voting procedures and social welfare functions. Journal of Economic Theory 10(2), 187–217 (1975)
10. Sertel, M.R., Sanver, M.: Strong equilibrium outcomes of voting games are the generalized Condorcet winners. Social Choice and Welfare 22, 331–347 (2004)
11. Thompson, D.R.M., Lev, O., Leyton-Brown, K., Rosenschein, J.: Empirical analysis of plurality election equilibria. In: The 12th International Conference on Autonomous Agents and Multiagent Systems (AAMAS), Saint Paul, Minnesota, USA (2013)
12. Xia, L., Conitzer, V.: Stackelberg voting games: Computational aspects and paradoxes. In: Proceedings of the 24th National Conference on Artificial Intelligence (AAAI), Atlanta, Georgia, USA, pp. 805–810 (2010)

Reliability Weighted Voting Games

Yoram Bachrach[1] and Nisarg Shah[2]

[1] Microsoft Research Cambridge, UK
yobach@microsoft.com
[2] Carnegie Mellon University, USA
nkshah@cs.cmu.edu

Abstract. We examine agent failures in weighted voting games. In our cooperative game model, R-WVG, each agent has a weight and a survival probability, and
the value of an agent coalition is the probability that its surviving members would
have a total weight exceeding a threshold. We propose algorithms for computing the value of a coalition, finding stable payoff allocations, and estimating the
power of agents. We provide simulation results showing that on average the stability level of a game increases as the failure probabilities of the agents increase.
This conforms to several recent results showing that failures increase stability in
cooperative games.

Keywords: Cooperative game theory, Weighted voting game, Reliability extension, Agent failures, Stability.

1 Introduction

Consider several firms collaborating to complete a joint project. The project requires
a threshold amount of a certain resource to be completed successfully, and each firm
owns a different amount of the resource that it can contribute to the project. If enough
firms commit their resources so that the total contributed amount is at least the threshold amount, the project is completed and generates a certain revenue. One important
question is how this revenue should be distributed among the participating firms. Traditionally, such domains were modelled as Weighted Voting Games (WVGs), and various
game theoretic solution concepts were used for revenue distribution (see [16]).

However, in the real world, a firm may promise to deliver resources but fail to do so
afterwards, or its delivered resources may fail during the execution of the project, due to
reasons beyond the firm's control. In this case, the project can only be completed if the
total amount of resources that did *not* fail exceeds the threshold. One might suggest to
model this as a WVG among the firms which successfully delivered working resources,
and distribute the revenue only among these firms. However, this might deter some
firms from participating in the first place, since due to such failures they may not get
paid even after exerting effort to deliver resources. One way to circumvent this is using
an ex-ante contract to divide the revenue (generated only if the project finishes) that is
independent of which firms eventually failed and which did not.

Another similar domain is the case of *lobbying agents* in parliamentary settings [26,9],
where the agents exert lobbying efforts to convince parties with different weights (e.g.,

B. Vöcking (Ed.): SAGT 2013, LNCS 8146, pp. 38–49, 2013.

the number of seats) to vote for a new legislation, but may fail to do so with a certain probability. Again, the agents might prefer an ex-ante contract for payoff division to avoid not being paid ex-post for their persuasion efforts in case they fail.

Clearly, such domains require explicit modeling of agent failures. Although failures were widely studied in *non cooperative game theory* [7,23,22], such analysis has surprisingly ignored the prominent WVGs model from cooperative game theory.

We study the effect of agent failures on the solutions of WVGs using the recently proposed reliability extension model [3]. The heart of a cooperative game is the *characteristic function* which maps every agent subset to the utility the agents achieve together. Under specified agent survival probabilities, and assuming such failures are independent, the reliability extension modifies the characteristic function and maps every agent subset to its *expected* value. We examine the reliability extension of WVGs, which we denote "R-WVGs" (Reliability Weighted Voting Games). We analyze how the reliability extension changes the outcome in WVGs, as captured by solutions such as the Shapley value [24] and the core [19], providing both theoretical and empirical results.

Our Contribution: The contribution of this paper is threefold. First, we contrast the computational hardness of various solution concepts in WVGs with that in R-WVGs. While the problems of computing the value of a coalition, testing emptiness of the core, and checking if a given imputation is in the core are in $\mathcal{P}$ for WVGs, we prove they are $\#\mathcal{P}$-hard and co$\mathcal{NP}$-hard for R-WVGs. For computing the value of a coalition, we provide an exact dynamic programming algorithm, as well as a polynomial time additive approximation method. We show that the latter two problems remain hard even if only one agent may fail. We develop an algorithm to compute a core imputation for R-WVGs with constantly many unreliable agents and small weights. Second, the hardness of computing power indices (the Shapley value and the Banzhaf index) in R-WVGs follows from the hardness in WVGs. We develop dynamic programming algorithms for computing these indices in restricted R-WVGs. Third, we provide simulation results for R-WVGs which indicate that, *on average*, lower reliability levels of agents lead to higher stability of the game, as measured by the probability of having a non-empty core, the least-core value, and the Cost of Stability.

1.1 Related Work

Computational aspects of cooperative game theory have recently received much attention. The problems of testing emptiness of the core and finding a core imputation have been investigated for many cooperative games, ranging from network games [4,1] through combinatorial games [14,10] to general representation languages [13]. The core is easy to analyze in simple games (including WVGs), as it is closely related to the existence of veto players (see [12]). However, R-WVGs are not simple games, and as our analysis shows, questions regarding the core are computationally harder in R-WVGs. While emptiness/non-emptiness of the core is a qualitative measure of stability, the least-core value [25] and the Cost of Stability (CoS) [5] serve as its quantitative generalizations; we use all three of them as stability measures. As weighted voting games (WVGs) also model decision making bodies [16,12], computing power indices (the Shapley value [24] and the Banzhaf index [8]) is a central question. Due to the hardness of computing them in WVGs [21], approximations have been proposed [17,2].

Failures were investigated in non-cooperative game theory, in congestion games [23,22], reliable network formation [7], Nash networks [11] and sensor networks [20], but received less attention in *cooperative* games. We used the recent reliability extension model of [3], and applied it to WVGs to obtain the class of R-WVGs.

Another recent paper [6] examines the core in the reliability extension of totally-balanced games, showing that in such games agent failures only help stability in terms of non-emptiness of the core. While the general theme is in accordance with our results, their analysis is irrelevant for R-WVGs as R-WVGs are not totally-balanced.

2   Preliminaries

A transferable utility cooperative game $G = (N, v)$ is composed of a set of agents $N = \{1, 2, \ldots, n\}$ and a characteristic function $v : 2^N \to \mathbb{R}$ mapping any coalition (agent subset) to the utility these agents achieve together. By convention, $v(\emptyset) = 0$. For an agent $i \in N$ and a coalition $S \subseteq N$, we denote $S \cup \{i\}$ by $S + i$ and $S \setminus \{i\}$ by $S - i$. A game is called *simple* if the characteristic function only takes values of 0 or 1, so $v : 2^N \to \{0, 1\}$. In simple games, a coalition $C \subseteq N$ is called winning if $v(C) = 1$, and losing otherwise.

Weighted Voting Games (WVGs): A WVG is a game where each agent $i \in N$ has a weight $w_i > 0$, and a coalition $C \subseteq N$ is winning iff its total weight exceeds a given threshold t: if $\sum_{i \in C} w_i \geq t$ then $v(C) = 1$, else $v(C) = 0$.

The Core: The characteristic function defines the value that a coalition achieves, but not how it should be *distributed* among its members. A payment vector $\mathbf{p} = (p_1, \ldots, p_n)$ is an *imputation* if $\sum_{i=1}^{n} p_i = v(N)$ (efficiency) and $p_i \geq v(\{i\})$ for every $i \in N$ (individual rationality). Here, p_i is the payoff to agent i, and the payoff to a coalition C is $p(C) = \sum_{i \in C} p_i$. The *core requirement* is that the payoff to every coalition is at least as much as it can gain on its own, so no coalition can gain by defecting from the grand coalition of all the agents. The *core* [19] is defined as the set of all imputations $\mathbf{p}$ such that $p(N) = v(N)$ and $p(S) \geq v(S)$ for all $S \subseteq N$.

The ϵ-Core: The definition of the core is quite demanding; many games of interest have empty core. One popular relaxation to circumvent this is the ϵ-core [25]. For any ϵ, the ϵ-core is the set of all payoff vectors $\mathbf{p}$ such that $p(N) = v(N)$ and $p(S) \geq v(S) - \epsilon$ for all $S \subseteq N$. One way to interpret this is that if a coalition incurs a cost of ϵ for deviating from the grand coalition, then the imputation is stable. Higher deviation cost makes it easier to find a stable imputation. For any game, the set $\{\epsilon \mid \text{the } \epsilon\text{-core is non-empty}\}$ has a minimum element $\epsilon_{\min}$, known as the *least core value* (LCV). The LCV is the minimal deviation cost admitting a stable enough payoff allocation. Higher LCV implies that the game is unstable.

The Cost of Stability: In games with an empty core, it is impossible to distribute the gains of the grand coalition N in a stable way. An external party can induce agent cooperation by offering a *supplemental payment* if the grand coalition is formed. Bachrach et. al. [5] formalized this as follows. Given a game $G = (N, v)$ and a supplemental payment $\Delta \in \mathbb{R}$, the *adjusted game* $G(\Delta) = (N, v')$, where the characteristic function is defined by: $v'(N) = v(N) + \Delta$ and $v'(S) = v(S)$ for all $S \neq N$. The *Cost of Stability* (CoS) of a game G, denoted $CoS(G)$, is the minimum supplement Δ^* for which the

core of the adjusted game $G(\Delta^*)$ is non-empty. The CoS quantifies the extent of instability by measuring the subsidy required to overcome agents' resistance to cooperation.
A higher CoS therefore indicates that the game is more unstable.

The Shapley Value: Power indices analyze the contributions of the agents to different
coalitions, proposing ways to divide the gains based on fairness criteria. The *marginal
contribution* of an agent $i \in N$ to a coalition $S \subseteq N - i$ is $v(S + i) - v(S)$. The
Shapley value is uniquely characterized by four important fairness axioms [15]. For
any permutation π of agents (i.e., $\pi : \{1, 2, \ldots, n\} \to \{1, 2, \ldots, n\}$ and π is onto),
let $\Gamma_i^\pi = \{j | \pi(j) < \pi(i)\}$ be the set of agents before i in π. The Shapley value is
the payoff vector $(\varphi_1, \ldots, \varphi_n)$, where φ_i is the Shapley value of agent i given by:
$\varphi(i) = \frac{1}{n!} \sum_{\pi \in S_n} (v(\Gamma_i^\pi + i) - v(\Gamma_i^\pi))$. For any coalition $S \subseteq N - i$, the number of
permutations $\pi \in S_n$ where $\Gamma_i^\pi = S$ is exactly $(|S|)! \cdot (n - |S| - 1)!$. Thus, summing
over coalitions, we get: $\varphi_i = \frac{1}{n!} \sum_{S \subseteq N-i} \left[(|S|)! \cdot (n - |S| - 1)! \cdot (v(S + i) - v(S)) \right]$.
The Banzhaf index is another prominent power index.

Reliability Games: A model for agent failures in cooperative games was proposed
in [3]. A *reliability game* $G = (N, v, \mathbf{r})$ consists of the set of agents $N = \{1, 2, \ldots, n\}$,
the *base characteristic function* $v : 2^N \to \mathbb{R}$ describing values in the absence of failures, and the reliability vector $\mathbf{r}$, where r_i is the probability of agent i surviving (i.e.,
not failing). The characteristic function $v^\mathbf{r}$ of the reliability game with failures now
considers the expected value of the survivors:

$$v^\mathbf{r}(S) = \sum_{S' \subseteq S} \Pr(S'|S) \cdot v(S') = \sum_{S' \subseteq S} \left(\prod_{i \in S'} r_i \cdot \prod_{j \in S \setminus S'} (1 - r_j) \right) \cdot v(S'). \quad (1)$$

Here, $\Pr[S'|S]$ is the probability that every agent in S' survives and every agent in $S \setminus S'$
fails. For the *base game* $G = (N, v)$, the game $G^\mathbf{r} = (N, v, \mathbf{r})$ is called the *reliability
extension* of G with the reliability vector $\mathbf{r}$. An agent is called fully reliable (or reliable)
if its reliability is 1, and unreliable otherwise.

3 Reliability Weighted Voting Games

In this paper, we examine the reliability extension of weighted voting games (R-WVGs).
Formally, an R-WVG $G^\mathbf{r} = (N, \mathbf{w}, t, \mathbf{r})$, where $N = \{1, \ldots, n\}$ is the set of agents,
$\mathbf{w} = (w_1, \ldots, w_n)$ is the vector of agent weights, t is the threshold (weight quota), and
$\mathbf{r}$ is the vector of agent survival probabilities. The characteristic function $v^\mathbf{r}$ is given by
Equation (1), where the base characteristic function follows $v(C) = 1$ if $\sum_{i \in C} w_i \geq t$
and $v(C) = 0$ otherwise.

We now discuss the complexity of computing various solutions in R-WVGs. We first
emphasize that R-WVGs are a generalization of WVGs: WVGs are recovered when all
agents are fully reliable, i.e., $\mathbf{r} = (1, \ldots, 1)$. Thus, solving R-WVGs is more demanding
than solving WVGs — any problem that is computationally hard for WVGs remains
hard in R-WVGs. For example, computing the Shapley value or the Banzhaf index in
WVGs is known to be $\#\mathcal{P}$-hard [21,16], and thus remains $\#\mathcal{P}$-hard even in R-WVGs.
However, certain prominent problems are easy for WVGs. For example, computing the

value of a coalition in a WVG is simple, as it only requires summing the weights of the members and testing whether the sum exceeds the threshold. Testing if the core is empty and checking if a given imputation is in the core are other examples of problems that are in $\mathcal{P}$ for WVGs. We show that all these problems become hard in R-WVGs.

Theorem 1. *Finding value of the grand coalition is #$\mathcal{P}$-hard in R-WVGs.*

Proof. We use a reduction from #$SUBSET$-SUM, the counting version of the subset sum problem. #$SUBSET$-SUM, which is known to be #$\mathcal{P}$-hard, requires counting the number of subsets of a given set S of positive integers that sum to another positive integer t. Take an instance (S,t) of #$SUBSET$-SUM. Let $|S| = n$. Create an R-WVG G_1 with n agents having elements of S as the weights, threshold t, and reliability vector $\mathbf{r} = (1/2, 1/2, \ldots, 1/2)$. Create another R-WVG G_2, which is identical to G_1 except the threshold in G_2 is $t + 1$. Let $v(G_1)$ and $v(G_2)$ denote the values of grand coalitions of G_1 and G_2 respectively.

Note that with the reliability vector $\mathbf{r} = (1/2, \ldots, 1/2)$, the value of the grand coalition is the fraction of coalitions having total weight at least as much as the threshold. Formally, let $\#^t S$ denote the number of subsets of S with total weight at least t. Then, $v(G_1) = (\#^t S)/2^n$ and $v(G_2) = (\#^{t+1} S)/2^n$. So if we can compute the value of the grand coalition in R-WVGs, we can compute $v(G_1)$ and $v(G_2)$, and obtain $\#^t S - \#^{t+1} S = 2^n \cdot (v(G_1) - v(G_2))$, which is the number of subsets of S that sum to exactly t, i.e., the answer to the #$SUBSET$-SUM instance (S,t).   □

Though it is hard to compute the value of a coalition, we can approximate it additively. Consider an R-WVG $G^{\mathbf{r}} = (N, \mathbf{w}, t, \mathbf{r})$ and any coalition $S \subseteq N$. To approximate $v^{\mathbf{r}}(S)$, run k iterations such that in each iteration, every agent $i \in S$ survives with probability r_i. Let C_t be the surviving sub-coalition in iteration t. Then, $v^{\mathbf{r}}(S) \approx \hat{V} = \frac{1}{k} \cdot \sum_{t=1}^{k} v(C_t)$. Using Hoeffding's inequality and Equation (1), it can be shown that $k = \frac{1}{2 \cdot \epsilon^2} \cdot \log(2/\delta)$ is sufficient to achieve $\Pr[|v^{\mathbf{r}}(S) - \hat{V}| > \epsilon] \leq \delta$.[1] Further, if the agent weights and the threshold are integers, we can use a dynamic programming approach to calculate the value exactly. For simplicity, consider the grand coalition.[2] Let $T(j, q)$ denote the value of the coalition $\{1, \ldots, j\}$ in the R-WVG where the threshold is changed to q. Then, $T(j, q) = r_j \cdot T(j-1, q - w_j) + (1 - r_j) \cdot T(j-1, q)$, where $T(0, q) = 0$ if $q > 0$ and for all j, $T(j, q) = 1$ if $q \leq 0$. Now $v(N) = T(n, t)$, which can be computed in time $O(t \cdot n)$.

4   The Core of R-WVGs

Next, we examine the complexity of finding core-related solutions. Checking if a given imputation is in the core, testing emptiness of the core, and finding a core imputation are computationally easy (in $\mathcal{P}$) for WVGs. We show that all of them are computationally hard for R-WVGs.

---

[1] This method works for the reliability extension of any cooperative game in general.

[2] The value of any coalition can be obtained in the same way by examining the restricted game where the other agents are removed.

4.1 Checking if a Given Imputation Is in the Core

Theorem 2. *Checking if a given imputation is in the core is co$\mathcal{NP}$-hard for R-WVGs, even with a single unreliable agent.*

Proof. We reduce $SUBSET$-SUM to the complement of our problem, i.e., checking if an imputation is *not* in the core. Given an instance (S, t) of $SUBSET$-SUM where $S = \{w_1, \ldots, w_n\}$, the question is to check if there is a subset of S whose elements sum to t. Define $W = \sum_{i=1}^{n} w_i$. If $W \leq t$, the reduction is trivial: reduce the case of $W < t$ to any NO instance, and the case of $W = t$ to any YES instance.

If $W > t$, construct an R-WVG G with $n + 1$ agents, where first n agents have reliability 1 and weights $w_1, \ldots, w_n$, agent $n + 1$ has reliability $(t + 1)/W$ and weight $W - t$, and threshold is W. Consider the payments $\mathbf{p} = \{p_1, \ldots, p_{n+1}\}$ where $p_i = w_i/W$ for $1 \leq i \leq n$ and $p_{n+1} = 0$. We show that $\mathbf{p}$ is not in the core of G iff the answer to the $SUBSET$-SUM instance is YES. First, $\mathbf{p}$ is an imputation since the value of every single agent is 0 and the sum of payoffs is 1 (the value of the grand coalition).

Next, $\mathbf{p}$ is not in the core iff there is a coalition with total payoff less than its value. It can be checked that any coalition not containing agent $n + 1$ or containing first n agents has value either 0 or 1, and receives total payoff no less than its value. Thus, a violating coalition must contain agent $n + 1$ and not all of the first n agents. Such a coalition has value $r_{n+1} = (t + 1)/W$ if the total weight of agents other than agent $n + 1$ in the coalition is at least t, and 0 otherwise. If this total weight is at least $t + 1$, the coalition receives at least $(t+1)/W$, which is its value. Thus, a violating coalition exists iff there is a subset of the first n agents whose weights sum to exactly t. $\qquad\square$

4.2 Testing Emptiness of the Core

Theorem 3. *Testing emptiness of the core in R-WVGs with a single unreliable agent (SUCORE) and testing emptiness of the ϵ-core in WVGs (EPSCORE) are polynomial-time reducible to each other.*

Proof. First, take an instance (G, ϵ) of EPSCORE where WVG $G = (N, \mathbf{w}, t)$ has n agents, weight vector $\mathbf{w}$ and threshold t, and $\epsilon \geq 0$. The task is to check if the ϵ-core of G is empty. Define $W = \sum_{i=1}^{n} w_i$. If $W \leq t$, the reduction is trivial: If $W < t$, the grand coalition has value 0 and a payoff of 0 to every agent is in the ϵ-core. If $W = t$, the grand coalition has value 1 but every other coalition has value 0, so a payoff of 1 to any single agent and 0 to the rest is in the ϵ-core. In both cases, the ϵ-core of G is not empty, so we reduce to any NO instance of SUCORE. If $W \geq t$, form an R-WVG $G' = (N', \mathbf{w}', t', \mathbf{r}')$ (instance of SUCORE) with $n + 1$ agents, weight vector $\mathbf{w}' = \{w_1, \ldots, w_n, W - t\}$, threshold $t' = W$, and reliability vector $\mathbf{r}' = \{1, \ldots, 1, 1 - \epsilon\}$. We show that the ϵ-core of G is empty iff the core of G' is empty.

Let v' be the characteristic function of G'. Since $v'(\{1, \ldots, n\}) = 1 = v'(N')$, agent $n + 1$ must receive zero payoff in any core imputation of G'. Thus, the core of G' is non-empty iff there exists a payoff vector $\mathbf{p} = \{p_1, \ldots, p_n, 0\}$ such that $\sum_{i=1}^{n} p_i = 1$ and the payoff to every coalition is at least its value. Any coalition not containing agent $n + 1$ or containing all of first n agents receives at least its value by construction. Any coalition containing agent $n + 1$ but not all of first n agents has value $r_{n+1} = 1 - \epsilon$

if the total weight of the reliable agents (among first n agents) in it is at least t, and 0 otherwise. Thus, G' has non-empty core iff there is a solution to: $\sum_{i=1}^{n} p_i = 1$ and $p(S) \geq 1 - \epsilon$ whenever $w(S) \geq t$. But this is the LP for checking emptiness of ϵ-core for G, so the core of G' is empty iff the ϵ-core of G is non-empty.

We show a reduction in the other direction. Take any instance $H = (N, \mathbf{w}, t, \mathbf{r})$ of SUCORE with n agents, weight vector $\mathbf{w}$, threshold t, reliability vector $\mathbf{r}$, and characteristic function v. Without loss of generality, let agent n be the unreliable agent with reliability $r_n = x$. Now, $v(\{1, 2, \ldots, n-1\}) \in \{0, 1\}$. If $v(\{1, 2, \ldots, n-1\}) = 0$, paying $v(N)$ to agent n and 0 to other agents is a core imputation. Hence, the core is not empty, and we reduce this to any NO instance of EPSCORE. If $v(\{1, 2, \ldots, n-1\}) = 1$, then agent n has zero payoff in any core imputation of H. Hence, the core of H is non-empty iff there exists a payoff vector $\mathbf{p} = \{p_1, \ldots, p_{n-1}, 0\}$ such that the payoff to any coalition is at least its value. Any coalition containing all first $n-1$ agents or not containing agent n receives at least as much as its value. Any coalition containing agent n but not all of first $n-1$ agents has value x if the total weight of the reliable agents (among first $n-1$ agents) in the coalition is at least $t - w_n$, and 0 otherwise. That is, the core of H is non-empty iff there is a solution to: $\sum_{i=1}^{n-1} p_i = 1$ and $p(S) \geq x$ whenever $w(S) \geq t - w_n$. However, this is exactly the LP for checking emptiness of ϵ-core for the WVG $H' = (N', \mathbf{w}', t - w_n)$ with the set of agents $N' = \{1, \ldots, n-1\}$, weight vector $\mathbf{w}' = \{w_1, \ldots, w_{n-1}\}$ and threshold $t - w_n$, with $\epsilon = 1 - x$.   □

Elkind et. al. [16] proved that testing emptiness of ϵ-core of WVGs is co$\mathcal{NP}$-hard. Further, they gave an algorithm to compute an ϵ-core imputation of a WVG using a separation oracle that runs in time pseudo-polynomial in agent weights. Theorem 3 and its constructive proof allow us to translate these results to the domain of R-WVGs.

Corollary 1. *Testing emptiness of the core in R-WVGs is co$\mathcal{NP}$-hard, even with a single unreliable agent.*

Corollary 2. *If all weights are represented in unary, finding a core imputation of an R-WVG with a single unreliable agent is in $\mathcal{P}$.*

4.3 Finding a Core Imputation

Finding a core imputation, if one exists, is computationally easy (in $\mathcal{P}$) for WVGs (see [12]). Theorem 1 shows that even computing the value of the grand coalition is #$\mathcal{P}$-hard for R-WVGs. Since total payoff in any core imputation equals the value of the grand coalition, finding a core imputation in R-WVGs is clearly #$\mathcal{P}$-hard as well.

Corollary 3. *Finding a core imputation is #$\mathcal{P}$-hard in R-WVGs.*

Corollary 2 gives us a pseudo-polynomial time algorithm to find a core imputation of an R-WVG with a single unreliable agent. We now extend this result to the case of a few (more than one) unreliable agents. The algorithm of [16] to find an ϵ-core imputation of a WVG works using a separation oracle (that runs in time pseudo-polynomial in weights) to solve the exponential sized LP of ϵ-core. It uses an important subroutine that finds, given any x, the minimum total payoff to any coalition with total weight at least x. We denote it MINPAY. So, MINPAY$(\mathbf{p}, x) = \min_{S \subseteq N, w(S) \geq x} p(S)$.

Consider an R-WVG $G^{\mathbf{r}} = (N, \mathbf{w}, t, \mathbf{r})$, where $|N| = n$. Without loss of generality, let UR be the set of unreliable agents and R be the set of reliable agents. Let $\mathbf{p} = \{p_1, \ldots, p_n\}$ denote a payoff vector. For any coalition S, $p(S) = \sum_{i \in S} p_i$ and $w(S) = \sum_{i \in S} w_i$. We aim to find a separation oracle for the LP of the core: $\sum_{i=1}^{n} p_i = v^{\mathbf{r}}(N)$, and $p(S) \geq v^{\mathbf{r}}(S)$ for all $S \subseteq N$.

Divide S into reliable and unreliable parts: $S = S_1 \cup S_2$, where $S_1 \subseteq UR$ and $S_2 \subseteq R$. Note that $p(S) = p(S_1 \cup S_2) = p(S_1) + p(S_2)$. We examine cases for $v^{\mathbf{r}}(S) = v^{\mathbf{r}}(S_1 \cup S_2)$. Consider the power set $\mathcal{P}(S_1) = \{T_1, \ldots, T_{2^{|S_1|}}\}$. Let $y_i = w(T_i)$, $q_i = \Pr[T_i | S_1]$ (the probability that exactly the agents in T_i survive out of agents in S_1), and $Q_i = \sum_{j \geq i} q_j$. Without loss of generality, let $y_i \leq y_{i+1}$ for all i. Note that $y_1 = 0$ as $T_1 = \emptyset$, and $Q_1 = 1$ as there must be a unique surviving sub-coalition. Now,

$$v^{\mathbf{r}}(S) = v^{\mathbf{r}}(S_1 \cup S_2) = \begin{cases} Q_1 = 1 & \text{if } w(S_2) \geq t - y_1 = t, \\ Q_i & \text{if } w(S_2) \in [t - y_i, t - y_{i-1}) \text{ where } i \in [2, 2^{|S_1|}], \\ 0 & \text{if } w(S_2) < t - y_{2^{|S_1|}}. \end{cases}$$

Using these observations, we can simplify the LP to:

$$\sum_{i=1}^{n} p_i = v^{\mathbf{r}}(N) \tag{2}$$

$$p(S_2) \geq Q_i - p(S_1), \forall (S_1 \subseteq UR, i \in [1, 2^{|S_1|}], S_2 \subseteq R) \text{ s.t. } w(S_2) \geq t - y_i.$$

Note that we have introduced additional constraints, but it is easy to check that they are redundant, and thus do not change the LP.[3] However, they enable us to use the subroutine MINPAY, for which we have a dynamic programming formulation.

Algorithm CORE-FEW-UNREL: Solve LP (2) using the following separation oracle.

ALGORITHM: SEPARATIONORACLE

Data: R-WVG $G^{\mathbf{r}} = (N, \mathbf{w}, t, \mathbf{r})$, payoff vector $\mathbf{p}$
Result: Either returns a violated constraint of LP (2) or returns SATISFIED
1. Compute $v^{\mathbf{r}}(N)$ from Equation (1): $v^{\mathbf{r}}(N) = \sum_{C \subseteq UR} \Pr[C | UR] \cdot v(C \cup R)$ (since agents in R always survive).
2. Check if $\sum_{i=1}^{n} p_i = v^{\mathbf{r}}(N)$. If not, then return this violated constraint.
3. For every $S_1 \subseteq UR$, compute y_i, q_i, and thus Q_i, for $i \in [1, 2^{|S_1|}]$. For all i, check if MINPAY$(t - y_i) \geq Q_i - p(S_1)$. If not, return the violated constraint.
4. If no violated constraints are found above, then return SATISFIED.

Running Time: The time required to compute $v^{\mathbf{r}}(N)$ is $O(2^{|UR|} \cdot n)$. For any $S_1 \subseteq UR$, the time required to compute y_i, q_i and Q_i is $O(2^{|S_1|} \cdot |S_1|)$. We make $O(2^{|S_1|})$ calls to MINPAY, each of which takes $O(n \cdot W)$ time, where $W = \sum_{i=1}^{n} w_i$. Thus, the total time required to check the constraints for any S_1 is $O(2^{|S_1|} \cdot n \cdot W)$. Summing over all $S_1 \subseteq UR$, the total running time is $O\left(\sum_{S_1 \subseteq UR} 2^{|S_1|} \cdot n \cdot W\right) =$

---

[3] The constraint $p(S_2) \geq Q_i - p(S_1)$ is required only when $w(S_2) \in [t - y_i, t - y_{i-1}]$, but we added it for all S_2 where $w(S_2) \geq t - y_i$. If $w(S_2) \in [t - y_j, t - y_{j-1}]$ for $j < i$ (or $w(S_2) \geq t - y_1$), then the constraint $p(S_2) \geq Q_j - p(S_1)$ (resp. $p(S_2) \geq Q_1 - p(S_1)$) strictly dominates the additional constraints added.

$O\left(\sum_{k=1}^{|UR|}\binom{|UR|}{k}\cdot 2^k\cdot n\cdot W\right)=O\left(3^{|UR|}\cdot n\cdot W\right)$. The last equation follows using binomial expansion. Thus, we have:

Theorem 4. *If all weights are represented in unary, finding a core imputation in an R-WVG with constantly many unreliable agents is in $\mathcal{P}$.*

Despite significant effort, we could not settle the question of existence of a pseudo-polynomial time algorithm for R-WVGs with arbitrarily many unreliable agents.

5 Power Indices

We now examine fair payoff divisions (power indices) in R-WVGs. Two prominent indices, the Shapley value and the Banzhaf index, are known to be #$\mathcal{P}$-hard even in WVGs [21,16], and thus also in R-WVGs. Bachrach et. al. [3] gave an algorithm to additively approximate the Shapley value, which can easily be adapted for the Banzhaf index. The algorithm works for reliability extensions of any cooperative game,[4] thus also for R-WVGs. Additionally, dynamic programming algorithms are known for computing both indices in WVGs [18,21] when the weights and the threshold are integers. We give non-trivial extensions of these algorithms for computing both indices in R-WVGs with identical agent reliabilities (uniform reliability case), and integer weights and threshold. We only give a sketch for the Shapley value due to lack of space. The details appear in the full version of the paper.[5]

Consider an R-WVG $G^{\mathbf{r}}=(N,\mathbf{w},t,\mathbf{r})$ where $r_j=p$ for all j (uniform reliability). Bachrach et. al. [3] showed that the Shapley value of agent i in the reliability extension of any cooperative game satisfies

$$\varphi_i=\frac{r_i}{|N|!}\cdot\sum_{\pi\in S_n}\left[\sum_{S\subseteq\Gamma_i^\pi}m_i(S)\cdot\Pr[S|\Gamma_i^\pi]\right],$$

where $m_i(S)=v(S+i)-v(S)$ is the marginal contribution of agent i to coalition S in the base game. Changing the order of summations, we get:

$$\varphi_i=\frac{r_i}{|N|!}\cdot\sum_{S\subseteq N-i}m_i(S)\left[\sum_{\Gamma\subseteq N-i\ s.t.\ S\subseteq\Gamma}\left(\Pr[S|\Gamma]\cdot\sum_{\pi\in S_n,\Gamma_i^\pi=\Gamma}1\right)\right].$$

Now, $\Pr[S|\Gamma]=p^{|S|}\cdot(1-p)^{|\Gamma|-|S|}$, and $\sum_{\pi\in S_n,\Gamma_i^\pi=\Gamma}1=|\Gamma|!\cdot(n-|\Gamma|-1)!$. Thus, all the required quantities except $m_i(S)$ depend only on $|S|$ and $|\Gamma|$. We break the summations of S and Γ further over the sizes of the coalitions, and show that the expression can be computed in pseudo-polynomial time. The overall running time of our methods[6] is $O(t\cdot n^2)$, where t is the threshold and n is the number of agents. We remark that this is identical to the running time of the known dynamic programming algorithms for these indices in WVGs. This is surprising, given our results of Sections 3 and 4 that moving from WVGs to R-WVGs raises the computational complexity of many questions significantly.

---

[4] Bachrach et. al. [3] consider network games, but their method works for any cooperative game.

[5] See http://www.cs.cmu.edu/~nkshah/papers.html for the full version.

[6] The running time is for both the Shapley value and the Banzhaf index.

6   The Relation between Reliability And Stability

We examine the relation between agent reliability and stability of the game in our R-WVG model. We randomly construct many R-WVGs using a *generation model* depending on a *reliability parameter*, quantify the degree of stability in each generated game according to some *stability measures*, and examine the *expected* degree of stability for each reliability parameter. We use three metrics as measures of stability. On the qualitative level, a game is *completely stable* if its core is non-empty, as there exists a fully stable payoff division. On the quantitative level, we use the least core value (LCV) and the Cost of Stability (CoS). The LCV is the minimal deviation cost that admits a stable imputation, so a low LCV indicates high stability. The CoS is the external subsidy required to make the grand coalition stable, so a low CoS also indicates high stability. These three measures are related: by definition, the core is non-empty iff the game has a non-positive LCV and iff the game has a non-positive CoS.

Bachrach et. al. [3] initiated the study of the relation between agent failures and stability. They showed that when starting with a simple game with *zero failure probabilities*, increasing failure probabilities can only *increase* stability of the game in terms of non-emptiness of the core, and thus under all our measures.[7] Later, it was demonstrated [6] that in simple games, increasing failure probabilities starting from *non-zero values* may not preserve non-emptiness of the core, i.e., might *reduce* stability under all our measures. These results apply to WVGs as they are simple games. However, it was proved [6] that non-emptiness of the core is *always* preserved when failure probabilities are increased, starting from possibly non-zero values, if the game is *totally balanced* (i.e., if every subgame has non-empty core). This discussion indicates that although there is evidence that failures help stability in other classes of cooperative games, the relation is not so clear-cut in WVGs; in some R-WVGs increasing failure probabilities increases stability, while in others it decreases stability. We empirically show that even in R-WVGs, *on average* increasing failure probabilities increases stability.

First, we analyze R-WVGs where reliabilities of all the agents are *equal*. For 100 values of uniform reliability from 0.01 to 1, we generated 10^6 R-WVGs with the number of agents drawn uniformly at random between 5 and 10. We had few agents since computing the stability measures is computationally hard, and we solve many games to compute the average stability level. Weights were sampled from various distributions: Gaussian, Uniform, Poisson and Exponential. Figures 1 (for Gaussian) and 2 (for Uniform) show that the average LCV, the average CoS, and the probability of having an empty core (measures of instability) increase with the uniform reliability. Thus, stability increases as the uniform failure probability increases, according to all our measures. The plots for Poisson and Exponential are omitted as they are very similar.

Next, we analyze games where only a few agents are unreliable. One such a domain is a decision making body where most decision makers are known to either support or object a legislation, and lobbying agents may convince the others to vote for it, but may fail with a certain probability. We built 10^4 WVGs with 30 agents, and weights uniformly chosen from 1 to 10.[8] We then made up to 5 agents unreliable *one by one*,

---

[7] Recall the quantitative and qualitative measure are linked.

[8] Our algorithms are pseudo-polynomial in the weights, so low weights are required.

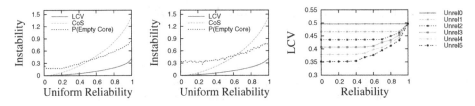

Fig. 1. Gaussian Weights   **Fig. 2.** Uniform Weights   **Fig. 3.** Few Unreliable Agents

changing their reliability to each of 10 values from 0.1 to 1, and measured the LCV using Algorithm CORE-FEW-UNREL.[9] The results, shown in Figure 3, indicate that *instability* (as measured by the average LCV) increases as agents become more reliable, so again increasing failure probabilities tends to increase stability on average. Further, we can see that the more agents we have that may fail, the more stable the game is.

All the above results reflect a similar pattern: Although there exist specific examples where making agents less reliable makes the game less stable, *on average* increasing failure probabilities in an R-WVG makes the game *more* stable. That is, failures help stabilize the game on average, which conforms to the results of [6].

7   Conclusion

We examined the impact of possible agents failures on the solutions to weighted voting games using the reliability extension model [3], which resulted in the class of R-WVGs. We contrasted the computational ease of calculating the value of a coalition and several core related questions in WVGs with hardness results for R-WVGs. We developed tractable tools for computing various solution concepts (core related or power indices) approximately, or exactly in restricted games. Using these tools, we explored the relation between agent reliability and stability, and empirically showed that on average higher failure probabilities make the game more stable.

Many questions are left open for future research. Could better computational tools be developed to solve R-WVGs, allowing us to handle larger games? Are there specific WVG domains that exhibit a different relation between agent reliabilities and stability? Does the general trend where introducing more failures causes the game to become more stable hold in other classes of cooperative games? Finally, how do failures affect other cooperative game solutions, such as coalition structures or the nucleolus?

References

1. Aziz, H., Lachish, O., Paterson, M., Savani, R.: Power indices in spanning connectivity games. In: Goldberg, A.V., Zhou, Y. (eds.) AAIM 2009. LNCS, vol. 5564, pp. 55–67. Springer, Heidelberg (2009)

---

[9] The algorithm finds a core imputation, but can easily be extended to compute the LCV, the CoS, and to test emptiness of the core.

2. Bachrach, Y., Markakis, E., Resnick, E., Procaccia, A., Rosenschein, J., Saberi, A.: Approximating power indices: theoretical and empirical analysis. In: AAMAS, pp. 105–122 (2010)
3. Bachrach, Y., Meir, R., Feldman, M., Tennenholtz, M.: Solving cooperative reliability games. In: UAI, pp. 27–34 (2011)
4. Bachrach, Y., Rosenschein, J.: Power in threshold network flow games. In: AAMAS, pp. 106–132 (2009)
5. Bachrach, Y., Elkind, E., Meir, R., Pasechnik, D., Zuckerman, M., Rothe, J., Rosenschein, J.S.: The cost of stability in coalitional games. In: Mavronicolas, M., Papadopoulou, V.G. (eds.) SAGT 2009. LNCS, vol. 5814, pp. 122–134. Springer, Heidelberg (2009)
6. Bachrach, Y., Kash, I., Shah, N.: Agent failures in totally balanced games and convex games. In: Goldberg, P.W. (ed.) WINE 2012. LNCS, vol. 7695, pp. 15–29. Springer, Heidelberg (2012)
7. Bala, V., Goyal, S.: A strategic analysis of network reliability. Review of Economic Design 5(3), 205–228 (2000)
8. Banzhaf, J.F.: Weighted voting doesn't work: A mathematical analysis. Rutgers Law Review 19, 317 (1964)
9. Baye, M.R., Kovenock, D., De Vries, C.G.: Rigging the lobbying process: an application of the all-pay auction. The American Economic Review 83(1), 289–294 (1993)
10. Bilbao, J.M.: Cooperative games on combinatorial structures. Kluwer Academic, Boston (2000)
11. Billand, P., Bravard, C., Sarangi, S.: Nash networks with imperfect reliability and heterogeous players. IGTR 13(2), 181–194 (2011)
12. Wooldridge, M.: Computational aspects of cooperative game theory. In: O'Shea, J., Nguyen, N.T., Crockett, K., Howlett, R.J., Jain, L.C. (eds.) KES-AMSTA 2011. LNCS (LNAI), vol. 6682, p. 1. Springer, Heidelberg (2011)
13. Conitzer, V., Sandholm, T.: Computing shapley values, manipulating value division schemes, and checking core membership in multi-issue domains. In: AAAI, pp. 219–225 (2004)
14. Deng, X., Papadimitriou, C.H.: On the complexity of cooperative solution concepts. MOR 19(2), 257–266 (1994)
15. Dubey, P.: On the uniqueness of the shapley value. IJGT 4(3), 131–139 (1975)
16. Elkind, E., Goldberg, L., Goldberg, P., Wooldridge, M.: On the computational complexity of weighted voting games. Ann. Math. Artif. Intell. 56(2), 109–131 (2009)
17. Fatima, S.S., Wooldridge, M., Jennings, N.R.: A randomized method for the Shapley value for the voting game. In: AAMAS, pp. 955–962 (2007)
18. Garey, M.R., Johnson, D.S.: Computers and Intractability: A Guide to the Theory of NP-Completeness. W. H. Freeman and Company (1979)
19. Gillies, D.B.: Some theorems on n-person games. PhD thesis, Princeton University (1953)
20. Kannan, R., Sarangi, S., Iyengar, S.: A simple model for reliable query reporting in sensor networks. In: Information Fusion, pp. 1070–1075 (2002)
21. Matsui, Y., Matsui, T.: A survey of algorithms for calculating power indices of weighted majority games. Journal of the Operations Research Society of Japan 43, 71–86 (2000)
22. Meir, R., Tennenholtz, M., Bachrach, Y., Key, P.: Congestion games with agent failures. In: AAAI, pp. 1401–1407 (2012)
23. Penn, M., Polukarov, M., Tennenholtz, M.: Congestion games with failures. In: EC, pp. 259–268 (2005)
24. Shapley, L.S.: A value for n-person games. In: Contrib. to the Theory of Games, pp. 31–40 (1953)
25. Shapley, L., Shubik, M.: Quasi-cores in a monetary economy with nonconvex preferences. Econometrica: Journal of the Econometric Society, 805–827 (1966)
26. Young, H.: The allocation of funds in lobbying and campaigning. Behavioral Science 23(1), 21–31 (1978)

The Power of Mediation in an Extended El Farol Game*

Dieter Mitsche[1], George Saad[2], and Jared Saia[2]

[1] Laboratoire Dieudonné, UMR CNRS 7351, Université de Nice
dmitsche@unice.fr
[2] Department of Computer Science, University of New Mexico
{george.saad,saia}@cs.unm.edu

Abstract. A mediator implements a correlated equilibrium when it proposes a strategy to each player confidentially such that the mediator's proposal is the best interest for every player to follow. In this paper, we present a mediator that implements the best correlated equilibrium for an extended El Farol game with symmetric players. The extended El Farol game we consider incorporates both negative and positive network effects.

We study the degree to which this type of mediator can decrease the overall social cost. In particular, we give an exact characterization of *Mediation Value (MV)* and *Enforcement Value (EV)* for this game. *MV* is the ratio of the minimum social cost over all Nash equilibria to the minimum social cost over all mediators of this type, and *EV* is the ratio of the minimum social cost over all mediators of this type to the optimal social cost. This sort of exact characterization is uncommon for games with both kinds of network effects. An interesting outcome of our results is that both the *MV* and *EV* values can be unbounded for our game.

Keywords: Nash Equilibria, Correlated Equilibria, Mediators and Network Effects.

1 Introduction

When players act selfishly to minimize their own costs, the outcome with respect to the total social cost may be poor. The Price of Anarchy [1] measures the impact of selfishness on the social cost and is defined as the ratio of the worst social cost over all Nash equilibria to the optimal social cost. In a game, with a high Price of Anarchy, one way to reduce social cost is to find a mediator of expected social cost less than the social cost of any Nash equilibrium.

In the literature, there are several types of mediators [2,3,4,5,6,7,8,9,10,11]. In this paper, we consider only the type of mediator that implements a correlated equilibrium (CE) [12].

---

* A full version with all the proofs is available at the authors' homepages.

B. Vöcking (Ed.): SAGT 2013, LNCS 8146, pp. 50–61, 2013.

A mediator is a trusted external party that suggests a strategy to every player separately and privately so that each player has no gain to choose another strategy assuming that the other players conform to the mediator's suggestion.

The algorithm that the mediator uses is known to all players. However, the mediator's random bits are unknown. We assume that the players are symmetric in the sense that they have the same utility function and the probability the mediator suggests a strategy to some player is independent of the identity of that player.

Ashlagi et al. [13] define two metrics to measure the quality of a mediator: the mediation value (MV) and the enforcement value (EV). In our paper, we compute these values, adapted for games where players seek to minimize the social cost. The *Mediation Value* is defined as the ratio of the minimum social cost over all Nash equilibria to the minimum social cost over all mediators. The *Enforcement Value* is the ratio of the minimum social cost over all mediators to the optimal social cost.

A mediator is optimal when its expected social cost is minimum over all mediators. Thus, the *Mediation Value* measures the quality of the optimal mediator with respect to the best Nash equilibrium; and the *Enforcement Value* measures the quality of the optimal mediator with respect to the optimal social cost.

1.1   El Farol Game

First we describe the traditional El Farol game [14,15,16,17]. El Farol is a tapas bar in Santa Fe. Every Friday night, a population of people decide whether or not to go to the bar. If too many people go, they will all have a worse time than if they stayed home, since the bar will be too crowded. That is a negative network effect [18].

Now we provide an extension of the traditional El Farol game, where both negative and positive network effects [18] are considered. The positive network effect is that if too few people go, those that go will also have a worse time than if they stayed home.

Motivation. Our motivation for studying this problem comes from the following discussion in [18].

"It's important to keep in mind, of course, that many real situations in fact display both kinds of [positive and negative] externalities - some level of participation by others is good, but too much is bad. For example, the El Farol Bar might be most enjoyable if a reasonable crowd shows up, provided it does not exceed 60. Similarly, an on-line social media site with limited infrastructure might be most enjoyable if it has a reasonably large audience, but not so large that connecting to the Web site becomes very slow due to the congestion."

We note that our El Farol extension is one of the simplest, non-trivial problems for which a mediator can improve the social cost. Thus, it is useful for studying the power of a mediation.

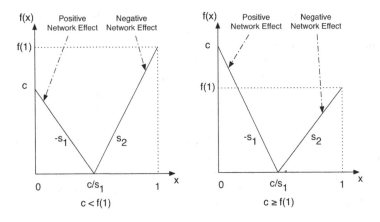

Fig. 1. The individual cost to go $f(x)$

Formal Definition of the Extended El Farol Game. We now formally define our game, which is non-atomic [19,20], in the sense that no individual player has significant influence on the outcome; moreover, the number of players is very large tending to infinity. The (c, s_1, s_2)-El Farol game has three parameters c, s_1 and s_2, where $0 < c < s_1$ and $s_2 > 0$. If x is the fraction of players to go, then the cost $f(x)$ for any player to go is as follows:

$$f(x) = \begin{cases} c - s_1 x & 0 \leq x \leq \frac{c}{s_1}, \\ s_2(x - \frac{c}{s_1}) & \frac{c}{s_1} \leq x \leq 1. \end{cases} \qquad (1)$$

and the cost to stay is 1. The function $f(x)$ is illustrated in the two plots of Figure 1.

Our Contributions. The main contributions of our paper are threefold:

- We design an optimal mediator, which implements the best correlated equilibrium for an extension of the El Farol game with symmetric players. Notably, this extension incorporates both negative and positive network effects.
- We give an exact characterization of the *Mediation Value* (*MV*) and the *Enforcement Value* (*EV*) for our game.
- We show that both the *MV* and *EV* values can be unbounded for our game.

Paper Organization. In Section 2, we discuss the related work. Section 3 states the definitions and notations that we use in the El Farol game. Our results are given in Section 4, where we show our main theorem that characterizes the best correlated equilibrium, and we compute accordingly the *Mediation Value* and the *Enforcement Value*. Finally, Section 5 concludes the paper and discusses some open problems.

2 Related Work

2.1 Mediation Metrics

Christodoulou and Koutsoupias [21] analyze the price of anarchy and the price of stability for Nash and correlated equilibria in linear congestion games. A consequence of their results is that the EV for these games is at least 1.577 and at most 1.6, and the MV is at most 1.015.

Brandt et al. [22] compute the mediation value and the enforcement value in ranking games. In a ranking game, every outcome is a ranking of the players, and each player strictly prefers high ranks over lower ones [23]. They show that for the ranking games with $n > 2$ players, $EV = n - 1$. They also show that $MV = n - 1$ for $n > 3$ players, and for $n = 3$ players where at least one player has more than two actions.

The authors of [3] design a mediator that implements a correlated equilibrium for a virus inoculation game [24,25]. In this game, there are n players, each corresponding to a node in a square grid. Every player has either to inoculate itself (at a cost of 1) or to do nothing and risk infection, which costs $L > 1$. After each node decides to inoculate or not, one node in the grid selected uniformly at random is infected with a virus. Any node, v, that chooses not to inoculate becomes infected if there is a path from the randomly selected node to v that traverses only uninoculated nodes. A consequence of their result is that EV is $\Theta(1)$ and MV is $\Theta((n/L)^{1/3})$ for this game.

Jiang et al. [26] analyze the price of miscoordination (PoM) and the price of sequential commitment (PoSC) in security games, which are defined to be a certain subclass of Stackelberg games. A consequence of their results is that MV is unbounded in general security games and it is at least $4/3$ and at most $\frac{e}{e-1} \approx 1.582$ in a certain subclass of security games.

We note that a poorly designed mediator can make the social cost worse than what is obtained from the Nash equilibria. Bradonjic et al. [27] describe the *Price of Mediation* (*PoM*) which is the ratio of the social cost of the worst correlated equilibrium to the social cost of the worst Nash equilibrium. They show that for a simple game with two players and two possible strategies, *PoM* can be as large as 2. Also, they show for games with more players or more strategies per player that *PoM* can be unbounded.

2.2 Finding and Simulating a Mediator

Papadimitriou and Roughgarden [28] develop polynomial time algorithms for finding correlated equilibria in a broad class of succinctly representable multi-player games. Unfortunately, their results do not extend to non-atomic games; moreover, they do not allow for direct computation of MV and EV, even when they can find the best correlated equilibrium.

Abraham et al. [29,30] describe a distributed algorithm that enables a group of players to simulate a mediator. This algorithm works robustly with up to linear size coalitions, and up to a constant fraction of adversarial players. The result suggests that the concept of mediation can be useful even in the absence of a trusted external party.

2.3   Other Types of Mediators

In all equilibria above, the mediator does not act on behalf of the players. However, a more powerful type of mediators is described in [2,4,5,6,7,8,9,10,11], where a mediator can act on behalf of the players that give that right to it.

For multistage games, the notion of the correlated equilibrium is generalized to the communication equilibrium in [31,32]. In a communication equilibrium, the mediator implements a multistage correlated equilibrium; in addition, it communicates with the players privately to receive their reports at every stage and selects the recommended strategy to each player accordingly.

3   Definitions and Notations

Now we state the definitions and notations that we use in the El Farol game.

Definition 1. A configuration $C(x)$ *characterizes that a fraction of players, x, is being advised to go; and the remaining fraction of players, $(1 - x)$, is being advised to stay.*

Definition 2. A configuration distribution $\mathcal{D}\{(C(x_1), p_1), .., (C(x_k), p_k)\}$ *is a probability distribution over $k \geq 2$ configurations, where $(C(x_i), p_i)$ represents that configuration $C(x_i)$ is selected with probability p_i, for $1 \leq i \leq k$. Note that $0 \leq x_i \leq 1$, $0 < p_i < 1$, $\sum_{i=1}^{k} p_i = 1$ and if $x_i = x_j$ then $i = j$ for $1 \leq i, j \leq k$.*

For any player i, let $\mathcal{E}_G^i$ be the event that player i is advised to go, and C_G^i be the cost for player i to go (when all other players conform to the advice). Also let $\mathcal{E}_S^i$ be the event that player i is advised to stay, and C_S^i be the cost for player i to stay. Since the players are symmetric, we will omit the index i.

A configuration distribution, $\mathcal{D}\{(C(x_1), p_1), .., (C(x_k), p_k)\}$, is a correlated equilibrium iff

$$\mathbf{E}\left[C_S | \mathcal{E}_G\right] \geq \mathbf{E}\left[C_G | \mathcal{E}_G\right],$$
$$\mathbf{E}\left[C_G | \mathcal{E}_S\right] \geq \mathbf{E}\left[C_S | \mathcal{E}_S\right].$$

Definition 3. A mediator *is a trusted external party that uses a configuration distribution to advise the players such that this configuration distribution is a correlated equilibrium. The set of configurations and the probability distribution are known to all players. The mediator selects a configuration according to the probability distribution. The advice the mediator sends to a particular player, based on the selected configuration, is known only to that player.*

Throughout the paper, we let n be the number of players.

4   Our Results

In our results, we assume that *the cost to stay* is 1; we justify this assumption at the end of this section. Our first results in Lemmas 1 and 2 are descriptions of the optimal social cost and the minimum social cost over all Nash equilibria for our extended El Farol game. We next state our main theorem which characterizes the best correlated equilibrium and determines the *Mediation Value* and *Enforcement Value*.

Lemma 1. *For any* (c, s_1, s_2)-*El Farol game, the optimal social cost is* $(y^* f(y^*) + (1 - y^*))n$, *where*

$$
y^* = \begin{cases}
\frac{1}{2}(\frac{c}{s_1} + \frac{1}{s_2}) & if \ \frac{c}{s_1} \le \frac{1}{2}(\frac{c}{s_1} + \frac{1}{s_2}) \le 1, \\
\frac{c}{s_1} & if \ \frac{1}{s_2} < \frac{c}{s_1}, \\
1 & otherwise.
\end{cases}
$$

Proof. By Equation (1), $f(x)$ has two cases. Let $f_1(x)$ be $f(x)$ for $x \in [0, \frac{c}{s_1}]$, and let $f_2(x)$ be $f(x)$ for $x \in [\frac{c}{s_1}, 1]$. Also let $h_1(x)$ be the social cost when $0 \le x \le \frac{c}{s_1}$, and let $h_2(x)$ be the social cost when $\frac{c}{s_1} \le x \le 1$. Thus, $h_1(x) = (xf_1(x) + (1 - x))n$ and $h_2(x) = (xf_2(x) + (1 - x))n$.

We know that $h_1(x)$ is minimized at $x = \frac{c}{s_1}$. In addition, we know that $h_2(x)$ is a quadratic function with respect to x, and thus it has one minimum over $x \in [\frac{c}{s_1}, 1]$ at $x = y^*$, where:

$$
y^* = \begin{cases}
\frac{1}{2}(\frac{c}{s_1} + \frac{1}{s_2}) & if \ \frac{c}{s_1} \le \frac{1}{2}(\frac{c}{s_1} + \frac{1}{s_2}) \le 1, \\
\frac{c}{s_1} & if \ \frac{1}{2}(\frac{c}{s_1} + \frac{1}{s_2}) < \frac{c}{s_1}, \\
1 & otherwise.
\end{cases}
$$

Let h^* be the optimal social cost. Then $h^* = min(h_1(\frac{c}{s_1}), h_2(y^*))$. Since $f_1(\frac{c}{s_1}) = f_2(\frac{c}{s_1})$, we have $h_1(\frac{c}{s_1}) = h_2(\frac{c}{s_1})$. Hence, $h^* = min(h_2(\frac{c}{s_1}), h_2(y^*))$. This implies that $h^* = h_2(y^*)$. □

Lemma 2. *For any* (c, s_1, s_2)-*El Farol game, if* $f(1) \ge 1$, *then the best Nash equilibrium is at which the cost to go in expectation is equal to the cost to stay; otherwise, the best Nash equilibrium is at which all players would rather go. The social cost of the best Nash equilibrium is* $min(n, f(1) \cdot n)$.

Proof. There are two cases for $f(1)$ to determine the best Nash equilibrium.
Case 1: $f(1) \ge 1$. Let N_y be a Nash equilibrium with the minimum social cost over all Nash equilibria and with a y-fraction of players that go in expectation. If $f(y) > 1$, then at least one player of the y-fraction of players would rather stay. Also if $f(y) < 1$, then at least one player of the $(1 - y)$-fraction of players would rather go. Thus, we must have $f(y) = 1$. Assume that each player has a mixed strategy, where player i goes with probability y_i. Recall that N_y has a y-fraction of players that go in expectation. Thus, $y = \frac{1}{n} \sum_{i=1}^{n} y_i$. Then the social cost is $\sum_{i=1}^{n} (y_i f(y) + (1 - y_i))$, or equivalently, n.
Case 2: $f(1) < 1$. In this case, the best Nash equilibrium is at which all players would rather go, with a social cost of $f(1) \cdot n$.

Therefore, the social cost of the best Nash equilibrium is $min(n, f(1) \cdot n)$. □

Theorem 1. *For any (c, s_1, s_2)-El Farol game , if $c \leq 1$, then the best correlated equilibrium is the best Nash equilibrium; otherwise, the best correlated equilibrium is $\mathcal{D}\{(C(0), p), (C(x^*), 1 - p)\}$, where $\lambda(c, s_1, s_2) = c(\frac{1}{s_1} + \frac{1}{s_2}) - \sqrt{\frac{c(\frac{1}{s_1} + \frac{1}{s_2})(c-1)}{s_2}}$,*

$$x^* = \begin{cases} \lambda(c, s_1, s_2) & \text{if } \frac{c}{s_1} \leq \lambda(c, s_1, s_2) < 1, \\ \frac{c}{s_1} & \text{if } \lambda(c, s_1, s_2) < \frac{c}{s_1}, \\ 1 & \text{otherwise.} \end{cases}$$

and $p = \frac{(1-x^*)(1-f(x^*))}{(1-x^*)(1-f(x^*))+c-1}$. Moreover,

1) the expected social cost is $(p + (1 - p)(x^* f(x^*) + (1 - x^*)))n$,
2) the Mediation Value (MV) is $\frac{\min(f(1),1)}{p+(1-p)(x^* f(x^*)+(1-x^*))}$ and
3) the Enforcement Value (EV) is $\frac{p+(1-p)(x^* f(x^*)+(1-x^*))}{y^* f(y^*)+(1-y^*)}$, where

$$y^* = \begin{cases} \frac{1}{2}(\frac{c}{s_1} + \frac{1}{s_2}) & \text{if } \frac{c}{s_1} \leq \frac{1}{2}(\frac{c}{s_1} + \frac{1}{s_2}) \leq 1, \\ \frac{c}{s_1} & \text{if } \frac{1}{s_2} < \frac{c}{s_1}, \\ 1 & \text{otherwise.} \end{cases}$$

Due to the space constraints, the proof of this theorem is not given here.

The following corollary shows that for $c > 1$, if $\lambda(c, s_1, s_2) \geq 1$, then the best correlated equilibrium is the best Nash equilibrium, where all players would rather go.

Corollary 1. *For any (c, s_1, s_2)-El Farol game, if $c > 1$ and $\lambda(c, s_1, s_2) \geq 1$ then MV $= 1$.*

Proof. By Theorem 1, when $\lambda(c, s_1, s_2) \geq 1$, $x^* = 1$ and $p = 0$. Now we prove that if $\lambda(c, s_1, s_2) \geq 1$, then the best correlated equilibrium is the best Nash equilibrium of the case $f(1) < 1$ in Lemma 2. To do so, we prove that $\lambda(c, s_1, s_2) \geq 1 \Rightarrow f(1) < 1$.

Now assume by way of contradiction that $\lambda(c, s_1, s_2) \geq 1 \Rightarrow f(1) \geq 1$. Recall that $f(1) = s_2(1 - \frac{c}{s_1})$. Then $\lambda(c, s_1, s_2) \geq 1 \Rightarrow \frac{c}{s_1} + \frac{1}{s_2} \leq 1$, or equivalently, $\lambda(c, s_1, s_2) \geq 1 \Rightarrow \frac{c}{s_1} + \frac{1}{s_2} \leq \lambda(c, s_1, s_2)$. Also recall that $\lambda(c, s_1, s_2) = c(\frac{1}{s_1} + \frac{1}{s_2}) - \sqrt{\frac{c(\frac{1}{s_1} + \frac{1}{s_2})(c-1)}{s_2}}$. Thus, we have:

$$\lambda(c, s_1, s_2) \geq 1 \Rightarrow \frac{c}{s_1} + \frac{1}{s_2} \leq c(\frac{1}{s_1} + \frac{1}{s_2}) - \sqrt{\frac{c(\frac{1}{s_1} + \frac{1}{s_2})(c-1)}{s_2}}$$

$$\Rightarrow s_2 \cdot \frac{c}{s_1} \leq -1,$$

which contradicts since s_1, s_2 and c are all positive. Therefore, for $c > 1$ and $\lambda(c, s_1, s_2) \geq 1$, MV must be equal to 1.   □

Now we show that MV and EV can be unbounded in the following corollaries.

Corollary 2. *For any $(2 + \epsilon, \frac{2+\epsilon}{1-\epsilon}, \frac{1}{\epsilon})$-El Farol game, as $\epsilon \to 0$, MV $\to \infty$.*

Proof. For any $(2 + \epsilon, \frac{2+\epsilon}{1-\epsilon}, \frac{1}{\epsilon})$-El Farol game, we have $f(1) = 1$. By Theorem 1, we obtain $x^* = 1 - \epsilon$, $f(x^*) = 0$ and $p = \frac{\epsilon}{1+2\epsilon}$ for $\epsilon \leq \frac{1}{2}(\sqrt{3} - 1)$. Thus we have

$$\lim_{\epsilon \to 0} MV = \lim_{\epsilon \to 0} \frac{\min(f(1), 1)}{\frac{\epsilon}{1+2\epsilon} + \epsilon(\frac{1+\epsilon}{1+2\epsilon})} = \infty.$$

$\square$

Corollary 3. *For any $(1 + \epsilon, \frac{1+\epsilon}{1-\epsilon}, \frac{1}{\epsilon})$-El Farol game, as $\epsilon \to 0$, EV $\to \infty$.*

Proof. For any $(1 + \epsilon, \frac{1+\epsilon}{1-\epsilon}, \frac{1}{\epsilon})$-El Farol game, by Theorem 1, we obtain $x^* = 1 + \epsilon^2 - \epsilon\sqrt{1+\epsilon^2}$ and $f(x^*) = 1 + \epsilon - \sqrt{1-\epsilon^2}$. Then we have

$$p = \frac{(1 - (1 + \epsilon^2 - \epsilon\sqrt{1+\epsilon^2}))(1 - (1 + \epsilon - \sqrt{1-\epsilon^2}))}{(1 - (1 + \epsilon^2 - \epsilon\sqrt{1+\epsilon^2}))(1 - (1 + \epsilon - \sqrt{1-\epsilon^2})) + \epsilon}.$$

Also we have $y^* = 1 - \epsilon$ and $f(y^*) = 0$ for $\epsilon \leq \frac{1}{2}$. Thus we have

$$\lim_{\epsilon \to 0} EV = \lim_{\epsilon \to 0} \frac{p + (1-p)(x^* f(x^*) + (1 - x^*))}{y^* f(y^*) + (1 - y^*)} = \infty.$$

$\square$

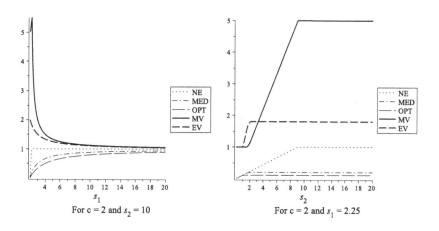

Fig. 2. NE, MED, OPT, MV and EV with respect to s_1 and s_2

Based on these results, we show in Figures 2 and 3 the social cost of the best Nash equilibrium (NE), the expected social cost of our optimal mediator (MED) and the optimal social cost (OPT), normalized by n, with respect to s_1, s_2 and

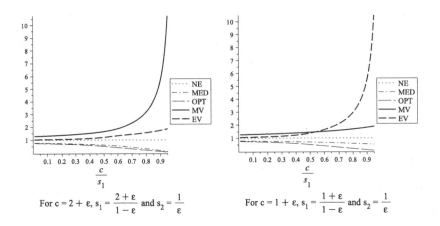

$$\text{For } c = 2 + \varepsilon, \, s_1 = \frac{2 + \varepsilon}{1 - \varepsilon} \text{ and } s_2 = \frac{1}{\varepsilon}$$ $$\text{For } c = 1 + \varepsilon, \, s_1 = \frac{1 + \varepsilon}{1 - \varepsilon} \text{ and } s_2 = \frac{1}{\varepsilon}$$

Fig. 3. NE, MED, OPT, MV and EV with respect to c/s_1

c/s_1. Also we show the corresponding *Mediation Value (MV)* and *Enforcement Value (EV)*.

In Figure 2, the left plot shows that for $c = 2$ and $s_2 = 10$, the values of NE, MED, OPT increase, each up to a certain point, when s_1 increases; however, the values of MV and EV decrease when s_1 increases. Moreover, MV reaches its peak at the point where the best Nash equilibrium starts to remain constant with respect to s_1. In the right plot, we set $c = 2$ and $s_1 = 2.25$; it shows that the values of NE, MED, OPT, MV and EV increase, each up to a certain point, when s_2 increases.

Figure 3 illustrates Corollaries 2 and 3, and it shows how fast MV and EV go to infinity with respect to c/s_1, where $c/s_1 = 1 - \epsilon$. The left plot shows that for any $(2 + \epsilon, \frac{2+\epsilon}{1-\epsilon}, \frac{1}{\epsilon})$-El Farol game, as $c/s_1 \to 1$ ($\epsilon \to 0$), $MV \to \infty$ and $EV \to 2$. In the right plot, for any $(1 + \epsilon, \frac{1+\epsilon}{1-\epsilon}, \frac{1}{\epsilon})$-El Farol game, as $c/s_1 \to 1$ ($\epsilon \to 0$), $EV \to \infty$ and $MV \to 2$.

Note that for any (c, s_1, s_2)-El Farol game, if $c/s_1 = 1$, then the best correlated equilibrium is at which all players would rather go with a social cost of 0, that is the best Nash equilibrium as well. Therefore, once c/s_1 is equal to 1, MV drops to 1.

The Cost to Stay Assumption

Now we justify our assumption that the cost to stay is unity. Let (c', s_1', s_2', t')-El Farol game be a variant of (c, s_1, s_2)-El Farol game, where $0 < c' < s_1'$, $s' > 0$ and the cost to stay is $t' > 0$. If x is the fraction of players to go, then the cost $f'(x)$ for any player to go is as follows:

$$f'(x) = \begin{cases} c' - s_1' x & 0 \le x \le \frac{c'}{s_1'}, \\ s_2'(x - \frac{c'}{s_1'}) & \frac{c'}{s_1'} \le x \le 1. \end{cases}$$

The following lemma shows that any (c', s_1', s_2', t')-El Farol game can be reduced to a (c, s_1, s_2)-El Farol game.

Lemma 3. *Any (c', s_1', s_2', t')-El Farol game can be reduced to a (c, s_1, s_2)-El Farol game that has the same Mediation Value and Enforcement Value, where $c = \frac{c'}{t'}, s_1 = \frac{s_1'}{t'}$ and $s_2 = \frac{s_2'}{t'}$.*

Proof. In a manner similar to Theorem (1), for any (c', s_1', s_2', t')-El Farol game, if $c > t'$, then the best correlated equilibrium is $\mathcal{D}\{(C(0), p'), (C(x'), 1 - p')\}$,

where $\lambda'(c', s_1', s_2', t') = c'(\frac{1}{s_1'} + \frac{1}{s_2'}) - \sqrt{\dfrac{c'(\frac{1}{s_1'} + \frac{1}{s_2'})(c' - t')}{s_2'}}$;

$$
x' = \begin{cases} \lambda'(c', s_1', s_2', t') & \text{if } \frac{c'}{s_1'} \le \lambda'(c', s_1', s_2', t') < 1, \\ \frac{c'}{s_1'} & \text{if } \lambda'(c', s_1', s_2', t') < \frac{c'}{s_1'}, \\ 1 & \text{otherwise.} \end{cases}
$$

and $p' = \frac{(1 - x')(t' - f(x'))}{(1 - x')(t' - f(x')) + c' - t'}$. Moreover,

1) the Mediation Value (MV') is $\dfrac{\min(f'(1), t')}{p't' + (1 - p')(x'f(x') + (1 - x')t')}$ and

2) the Enforcement Value (EV') is $\dfrac{p't' + (1 - p')(x'f(x') + (1 - x')t')}{y'f(y') + (1 - y')t'}$, where

$$
y' = \begin{cases} \frac{1}{2}(\frac{c'}{s_1'} + \frac{t'}{s_2'}) & \text{if } \frac{c'}{s_1'} \le \frac{1}{2}(\frac{c'}{s_1'} + \frac{t'}{s_2'}) \le 1, \\ \frac{c'}{s_1'} & \text{if } \frac{t'}{s_2'} < \frac{c'}{s_1'}, \\ 1 & \text{otherwise.} \end{cases}
$$

Similarly, for $c \le t'$, we have $MV' = 1$ and $EV' = \frac{\min(f'(1), t')}{y'f(y') + (1 - y')t'}$.

For both cases, by Theorem 1, if we set $c = c'/t'$, $s_1 = s_1'/t'$ and $s_2 = s_2'/t'$, then we have $f'(1) = f(1) \cdot t'$; also we get $y' = y^*$ and $\lambda'(c', s_1', s_2', t') = \lambda(c, s_1, s_2)$. This implies that $f'(y') = f(y^*) \cdot t'$ and $x' = x^*$; which in turn $f'(x') = f(x^*) \cdot t'$ and $p' = p$. Thus, we obtain $MV' = MV$ and $EV' = EV$. $\square$

5   Conclusion

We have extended the traditional El Farol game to have both negative and positive network effects. We have described an optimal mediator, and we have measured the *Mediation Value* and the *Enforcement Value* to completely characterize the benefit of our mediator with respect to the best Nash equilibrium and the optimal social cost.

Several open questions remain including the following: can we generalize our results for our game where the players choose among $k > 2$ actions? How many configurations are required to design an optimal mediator when there are $k > 2$ actions? Another problem is characterizing the MV and EV values for our game with the more powerful mediators in [2,4,5,6,7,8,9,10,11]. How much would these more powerful mediators reduce the social cost over our type of weaker mediator?

References

1. Koutsoupias, E., Papadimitriou, C.: Worst-case equilibria. In: Meinel, C., Tison, S. (eds.) STACS 1999. LNCS, vol. 1563, pp. 404–413. Springer, Heidelberg (1999)
2. Ashlagi, I., Monderer, D., Tennenholtz, M.: Mediators in position auctions. In: Proceedings of the 8th ACM Conference on Electronic Commerce, EC 2007, pp. 279–287. ACM, New York (2007)
3. Díaz, J., Mitsche, D., Rustagi, N., Saia, J.: On the power of mediators. In: Leonardi, S. (ed.) WINE 2009. LNCS, vol. 5929, pp. 455–462. Springer, Heidelberg (2009)
4. Forgó, F.: A generalization of correlated equilibrium: A new protocol. Mathematical Social Sciences 60(3), 186–190 (2010)
5. Forgó, F.: Measuring the power of soft correlated equilibrium in 2-facility simple non-increasing linear congestion games. Central European Journal of Operations Research, 1–17 (2012)
6. Monderer, D., Tennenholtz, M.: Strong mediated equilibrium. Artif. Intell. 173(1), 180–195 (2009)
7. Peleg, B., Procaccia, A.D.: Implementation by mediated equilibrium. International Journal of Game Theory 39(1-2), 191–207 (2010)
8. Rozenfeld, O., Tennenholtz, M.: Strong and correlated strong equilibria in monotone congestion games. In: Spirakis, P.G., Mavronicolas, M., Kontogiannis, S.C. (eds.) WINE 2006. LNCS, vol. 4286, pp. 74–86. Springer, Heidelberg (2006)
9. Rozenfeld, O., Tennenholtz, M.: Group dominant strategies. In: Deng, X., Graham, F.C. (eds.) WINE 2007. LNCS, vol. 4858, pp. 457–468. Springer, Heidelberg (2007)
10. Rozenfeld, O., Tennenholtz, M.: Routing mediators. In: Proceedings of the 20th International Joint Conference on Artificial Intelligence, IJCAI 2007, pp. 1488–1493. Morgan Kaufmann Publishers Inc., San Francisco (2007)
11. Tennenholtz, M.: Game-theoretic recommendations: some progress in an uphill battle. In: Proceedings of the 7th International Joint Conference on Autonomous Agents and Multiagent Systems, AAMAS 2008, Richland, SC, pp. 10–16 (2008)
12. Aumann, R.: Subjectivity and correlation in randomized games. Mathematical Economics 1, 67–96 (1974)
13. Ashlagi, I., Monderer, D., Tennenholtz, M.: On the value of correlation. Journal of Artificial Intelligence Research (JAIR) 33, 575–613 (2008)
14. Arthur, B.: Bounded rationality and inductive behavior (the el farol problem). American Economic Review 84, 406–411 (1994)
15. De Cara, M., Pla, O., Guinea, F.: Competition, efficiency and collective behavior in the "el farol" bar model. The European Physical Journal B - Condensed Matter and Complex Systems 10(1), 187–191 (1999)
16. Challet, D., Marsili, M., Ottino, G.: Shedding light on el farol. Physica A: Statistical Mechanics and Its Applications 332, 469–482 (2004)
17. Lus, H., Aydin, C., Keten, S., Unsal, H., Atiligan, A.: El farol revisited. Physica A: Statistical Mechanics and Its Applications 346(3-4), 651–656 (2005)
18. Easley, D., Kleinberg, J.: Networks, Crowds, and Markets: Reasoning About a Highly Connected World. Cambridge University Press, New York (2010)
19. Aumann, R., Shapley, L.: Values of Non-Atomic Games. A Rand Corporation Research Study Series. Princeton University Press (1974)
20. Schmeidler, D.: Equilibrium points of nonatomic games. Journal of Statistical Physics 7(4), 295–300 (1973)
21. Christodoulou, G., Koutsoupias, E.: On the price of anarchy and stability of correlated equilibria of linear congestion games. In: Brodal, G.S., Leonardi, S. (eds.) ESA 2005. LNCS, vol. 3669, pp. 59–70. Springer, Heidelberg (2005)

22. Brandt, F., Fischer, F., Harrenstein, P., Shoham, Y.: A game-theoretic analysis of strictly competitive multiagent scenarios. In: Proceedings of the 20th International Joint Conference on Artifical Intelligence, IJCAI 2007, pp. 1199–1206. Morgan Kaufmann Publishers Inc., San Francisco (2007)

23. Brandt, F., Fischer, F., Shoham, Y.: On strictly competitive multi-player games. In: Proceedings of the 21st National Conference on Artificial Intelligence, AAAI 2006, vol. 1, pp. 605–612. AAAI Press (2006)

24. Aspnes, J., Chang, K., Yampolskiy, A.: Inoculation strategies for victims of viruses and the sum-of-squares partition problem. Journal of Computer and System Science 72(6), 1077–1093 (2006)

25. Moscibroda, T., Schmid, S., Wattenhofer, R.: When selfish meets evil: byzantine players in a virus inoculation game. In: Proceedings of the Twenty-Fifth Annual ACM Symposium on Principles of Distributed Computing, PODC 2006, pp. 35–44. ACM, New York (2006)

26. Jiang, A.X., Procaccia, A.D., Qian, Y., Shah, N., Tambe, M.: Defender (mis)coordination in security games. In: International Joint Conference on Artificial Intelligence, IJCAI (2013)

27. Bradonjic, M., Ercal-Ozkaya, G., Meyerson, A., Roytman, A.: On the price of mediation. In: Proceedings of the 10th ACM Conference on Electronic Commerce, EC 2009, pp. 315–324. ACM, New York (2009)

28. Papadimitriou, C.H., Roughgarden, T.: Computing correlated equilibria in multi-player games. J. ACM 55(3), 1–29 (2008)

29. Abraham, I., Dolev, D., Gonen, R., Halpern, J.: Distributed computing meets game theory: Robust mechanisms for rational secret sharing and multiparty computation. In: Proceedings of the 25th Annual ACM Symposium on Principles of Distributed Computing, PODC 2006, pp. 53–62. ACM, New York (2006)

30. Abraham, I., Dolev, D., Halpern, J.Y.: Lower bounds on implementing robust and resilient mediators. In: Canetti, R. (ed.) TCC 2008. LNCS, vol. 4948, pp. 302–319. Springer, Heidelberg (2008)

31. Forges, F.: An approach to communication equilibria. Econometrica 54(6), 1375–1385 (1986)

32. Myerson, R.B.: Multistage games with communication. Econometrica 54(2), 323–358 (1986)

Friend of My Friend:
Network Formation with Two-Hop Benefit*

Elliot Anshelevich, Onkar Bhardwaj, and Michael Usher

Rensselaer Polytechnic Institute, Troy NY, USA

Abstract. How and why people form ties is a critical issue for understanding the fabric of social networks. In contrast to most existing work,
we are interested in settings where agents are neither so myopic as to
consider only the benefit they derive from their immediate neighbors,
nor do they consider the effects on the entire network when forming connections. Instead, we consider games on networks where a node tries to
maximize its utility taking into account the benefit it gets from the nodes
it is directly connected to (called *direct benefit*), as well as the benefit it
gets from the nodes it is acquainted with via a two-hop connection (called
two-hop benefit). We call such games *Two-Hop* Games. The decision to
consider only two hops stems from the observation that human agents
rarely consider "contacts of a contact of a contact" (3-hop contacts) or
further while forming their relationships. We consider several versions
of Two-Hop games which are extensions of well-studied games. While
the addition of two-hop benefit changes the properties of these games
significantly, we prove that in many important cases good equilibrium
solutions still exist, and bound the change in the price of anarchy due to
two-hop benefit both theoretically and in simulation.

1 Introduction

How and why people form ties is a critical issue for understanding the fabric of social networks. In various models, including public good games (see
e.g., [8, 11, 15] and the references therein), stable matching (see e.g., [6, 14]),
and others [5], it is often assumed that people make strategic decisions or form
friendships/partnerships based on the benefit they derive from their immediate
neighbors, independent of the rest of the network. On the opposite end of the
spectrum, many game-theoretic models such as [13] and its many extensions
(see [7] and references therein) consider players that form a network with the
goal of maximizing their influence over nodes that can be far away from them,
i.e., caring not just about their local neighborhood but about their position in
the entire network. In many settings, however, agents are neither so myopic as
to consider only the benefit they get from their immediate connections alone,
nor do they form relations considering the effects on the whole network. For

---

* This work was supported in part by NSF awards CCF- 0914782, CCF-1101495, and
CNS-1017932.

B. Vöcking (Ed.): SAGT 2013, LNCS 8146, pp. 62–73, 2013.

example, one of the aspects people consider when forming a relationship is the *two-hop* benefit they can get from the friends of such a friend. This is especially important in the world of business, but also occurs naturally when forming every-day friendships and collaborations: we judge people by the company they keep, and become better friends with those whose friends we like as well. Inspired by such settings, we consider games on networks where a node tries to maximize its utility taking into account the benefit it gets from the nodes it is directly connected to (called *direct benefit*), as well as the benefit it gets from the nodes it is acquainted with via a two-hop connection (called *two-hop benefit*). We will call such games *Two-Hop* Games.

Before formally defining Two-Hop games, we point out a difference between two concepts - one being the ability to *form* a relationship with someone, and an-other being the ability to *extract benefit* out of a direct or two-hop acquaintance. The ability to form a relationship indicates whether two agents can interact directly with each other (due to geographical proximity, etc). The ability to ex-tract benefit out of a direct or two-hop acquaintance instead tells us about how compatible the agents are with each other. We distinguish between these two concepts by having two graphs:

- Connection Graph(G_C): The edges in this graph denote which pairs of agents are able to form connections/relationships with each other.
- Friendship Graph(G_F): The edges in this graph indicate whether two agents are compatible with each other. If they are compatible, then they can derive benefit if they are connected either directly or via a two-hop connection. Thus G_F governs the utility extracted from acquaintances (the formation of which is governed by G_C).

Two-Hop Games. Now we will formally define Two-Hop games and discuss how some well-known games can be naturally extended to their two-hop versions. Each game is specified by a triple (G_C, G_F, k), with G_C and G_F having the same node set and $k \geq 0$. These nodes are the players of the game. We want to model the case where different friendships and relationships can be of different strength. Thus the strategy of a node, say u, consists of choosing to contribute to each of its adjacent edges (uv) in G_C, with an amount $0 \leq x_u^{uv} \leq 1$. The contribution x_u^{uv} represents the effort u puts into its relationship with v. Note that we restrict the contributions of u to edges adjoining u in G_C, as those are the nodes that u can connect to directly. We can represent the strategy of u in a compact way using a vector $\mathbf{x_u} = (x_u^{uv})$ with number of components equal to the degree of u in G_C. As usual, $\mathbf{x_{-u}} = (\mathbf{x_1}, \cdots, \mathbf{x_{u-1}}, \mathbf{x_{u+1}}, \cdots, \mathbf{x_n})$ denotes the strategies of all other nodes except u. We restrict $\sum_{(uv) \ni u} x_u^{uv} = \min(k, d_u)$ where d_u is the degree of u in G_C. The limit of k represents the fact that any person has only finite time/resources at his disposal to form acquaintances, and thus can contribute at most k effort in total. The objective of a node u is to maximize its utility given by

$$U_u(\mathbf{x_u}, \mathbf{x_{-u}}) = \sum_{\substack{(uw) \in G_F \cap G_C}} r_{uw}(x_u^{uw}, x_w^{uw}) + \sum_{\substack{(uw) \in G_F \\ (uv), (vw) \in G_C}} s_{uvw}(x_u^{uv}, x_v^{uv}, x_v^{vw}, x_w^{vw}) (1)$$

The function $r_{uw}(x_u^{uw}, x_w^{uw})$ represents the strength of the direct relationship between u and w: this depends only on the effort that u and w put into the relationship. The function $s_{uvw}(x_u^{uv}, x_v^{uv}, x_v^{vw}, x_w^{vw})$ represents the strength of a bond between u and w formed due to a mutual friend v. The strength of such a two-hop acquaintance can potentially depend on all the intermediate efforts on the 2-link path. Thus the utility of a node u is the total strength of its (direct and 2-hop) relationships with all of its neighbors in G_F, i.e., the nodes who actually benefit node u. Note that the two-hop benefit over all two-hop paths between u and w adds up: a larger number of mutual friends increases how much people can influence each other, a larger number of internal referrals increases the chances that a job-seeker gets an interview, etc. We are interested in the following two types of Two-Hop games.

– *Sum Two-Hop Games (S2H Games)*:

$$r_{uv}(x_u^{uv}, x_v^{uv}) = x_u^{uv} + x_v^{uv} \tag{2}$$

$$s_{uvw}(x_u^{uv}, x_v^{uv}, x_v^{vw}, x_w^{vw}) = \alpha \cdot (x_u^{uv} \cdot x_v^{uv} + x_v^{vw} \cdot x_w^{uw}) \tag{3}$$

We call $0 \leq \alpha \leq 1$ the two-hop benefit factor. It represents the intuitive notion that a two-hop acquaintance between u and w via v should yield less benefit than a direct acquaintance. Eqn 2 defines the strength of a relationship as the addition of strengths in each direction: strength in the direction $u \to v$ is given by x_u^{uv}, and in the reverse direction given by x_v^{uv}. Similarly the term $x_u^{uv} \cdot x_v^{uv}$ in Eqn 3 represents the strength of the two-hop acquaintance between u and w via v in the direction $u \to v \to w$ (e.g., how likely information is to pass from u to w via v). The term $x_v^{vw} \cdot x_w^{uw}$ is the strength of this indirect relationship in the other direction.

If not for the 2-hop effect, S2H Games would be a simple variation of network contribution games [5], in which nodes decide how much to contribute to adjacent edges, and receive the sum of contributions to these edges as their utility. It is easy to see that in such a game, a Nash Equilibrium always exists and its Price of Anarchy is 1. As we show in this paper, however, the addition of 2-hop benefit changes the properties of this game.

– *Min Two-Hop Games (M2H Games)*:

$$r_{uv}(x_u^{uv}, x_v^{uv}) = \min(x_u^{uv}, x_v^{uv}) \tag{4}$$

$$s_{uvw}(x_u^{uv}, x_v^{uv}, x_v^{vw}, x_w^{uw}) = \alpha \cdot \min(x_u^{uv}, x_v^{uv}) \cdot \min(x_v^{vw}, x_w^{uw}) \tag{5}$$

In M2H games, a relationship is only strong if *both* participants contribute a lot of effort. As before, the strength of a 2-hop effect is the product of the strengths of the two relationships in the 2-link path, attenuated by a factor $\alpha \in [0, 1]$.

Without the 2-hop effect, this game is essentially a fractional version of k-stable matching (see the full version [4] for details), in which every node u chooses k nodes among its neighbors to be matched with, and only gets utility from an edge (u, v) if v chooses u as well. As discussed below, existing work on stable matching immediately implies various results about the existence and quality of equilibrium for such a game. However, just as with S2H games, the addition of 2-hop benefit greatly changes the properties of this game.

To assess the quality of a solution M in S2H and M2H games, we will use social welfare, given by $U(M) = \sum_u U_u(\mathbf{x_u}, \mathbf{x_{-u}})$. For S2H games, we will focus on the existence and the quality of Nash Equilibria (NE's). For M2H games, however, using the concept of 2-strong Nash Equilibrium, also called pairwise equilibrium [17], makes more sense to consider than the concept of Nash equilibrium. A pairwise equilibrium (PE) is a solution stable with respect to deviations by any pair of players, as well as any single player. This is consistent with previous work on such games: if we think of integral versions of these games (where x_u^{uv} is constrained to be in $\{0, 1\}$) as network formation games, then S2H corresponds to a game where a node can unilaterally form a link and reap the benefits of this link, while M2H corresponds to a game in which both endpoints of a link are needed to form this link. Traditionally pairwise equilibria have been used to analyze the latter types of games [5, 12] simply due to the fact that any single-player deviation would not be able to create a new link. Similarly, in our fractional version of M2H, it is reasonable to expect for a pair of people (u, v) to increase the level of their friendship at the same time, thus increasing $\min(x_u^{uv}, x_v^{uv})$. Thus for M2H games, we study pairwise equilibria and investigate their quality compared to the optimal solution. We call the ratio between the quality of the worst pairwise equilibrium and the optimal solution 2-PoA to differentiate it from the PoA (price of anarchy) with respect to Nash Equilibria.

Our Contribution. We define Two-Hop games, which are natural generalizations of well-studied games. As mentioned above, S2H games without any two-hop benefit reduce to simple network contribution games; thus they are potential games, an integral NE always exists for them, and they have Price of Anarchy (PoA) of 1. As we show in Sec 2, despite the introduction of two-hop benefit, a NE always exists for general S2H games. However, an *integral* NE may no longer exist, and S2H games are not potential games (except for some special cases: see Theorem 4).

The introduction of two-hop benefit also changes the behavior of PoA. For the important special cases of $G_F \subseteq G_C$ (I can connect to all of my friends) and $G_C \subseteq G_F$ (I can only connect to friends), we show a tight PoA bound of $1 + \alpha k$, and in the very nice case when G_F and G_C are complete graphs, the PoA is $\frac{1+\alpha k}{1+\alpha(k-1)}$. As we show in Theorem 5, in general for S2H games PoA decreases as the overlap between G_F and G_C increases, i.e., PoA decreases as nodes get more opportunities to form acquaintances with nodes they are compatible with. For example, if every node has at least $k/2$ nodes which are its neighbors in both G_F and G_C, then the PoA is at most $1 + 2\alpha k$. Note that for the most reasonable values of α the PoA bounds above are rather small. For example, we can often assume that a single direct friendship brings more benefit than connecting to someone solely because of the 2-hop contacts being made; this is quantified by assuming that $\alpha \leq \frac{1}{k}$ since any node can have at most k friends. For this range of α, the above PoA bounds become merely 2 and 3. We further consider weighted

S2H games (See Sec 2.2) in which different acquaintances can potentially yield different intrinsic benefit, and show that the results obtained for S2H games also hold for weighted games when $G_F \subseteq G_C$.

Because of its connection to many-to-many stable matching, it is not difficult to show that for M2H games without 2-hop benefit an integral pairwise equilibrium (PE) always exists, and 2-PoA is at most 2. For general M2H games with 2-hop benefit, however, we show that a integral PE may not exist (existence of a fractional PE for this and related games is an important open question). For the cases when PE does exist, our main result for M2H games proves that 2-PoA for the important case of $G_C \subseteq G_F$ is at most $2 + 2\alpha k$, which for the "reasonable" range of $\alpha \in [0, 1/k]$ mentioned above evaluates to at most 4.

For weighted versions, we also carried out simulations by scattering nodes uniformly in a unit square and experimented with different classes of weight functions which depend on the distance between the nodes. We found that although the worst-case PoA bounds could be quite high, the average quality of equilibria was very close to the optimum. We also found that although integral NE may not exist for S2H games, in our simulations for majority of the instances it did exist, and simple dynamics converged to it extremely quickly in almost all instances. Our simulations also showed that as two-hop benefit decreases, nodes transition from forming small interconnected clusters to forming more of a "backbone" tree-like network.

Related Work. Network formation games, and games on networks more generally, have been studied extensively. In many network formation games, nodes connect to each other with the goal of maximizing their utility, which depends on their position in the network. For example, it may depend on the average distance to the rest of the nodes, or on various other notions of centrality and node "importance", see e.g., [13] and its many extensions (see [7] and the references therein). On the other hand, in many models the agents are concerned only about the direct benefit they derive from their immediate neighbors. Stable matching games (See e.g., [1, 3, 6, 14] etc.) and network contribution games [5] are some examples, as well as many others [2, 8, 11, 15]. As discussed in Sec 1, however, we are interested in settings where agents may neither be concerned about the actions of remote nodes nor be so myopic as to consider only the benefit they derive from their immediate neighbors. Two-Hop games fall under this category. The decision to consider only two hops stems from the observation that human agents rarely consider "contacts of a contact of a contact" (3-hop contacts) or further while forming their relationships.

As mentioned above, we distinguish between the ability of a pair of agents to interact directly (represented by the connection graph G_C) and their capability of being able to derive benefit if they are connected directly or via a two-hop connection (represented by the friendship graph G_F). The friendship graph G_F can be seen as a social context which dictates the benefits obtained by the nodes by playing a game on the connection graph G_C. Some other work which explores different forms of social context are [9] and [16]. In this work, the cost of a node in a resource-sharing game depends on its own cost and the costs of its "friends", where friend nodes are its neighbors in an underlying social network.

Finally, as we discuss in detail in the full vesion [4], M2H games without two-hop benefit reduce to a fractional, many-to-many version of "correlated" stable matching [1] for which a stable matching is known to exist for arbitrary graphs, and the 2-PoA (quality of stable matching compared to the optimum one) is at most 2 [3]. To the best of our knowledge the correlated version of many-to-many stable matching has not been studied before; however it is easy to see that existence of integral stable matching and the same bound on 2-PoA still holds.

2   Sum Two-Hop Games

Recall that in S2H games, the utility $U_u(\mathbf{x_u}, \mathbf{x_{-u}})$ of a node u is obtained by substituting Eqn (2) and (3) into Eqn (1) which gives us

$$U_u(\mathbf{x_u}, \mathbf{x_{-u}}) = \sum_{(uv) \in G_C \cap G_F} (x_u^{uv} + x_v^{uv}) + \alpha \cdot \sum_{\substack{(uv),(vw) \in G_C \\ s.t.\ (uw) \in G_F}} (x_u^{uv} \cdot x_v^{uv} + x_v^{vw} \cdot x_w^{uw}) \quad (6)$$

Without any two-hop benefit, there always exists an integral pure NE for S2H games (i.e., a NE in which all the contributions are either 0 or 1) and they are exact potential games. Even after introducing two-hop benefit (i.e., $\alpha > 0$), we can prove that a NE always exists for S2H games using Proposition 20.3 of [18]. However, all NE may be fractional, and this seizes to be a potential game. All of the missing proofs appear in the full version [4].

Theorem 1. *For S2H games, a pure Nash Equilibrium always exists.*

Theorem 2. *There are instances of the general S2H game which do not admit any integral pure Nash equilibrium.*

Theorem 3. *The general S2H game is not a potential game.*

However, we now give a family of instances for which S2H games are exact potential games and an integral NE exists. Let $d_C(u, v)$ denote the distance between u and v in G_C. Then the following theorem holds.

Theorem 4. *If $d_u \geq k$ for all nodes and if the condition $d_C(u, v) \leq 2$ implies $(uv) \in G_F$ for all the pairs of nodes then the S2H game is an exact potential game and an integral NE exists.*

2.1   Price of Anarchy

We know that without any two-hop benefit, the PoA of S2H games is 1. With two-hop benefit, PoA can become unbounded for arbitrary G_F and G_C, if $G_F \cap G_C = \emptyset$. However, as the overlap between G_F and G_C increases then the PoA for S2H games decreases and for the interesting cases of $G_F \subseteq G_C$ and $G_C \subseteq G_F$, PoA becomes $1 + \alpha k$. Increasing the overlap between G_F and G_C can be interpreted as nodes getting more opportunities to become directly acquainted with the nodes they are compatible with.

We formally quantify what we mean by overlap between G_F and G_C. Let F_v denote the degree of v in $G_F \cap G_C$. We define overlap between G_F and G_C as $\rho(G_F, G_C) = \min_v F_v$. We now give PoA bounds for several interesting cases:

Theorem 5. *For the S2H game,*

1. *For arbitrary G_F and G_C, PoA $\leq 1 + \alpha k \cdot \frac{k}{\min(k, \rho(G_F, G_C))}$. Thus when there is large overlap between G_F and G_C, say $\rho(G_F, G_C) \geq k/2$ then we have PoA $\leq 1 + 2\alpha k$.*
2. *Furthermore, if $G_C \subseteq G_F$ or $G_F \subseteq G_C$ then PoA $\leq 1 + \alpha k$.*
3. *For the special case of $G_F = G_C = K_n$, we have PoA $= \frac{1 + \alpha k}{1 + \alpha(k-1)}$.*

[Proof sketch:] For arbitrary G_F and G_C, the PoA bound follows by simple observations on the minimum utility obtained by a node in a NE and its maximum attainable utility. The PoA can be large for a small overlap because even if a node is capable of getting little direct benefit because of its small degree in $G_F \cap G_C$ (which is a lower bound on minimum utility obtained by it in a NE), it can still get a large two-hop benefit (hence large maximum attainable utility) by connecting to a lot of its friends via $G_C \setminus G_F$. However this changes in $G_C \subseteq G_F$ because $G_C \setminus G_F = \emptyset$. This also changes with $G_F \subseteq G_C$ because here if a node can get a large two-hop benefit by connecting to a lot of friends in G_C then it must have a lot of friends, and therefore its degree in $G_F \cap G_C$ must be high. Thus these cases result in a much improved bound on PoA regardless of the overlap size. For details of the proof, see the full version [4].

Theorem 6. *The bounds on the price of anarchy in Theorem 5 are asymptotically tight.*

Fig 1 shows the scheme of an instance for proving asymptotic tightness of PoA bound for arbitrary G_F and G_C. It contains five sets of nodes, given by

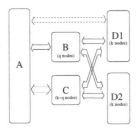

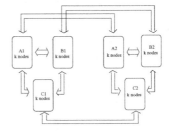

Fig. 1. Tight example for PoA of S2H games for arbitrary G_F and G_C

Fig. 2. Tight example for PoA for the S2H game when $G_F = G_C$

A, B, C, D_1, D_2. No edges exist between the nodes belonging to the same set. Solid bidirectional arrow between two sets denotes that all the nodes in one set are connected with all the nodes in the other set in $G_F \cap G_C$. Similarly semi-solid bidirectional arrow between two sets denotes that all the nodes in one set are connected with all the nodes in the other set in $G_F \setminus G_C$, whereas dotted bidirectional arrow means them being connected in $G_C \setminus G_F$.

It is sufficient to describe an instance with $G_F = G_C$ to show the tightness of the PoA bound for both $G_F \subseteq G_C$ and $G_C \subseteq G_F$. See Fig 2 for the scheme of such an instance. It consists of six sets A_1, B_1, C_1, A_2, B_2, C_2, each containing k nodes. No edges exist between the nodes belonging to the same set. Solid bidirectional arrow between two sets denotes that all the nodes in one set are connected with all the nodes in the other set.

2.2 Weighted S2H Games

Sometimes a person can have different levels of intrinsic interest in different acquaintances. We incorporate this scenario into S2H games by having a positive weight f^{uv} on each edge $(uv) \in G_F$. We call this extension as *Weighted S2H Games*. The utility of a node $U_u(\mathbf{x_u}, \mathbf{x_{-u}})$ in Weighted S2H games is given by:

$$U_u(\mathbf{x_u}, \mathbf{x_{-u}}) = \sum_{\substack{(uv) \in G_C \cap G_F}} (x_u^{uv} + x_v^{uv}) f^{uv} + \alpha \sum_{\substack{(uw) \in G_F \\ (uv),(vw) \in G_C}} (x_u^{uv} x_v^{vw} + x_w^{vw} x_v^{uv}) f^{uw} \qquad (7)$$

It is not difficult to see that the argument for the existence of NE for Weighted S2H games is the same as the argument for the existence of NE of S2H games. Also, despite having arbitrary weights on the edges of G_F, whenever we have $G_F \subseteq G_C$ the PoA proves to be at most $1 + \alpha k$ as it was in the absence of weights. Here by $G_F \subseteq G_C$, we mean that the unweighted version of G_F is a subset of G_C. Because of having arbitrary positive weights on the edges of G_F we do not treat the case of $G_F = G_C = K_n$ (i.e., the unweighted G_F is equal to K_n) separately but view it as a special case of $G_F \subseteq G_C$. Thus we get the following results (See the full version [4] for the proofs):

Theorem 7. *For Weighted S2H games, a Nash Equilibrium always exists.*

Theorem 8. *For Weighted S2H games, whenever $G_F \subseteq G_C$ we have PoA $\leq 1 + \alpha k$.*

3 Min Two-Hop (M2H) Games

Recall that M2H games are a natural extension of fractional stable matching games obtained by introducing two-hop benefit. Denoting $\min(x_u^{uv}, x_v^{uv})$ by x^{uv}, the utility $U_u(\mathbf{x_u}, \mathbf{x_{-u}})$ of a node u in M2H games can be written as:

$$U_u(\mathbf{x_u}, \mathbf{x_{-u}}) = \sum_{\substack{(uv) \in G_C \cap G_F}} x^{uv} + \alpha \cdot \sum_{\substack{(uv),(vw) \in G_C \\ (uw) \in G_F}} x^{uv} \cdot x^{vw} \qquad (8)$$

We will call the first summation in Eqn 8 as direct benefit of node u and the term with the coefficient of α in Eqn (8) as two-hop benefit of node u. Recall from Sec 1 that we use the concept of pairwise equilibria (PE's) to assess the quality of a solution in M2H games, denoting the ratio between the worst PE and the optimal solution as 2-PoA.

Recall that in an integral PE all the contributions are either 0 or 1. An integral PE exists for M2H games without two-hop benefit (See Related Work). With two-hop benefit, we can construct an instance of M2H games which does not admit any integral pure NE by adapting Example 1 from [10] which is an instance of stable roommates problem such that no stable matchings exist. However, we give 2-PoA bounds in Thm 10 for some important cases to assess the quality of PE's whenever they exist.

Theorem 9. *There exist instances of M2H games that do not admit any integral pairwise equilibrium.*

Theorem 10. *For the M2H game:*

1. *If $G_C \subseteq G_F$ then 2-PoA $\leq 2 + 2\alpha k$.*
2. *For the special case of $G_F = G_C = K_n$, a PE always exists and 2-PoA tends to $\frac{1+\alpha k}{1+\alpha(k-1)}$ as $n \to \infty$.*

[Proof sketch of Thm 10] Consider case $G_C \subseteq G_F$. Let M^* denote an optimum solution and let M denote a PE. Let y_u^{uv}'s (or r_u^{uv}'s) denote the contributions of u in M^* (or in M). Let $y^{uv} = \min(y_u^{uv}, y_v^{uv})$ and $r^{uv} = \min(r_u^{uv}, r_v^{uv})$.

We prove the 2-PoA bound for $G_C \subseteq G_F$ in two steps. First, we show that the direct benefit component of $U(M^*)$ is at most two times $U(M)$. Recall that while proving that a maximum matching μ^* is at most twice the size of any maximal matching μ, we use a property that every edge in $\mu^* \setminus \mu$ has at least one of its endpoints matched in μ. Analogously, we show that for M2H games with $G_C \subseteq G_F$, each edge (uv) in $G_F \cap G_C$ s.t. $y^{uv} > r^{uv}$ (i.e., an edge contributing more direct-benefit to M^* than M) has at least one of the endpoints, say u, such that in PE M, node u attains the maximum direct benefit that it can attain in any solution. Using this, we further prove that the direct benefit component of $U(M^*)$ is at most two times $U(M)$. Next, we bound the two-hop benefit component of $U(M^*)$ as

$$\sum_{\substack{(uv) \ni u, (vw) \ni v \\ (uw) \in G_F, w \neq u}} \alpha y^{uv} y^{vw} \leq \left(\sum_{(uv) \ni u} \alpha y^{uv} \right) \sum_{(vw) \ni v} y^{vw} \leq \left(\sum_{(uv) \ni u} y^{uv} \right) \cdot \alpha k \qquad (9)$$

For $G_C \subseteq G_F$, the term $\sum_{(uv) \ni u} y^{uv}$ is at most the direct benefit obtained by u in M^*, thus the two-hop benefit component of $U(M^*)$ is at most αk times the direct benefit component of $U(M^*)$. Combining it with the bound derived above on the direct benefit component of $U(M^*)$, we get the desired 2-PoA bound.

For the case of $G_F = G_C = K_n$, it can be verified that every node making a contribution of $k/(n-1)$ to every adjacent link is a PE. Now let us give an outline of how to prove the 2-PoA bound for this case. Let us define a set T as the set of nodes which obtain a direct benefit of strictly less than k in a PE M, i.e. $u \in T$ whenever $\sum_{(uv) \ni u} x^{uv} < k$ in M. Let $\bar{T}$ denote the complement of T.

In the full version [4], we prove that T contains at most k nodes. Thus for $n \gg k$, almost all the nodes belong to $\bar{T}$. Notice that making $x^{uv} = \delta$ s.t. $v \in \bar{T}$ brings node u an utility of at least $\delta(1 + \alpha(k-1))$. It is not difficult to see that if we combine the condition that $\sum_{(uv) \ni u} x^{uv} = k$ for $u \in \bar{T}$ with the observation that almost all the nodes belong to $\bar{T}$, then we have that most of the contributions of nodes in $\bar{T}$ are made to the edges with another endpoint also in $\bar{T}$. Since almost all the nodes in K_n are in $\bar{T}$ for $n \gg k$, this implies that a lower bound on $U(M)$ is approximately $n(k + \alpha k(k-1))$ as $n \to \infty$. Combining this with the observation that $U(M^*) \leq n(k + \alpha k^2)$, we obtain the desired result.

4   Empirical Findings

We performed a number of simulations on S2H and M2H games. Specifically, we investigated Weighted S2H and Weighted M2H games with 100 nodes scattered uniformly randomly inside a unit square, with $G_F = G_C$ being complete graphs, and $d(u,v)$ being the distance between nodes u and v. We considered the following three kinds of weight functions:

- Inverse: For the inverse weight function, we set $f^{uv} = 1/d(u,v)$.
- Exponential: For the exponential weight function, we set $f^{uv} = \frac{e^{-d(u,v)} - e^{-\sqrt{2}}}{1 - e^{-\sqrt{2}}}$.
  The weight function has been normalized to take value 1 when $d(u,v) = 0$ and 0 when $d(u,v) = \sqrt{2}$, with $\sqrt{2}$ being the largest distance between any two nodes located in a unit square.
- Linear: For the linear weight function, we set $f^{uv} = 1 - d(u,v)/\sqrt{2}$. Again, the weight function has been normalized to take value 1 when $d(u,v) = 0$ and 0 when $d(u,v) = \sqrt{2}$.

The attenuation with distance becomes steeper as the weight functions change from linear to exponential to inverse. In our simulations, we consider only integral strategies, with $k = 3$ and $\alpha = 1/6$. We performed natural better-response dynamics (see the full version [4] for details), with each player able to unilaterally deviate in S2H games, but with pairwise deviations allowed in M2H games.

Although in Section 2 we gave an instance of S2H games where an integral NE does not exist, we found that in simulation better-response dynamics converge to an integral NE almost all of the time. To give specific numbers, we found that for linear, exponential, and inverse weight functions, we have convergence for 99%, 97% and 73% of the simulation instances for Weighted S2H games. Moreover, the convergence, when it occurs, is *extremely* fast: most of the instances either converged within 8 or 9 rounds, or did not converge within 500 rounds, as demonstrated in the full version [4]. (A round is a series of better-responses in which every player is given a chance to change their strategy at least once.) The same is not true for Weighted M2H games: over 65 percent of our instances did not converge to a PE even after 5000 rounds. Thus, in the settings that we

examine, we found that we are much more likely to have an (integral) NE in
S2H games (with extremely fast convergence to a stable solution) compared to
Weighted M2H games.

The quality of NE that our simulations converged to was extremely close
to optimal, usually within a few percent of the centralized optimal solution,
indicating that our theoretical bounds are truly only for the worst case, not
average case. Table 1 shows the average values of the equilibria obtained for our
simulations when the weight functions are linear. The values for exponential and
inverse utilities are similar, with the highest numbers of 1.15 obtained for inverse
utilities. We can see that on average, the value of the equilibria obtained is very
close to the value of an optimum solution. The values are also consistent with
worst-case PoA being $1 + \alpha k$, which decreases as α decreases. As α decreases,
the NEs and optimum solutions both converge to nodes following the strategy
of contributing to edges leading to their closest neighbors. Thus there is less
tendency to deviate from the strategy in OPT, resulting in better equilibria.

We also saw that as α decreases or as the weight functions go from linear to
exponential to inverse, the nodes stop forming "clusters" and instead form a
"backbone"-type network resulting from their connecting to the closest possible
nodes. As α decreases or the weight functions become steeper, nodes naturally
begin to connect only to their closest neighbors instead of connecting to slightly
farther nodes that give them two-hop benefit. When α is large, it is advantageous
to form small cliques, since this maximizes two-hop benefit. When α is small,
however, nodes simply connect to their closest k neighbors, and little clustering
takes place. This effect is especially pronounced in M2H games due to bilateral

Table 1. Average value of equilibria obtained with α for $n = 100$, $k = 3$. PoA here
refers to the ratio between the computed NE and the optimum solution.

α	Average PoA (with std dev) Weighted S2H games	Average PoA (with std dev) Weighted M2H games
1/2	1.033 (0.0021)	1.043 (0.0063)
1/4	1.031 (0.0018)	1.037 (0.0049)
1/6	1.029 (0.0016)	1.032 (0.0040)
1/8	1.028 (0.0025)	1.030 (0.0043)

Fig. 3. Typical pairwise equilibria computed for the Weighted M2H games with linear
weight function and (Left) $\alpha = 1/2$ (Middle) $\alpha = 1/8$ (Right) $\alpha = 1/16$

nature of acquaintance (see Fig 3). It would be interesting to investigate further if a phase transition occurs as α decreases, where the clustering effect suddenly disappears, or whether this effect occurs gradually. For more discussion on this topic, see the full version [4].

Acknowledgements. We thank Martin Hoefer for many useful discussions about Two-Hop games.

References

[1] Abraham, D.J., Levavi, A., Manlove, D.F., O'Malley, G.: The stable roommates problem with globally ranked pairs. Internet Mathematics 5(4), 493–515 (2008)

[2] Abramson, G., Kuperman, M.: Social games in a social network. Physical Review E 63(3), 030901 (2001)

[3] Anshelevich, E., Das, S., Naamad, Y.: Anarchy, stability, and utopia: Creating better matchings. In: Mavronicolas, M., Papadopoulou, V.G. (eds.) SAGT 2009. LNCS, vol. 5814, pp. 159–170. Springer, Heidelberg (2009)

[4] Anshelevich, E., Bhardwaj, O., Usher, M.: Friend of my friend: Network formation with two-hop benefit (2013), http://www.cs.rpi.edu/~eanshel/

[5] Anshelevich, E., Hoefer, M.: Contribution games in networks. Algorithmica 63(1-2), 51–90 (2012)

[6] Baiou, M., Balinski, M.: Many-to-many matching: stable polyandrous polygamy (or polygamous polyandry). Discrete Applied Mathematics 101(13), 1–12 (2000)

[7] Bei, X., Chen, W., Teng, S., Zhang, J., Zhu, J.: Bounded budget betweenness centrality game for strategic network formations. Theoretical Computer Science 412(52), 7147–7168 (2011)

[8] Bramoullé, Y., Kranton, R.: Public goods in networks. Journal of Economic Theory 135(1), 478–494 (2007)

[9] Buehler, R., et al.: The price of civil society. In: Chen, N., Elkind, E., Koutsoupias, E. (eds.) WINE 2011. LNCS, vol. 7090, pp. 375–382. Springer, Heidelberg (2011)

[10] Chung, K.S.: On the existence of stable roommate matchings. Games and Economic Behavior 33(2), 206–230 (2000)

[11] Corbo, J., Calvó-Armengol, A., Parkes, D.C.: The importance of network topology in local contribution games. In: Deng, X., Graham, F.C. (eds.) WINE 2007. LNCS, vol. 4858, pp. 388–395. Springer, Heidelberg (2007)

[12] Corbo, J., Parkes, D.: The price of selfish behavior in bilateral network formation. In: Proceedings of PODC, pp. 99–107. ACM (2005)

[13] Fabrikant, A., Luthra, A., Maneva, E., Papadimitriou, C., Shenker, S.: On a network creation game. In: Proceedings of PODC, pp. 347–351. ACM (2003)

[14] Gale, D., Shapley, L.S.: College admissions and the stability of marriage. In: American Mathematical Monthly, pp. 9–15 (1962)

[15] Galeotti, A., Goyal, S., Jackson, M.O., Vega-Redondo, F., Yariv, L.: Network games. The Review of Economic Studies 77(1), 218–244 (2010)

[16] Hoefer, M., Skopalik, A.: Social context in potential games. In: Goldberg, P.W. (ed.) WINE 2012. LNCS, vol. 7695, pp. 364–377. Springer, Heidelberg (2012)

[17] Jackson, M.O., Wolinsky, A.: A strategic model of social and economic networks. Journal of Economic Theory 71(1), 44–74 (1996)

[18] Osborne, M.J., Rubinstein, A.: A course in game theory. MIT Press (1994)

Load Rebalancing Games in Dynamic Systems with Migration Costs

Sofia Belikovetsky and Tami Tamir

School of Computer Science, The Interdisciplinary Center (IDC), Herzliya, Israel

Abstract. We consider the following dynamic load balancing game: Given an initial assignment of jobs to identical parallel machines, the system is modified; specifically, some machines are added or removed. Each job's cost is the load on the machine it is assigned to; thus, when machines are added, jobs have an incentive to migrate to the new unloaded machines. When machines are removed, the jobs assigned to them must be reassigned. Consequently, other jobs might also benefit from migrations. In our *job-extension penalty* model, for a given *extension parameter* $\delta \geq 0$, if the machine on which a job is assigned to in the modified schedule is different from its initial machine, then the job's processing time is extended by δ.

We provide answers to the basic questions arising in this model. Namely, the existence and calculation of a Nash Equilibrium and a Strong Nash Equilibrium, and their inefficiency compared to an optimal schedule. Our results show that the existence of job-migration penalties might lead to poor stable schedules; however, if the modification is a result of a sequence of improvement steps or, better, if the sequence of improvement steps can be supervised in some way (by forcing the jobs to play in a specific order) then any stable modified schedule approximates well an optimal one.

Our work adds two realistic considerations to the study of job scheduling games: the analysis of the common situation in which systems are upgraded or suffer from failures, and the practical fact according to which job migrations are associated with a cost.

1   Introduction

The well-studied *load balancing* problem considers a scenario in which a set of jobs needs to be assigned on a set of identical parallel machines. Each job j, is associated with a processing time p_j and the goal is to balance the load on the machines. In contrast to the traditional load balancing problem, where a central designer determines the allocation of jobs to machines and all the participating entities are assumed to obey the protocol, in the *load balancing game*, each job is owned by a selfish agent who wishes to optimize its own objective.

Given an assignment, each job incurs a cost which is equal to the total load on the machine it is assigned to. This cost function characterizes systems in which jobs are processed in parallel, or when all jobs on a particular machine have the same single pick-up time, or need to share some resource simultaneously.

B. Vöcking (Ed.): SAGT 2013, LNCS 8146, pp. 74–85, 2013.

This problem has been widely studied in recent years from a game theoretic perspective, see [14,2,6,9], and a survey in [17].

In this work, we consider a dynamic variant of this game. Specifically, we are given an assignment, s_0, of n jobs on m_0 machines. The system is modified, namely, m' machines are added or removed. When machines are added, jobs will naturally have an incentive to migrate to the new unloaded machines. When machines are removed, the jobs assigned to the removed machine must be reassigned. As a result of these migrations, other jobs might also benefit from migrations. The goal is to find a pure *Nash Equilibrium* (NE) assignment, s, in the modified system. In such an assignment, no job can reduce its cost by migrating to a different machine. Apparently, this can be viewed as a new instance of the load balancing game. However, in the model we consider, a deviation from the initial assignment is associated with a penalty. We introduce and study the job-extension penalty model. In this model, we are given *an extension parameter* $\delta \geq 0$. If the machine on which job j is scheduled in s is different from its initial machine in s_0, then the processing time of j is extended to be $p_j + \delta$. Practically, this penalty is justified since the reassignment of j causes some extra work on the system, for example, if some preprocessing or configuration set-up was already performed according to the initial assignment.

We distinguish between the following scenarios:

1. The initial schedule s_0 might be a pure NE, or not.
2. The system's modification might be addition or removal of machines.
3. The modified schedule is achieved by performing a sequence of improvement steps, a sequence of best-improvement steps, or arbitrarily.
4. Improvement steps are done unilaterally or by coalitions.

Applications: Traditional analysis of job scheduling assumes a central utility that determines the allocation of jobs to machines and all the participating entities are assumed to obey the protocol. However, in practice, many systems are used by heterogeneous, autonomous agents, which often display selfish behavior and attempt to optimize their own objective rather than the global objective. Game theoretic analysis provides us with the mathematical tools to study such situations, and indeed has been extensively used recently to analyze multiagent systems. This trend is motivated in part by the emergence of the Internet, which is composed of distributed computer networks managed by multiple administrative authorities and shared by users with competing interests [15].

Our work adds two realistic considerations to the study of job scheduling using game theoretic analysis. First, we assume that the system is dynamic and resources might be added or removed - this reflects the common situation in which systems are upgraded or suffer from failures. Second, we assume that job migrations are associated with a cost. Indeed, in real systems, migrations do incur some cost.

The added cost might be due to the transferring overhead or due to set-up time that should be added to the job's processing time in its new location. Consider for example an initial allocation of clients to download servers. Assume that some preprocessing is done at the time a client is assigned to a server, before the download actually begins (e.g., locating the required file, format conversion, etc.). Clients might choose to switch to a mirror server. Such a change would require repeating the preprocessing work on the new server.

Another example of a system in which extension penalty occurs is of an RPC (Remote Procedure Call) service. In this service, a cloud of servers enables service to simultaneous users. When the system is upgraded, more virtual servers are added. Users might switch to the new servers and get a better service (with less congestion), however, some set-up time and configuration tuning is required for each new user.

Note that in all the above applications, the delay caused due to a migration is independent of the migrating job. A similar, low-tech, application is freight transport, in which the whole cargo ship is delayed when items are added. The registration and handling of a new item takes fixed time, independent of the item's size and weight.

1.1   Model and Preliminaries

A job rescheduling setting is defined by the tuple $G = \langle M_0, M', N, p_j, \delta \rangle$, where M_0 is a set of initial identical machines and M' is a set of added or removed machines. If the modification is machines' addition, then M' is a set of new machines, all identical to the machines in M_0. If the modification is machines' removal then $M' \subseteq M_0$. We denote by m_0, m' the number of machines in M_0, M', respectively. $N = \{1, \ldots, n\}$ is the set of jobs. For each job $j \in N$, p_j denotes the processing time of job j. $\delta > 0$ is the extension parameter, i.e, the time penalty that is added to the processing time of a migrating job. An assignment method produces an assignment $s = (s(1), \ldots, s(n))$ of jobs to machines, where $s(j)$ is the machine to which job j is assigned. The assignment is referred to as a schedule. We use s_0, s to denote the initial and the modified schedules, respectively. In s, the processing time of a job $j \in N$ on machine $i \in M_0 \bigcup M'$ is p_j if $i = s_0(j)$ and $p_j + \delta$ otherwise. The load on a machine i in a schedule s is the sum of the processing times (including the extension penalty) of the jobs assigned to i, that is, $L_i(s) = \sum_{j:s(j)=i} p_j + \delta_{i,j}$ where $\delta_{i,j} = 0$ if $s_0(j) = i$ and δ otherwise. For a job $j \in N$, let $c_j(s)$ be the cost of job j in the schedule s, then $c_j(s) = L_{s(j)}$.

An assignment s is a pure Nash equilibrium (NE) if no job $j \in N$ can benefit from unilaterally deviating from its machine to another machine; i.e., for every $j \in N$ and every machined $i \neq s(j)$, $L_i + p_j + \delta_{i,j} \geq L_{s(j)}$.

Some of our results refer to outcomes of a sequence of improvement steps. Better-Response Dynamic (RD) is a local-search method where in each step some player is chosen and is allowed to change his assignment, given the assignment of the others. In better-RD, any improvement step can be performed. In best-RD, players select their best possible response. When best-RD or better-RD are performed, one job might migrate several times. The extension penalty is independent of the number of steps and only the final assignment matters. In particular, if j leaves its machine in s_0 and returns to it later, then j is not extended. This is justified by the applications motivating our work - in which the penalty is not due to physical migration cost but due to the adjustment of the job's processing to a new machine.

It is well known that decentralized decision-making may lead to sub-optimal solutions from the point of view of society as a whole. We quantify the inefficiency incurred due to self-interested behavior according to the *price of anarchy* (PoA) [14,15] and *price of stability* (PoS) [1] measures. The *PoA* is the worst-case inefficiency of a Nash equilibrium, compared to the social optimum (SO), while the *PoS* measures the best-case inefficiency of a Nash equilibrium. The social objective function we consider is the egalitarian one, i.e., we wish to minimize the cost of the job with the highest cost. In scheduling terms, this is equivalent to minimizing the maximal load on some machine (also known as *makespan*). For a schedule s, $makespan(s) = max_j c_j(s) = max_i L_i(s) = L_{max}(s)$. Formally,

Definition 1. *Let $\mathcal{G}$ be a family of games, and let $G \in \mathcal{G}$ be some game in this family. Let $\Phi(G)$ be the set of Nash equilibria of the game G. If $\Phi(G) \neq \emptyset$:*

- *The* price of anarchy *of the game G is the ratio between the* maximal cost *of a NE and the SO of G. That is $PoA(G) = max_{s \in \Phi(G)} L_{max}(s)/OPT(G)$. The* price of anarchy *of the family of games $\mathcal{G}$ is $PoA(\mathcal{G}) = Sup_{G \in \mathcal{G}} PoA(G)$.*
- *The* price of stability *of the game G is the ratio between the* minimal cost of a *NE and the SO of G. That is, $PoS(G) = min_{s \in \Phi(G)} L_{max}(s)/OPT(G)$, and the* price of stability *of the family of games $\mathcal{G}$ is $PoS(\mathcal{G}) = Sup_{G \in \mathcal{G}} PoS(G)$.*

In section 4 we study coordinated deviations. A set of players $\Gamma \subseteq N$ forms a *coalition* if there exists a move where each job $j \in \Gamma$ strictly reduces its cost. An assignment s is a *strong equilibrium* (SE) if there is no coalition $\Gamma \subseteq N$ that has a beneficial move from s. The *strong PoA* and the *strong PoS* are defined similarly, where $\Phi(G)$ refers to the set of strong Nash equilibria.

1.2   Related Work

The minimum makespan problem corresponds to the centralized version of our game in which all jobs obey the decisions of one utility. This is a well-studied NP-hard problem, having a simple greedy $(\frac{4}{3} - \frac{1}{3m})$-approximation algorithm (LPT) [12], and a PTAS [13].

In the associated load balancing game, each job is controlled by a selfish agent who aims to minimize its cost - given by the load on the machine it is assigned to. Fotakis et al. showed that LPT-schedules are NE schedules [11]. In [7], Even-dar et al. analyzed the convergence time of BRD on unrelated machines. Note that our model can be seen as a restricted case of scheduling on unrelated machines. For every job j and machine M_i, the processing time of j on M_i is p_j if $i = s_0(j)$ and $p_j + \delta$ otherwise. Our analysis provides tighter results than those known for unrelated machines [6].

The concept of the price of anarchy (PoA) was introduced by Koutsoupias and Papadimitriou in [14]. They proved that the price of anarchy of job scheduling games is $2 - \frac{1}{m}$. In [10], Finn and Horowitz presented an upper bound of

$2 - \frac{2}{m+1}$ for the price of anarchy in load balancing games with identical machines. Note that in this game, the PoA is equivalent to the makespan approximation.

Other related work deal with cost functions that depend on the internal order of jobs, (e.g., [5,3]), or a cost function that is based on both the congestion on the machine and its activation cost [8]. Some of our results bound the inefficiency of a NE produced by a sequence of improvement steps (best or better-RD). Analysis of such *sequential* NE was initiated in [16] for several other games. Coordinated deviations were studied by Andelman et al. in [2]. A survey of results on selfish load balancing appears in [17].

1.3   Our Results

We study the problem of equilibrium existence, calculation, and inefficiency in the load rebalancing game with uniform extension penalty. We show that any job scheduling game with added or removed machines possesses at least one Nash equilibrium schedule. Moreover, some optimal solution is also a Nash equilibrium, and thus, the PoS is 1. We show that in general, the PoA is unbounded when machines are either added or removed. The PoA can be bounded if the modified schedule is achieved by performing improvement steps. Specifically, for a NE that is achieved by performing improvement steps, we show that the PoA is

1. $\frac{m'-1}{m_0} + 2$ when machines are added and s_0 is a NE.
2. $m_0 + m'$ when machines are added and s_0 is not a NE.
3. $m_0 - m'$ when machines are removed (and s_0 is either a NE or not a NE).
4. $2 - \frac{1}{m_0-m'}$ when machines are removed, s_0 is a NE, and jobs are activated in a specific order, denoted *two-phase better-RD*.

For all the above cases we prove the upper bound and provide matching lower bounds. The lower bounds are tight for some values of m_0, m' and almost tight for other values.

We also analyze the load rebalancing game assuming coordinated deviations are allowed. We prove that a strong equilibrium exists for all system modifications and that the $SPoS$ is 1. We show that the $SPoA$ is 3 for both the adding and removing machines scenarios and that this bound is tight. Moreover, we provide a closer analysis of the $SPoA$, and bound this value as a function of the ratio between δ and OPT. Specifically, we show that the $SPoA$ is $2 + \frac{\delta}{OPT}$ and that this bound is tight. Moreover, it is achieved even when the SE is reached by a sequence of coalitional improvement steps. Finally, we show that for any value of $\delta > 0$, it is NP-hard to determine whether a given modified schedule is a SE.

We note that in a dynamic setting in which machines are added or removed and migrations are free of cost (i.e., when $\delta = 0$), then the results known for classic load balancing games apply. In particular, the PoA assuming $\delta = 0$ is $2 - \frac{2}{m+1}$ for a game with m machines in the modified systems. The proofs are identical to the proofs for a fixed number of machines. Thus, the difference between our results and the results for the classic load balancing game are due to the migration penalty.

Due to space constraints most of the proofs are given in the full version [4].

2   Machines' Addition

In this section we study the scenario in which the system's modification involves an addition of machines and uniform extension penalty is applied. Specifically, for a given parameter $\delta > 0$, if a job is assigned to a machine different than its original machine then its processing time is extended to be $p_j + \delta$. Recall that m_0, m' denote the initial and added number of machines, respectively.

2.1   Equilibrium Existence and Computation

Every instance of the load rebalancing game with added machines and uniform extension penalty admits at least one pure Nash equilibrium. This follows from the fact that the sorted machines' load-vector corresponding to the schedule is decreasing lexicographically with any beneficial move. Thus, any better-RD process converges to a NE.

The next question we consider is how many moves are required to reach a NE. The following result shows that, for any given initial assignment, there exists a short sequence of beneficial moves that leads to a NE. Assume that the jobs are sorted according to their processing length, that is, $p_1 \geq p_2 \geq \ldots \geq p_n$. *Max-length best-RD* activates the jobs one after the other according to the sorted order. An activated job j plays a best response, i.e., it moves to a machine that minimizes its cost (or remain on $s_0(j)$ if no beneficial move exists).

We show that after a single phase of *max-length best-RD*, the system reaches a NE. While this result is valid also for the classic load balancing game [17], its proof for the load rebalancing game is more involved, since migrating jobs have stronger incentive to return to their initial machine (and get rid of the penalty).

Theorem 1. *Let s_0 be any initial schedule of n jobs on m_0 machines. Assume that m' machines are added. Starting from s_0, max-length best-RD reaches a pure Nash equilibrium after each job is activated once.*

2.2   Equilibrium Inefficiency

In this section we bound the *price of stability* and the *price of anarchy* of our game, distinguishing between various initial states and convergence methods. For the classic load balancing game, with no extension penalty, it is known that $PoS = 1$ and $PoA = 2 - \frac{2}{m+1}$. We show that in our model $PoS = 1$ and the PoA is not bounded by a constant. It can be arbitrary large if the schedule is not achieved by a sequence of improvement steps and bounded by $\frac{m'-1}{m_0} + 2$ if the schedule is achieved by a sequence of improvement steps. We also show that if the initial schedule is not a NE but the schedule is achieved by performing a sequence of improvement steps, the PoA is bounded by $m_0 + m'$.

It is easy to see that a beneficial move does not increase the makespan. Therefore, by performing best-RD starting from any optimal assignment, we reach a NE whose makespan is equal to the optimum. This implies that the PoS equals 1. We turn to analyze the PoA. We first mention (see [4]) that when the NE is not necessarily achieved by a sequence of beneficial moves, the PoA is unbounded. The bound is valid even if $m' = 0$ and the initial schedule is a NE.

The more realistic scenario is when the NE is reached by performing beneficial moves, starting from a NE schedule s_0. We provide an upper bound, which is tight when $m' \bmod m_0 = 1$, and almost tight for any other case.

Theorem 2. *When the NE is reached by better-RD, $PoA \leq \frac{m'-1}{m_0} + 2$.*

Let $m' = km_0 + \alpha$ for integers k and $\alpha < m_0$. By Theorem 2, we have that the PoA is at most $\frac{m'-1}{m_0} + 2 = \frac{m_0 k + \alpha - 1}{m_0} + 2 = k + \frac{\alpha - 1}{m_0} + 2$. We show that for $\alpha = 1$ and any k, the bound is tight. Almost tight analysis for other values of k, α is given in [4].

Theorem 3. *For any number of machines m_0, for any integer $k > 0$, and for any $\rho > 0$, there exists an input with $m' = km_0 + 1$ added machines, for which $PoA > 2 + \frac{m'-1}{m_0} - \rho$, and the NE is reached by better-RD.*

Proof. Given ρ, m_0, k, let $m' = km_0 + 1$. Let B be an integer such that $\rho \geq \frac{k+2}{B+1}$. In addition, let $\varepsilon = \frac{1}{(k+1)m'B}$ and $\delta = 1 - \varepsilon$.

The set of jobs includes $m' + m_0 = (k+1)m_0 + 1$ jobs of length B, and $1/\varepsilon = (k+1)m'B$ jobs of length ε. In the initial assignment, a single machine is assigned $k+2$ jobs of length B and each of the other $m_0 - 1$ machines is assigned $k+1$ jobs of length B, as well as some jobs of length ε, such that the ε-jobs are assigned in a balanced way and the assignment is a NE. Note that the load on the first machine is $(k+2)B$ and the load on each of the other $m_0 - 1$ machines is between $(k+1)B$ and $(k+1)B + 1$.

We present the construction of the lower bound in Fig. 1, where $m_0 = 3$ and $k = 1$ (implying $m' = 4$). The initial assignment is given in Fig. 1(a).

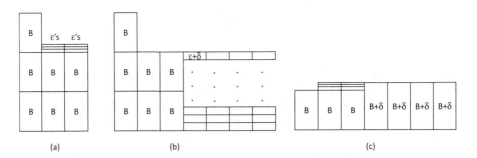

Fig. 1. An instance achieving the maximal possible PoA. (a) the initial assignment, (b) the worst NE, and (c) the best NE.

Assume that m' machines are added and improving steps are performed. A possible NE (see Fig. 1(b)) is a one in which the long jobs remain on M_0 and every new machine is assigned $(k+1)B$ jobs of length ε. The load on the first machine remains $(k+2)B$. The load on each of the other $m_0 - 1$ machines of M_0 is $(k+1)B$. The load on every new machine is $(k+1)B(\delta + \varepsilon) = (k+1)B$.

The maximum load is $(k+2)B$ - achieved on the first machine. This assignment is a NE as the shortest job on the most loaded machine has length B - which is exactly the gap from the load on all other machines. Also, the other machines are perfectly balanced, therefore no migrations are beneficial.

On the other hand, the following is an optimal assignment (see Fig. 1(c)): One job of length B migrates to each of the new machines. The other m_0 jobs of length B as well as all jobs of length ε remain on the original machines M_0. The maximal load on M_0 is at most $B + 1$. The load on every new machine is $B + \delta < B + 1$.

The ratio between the maximal loads of the two assignments is $\frac{(k+2)B}{B+1}$. The value of B was selected such that this is more than $2 + k - \rho = 2 + \frac{m'-1}{m_0} - \rho$. □

Finally, we analyze the *PoA* for arbitrary initial assignment. The analysis is tight for any number of machines m_0, m' and for any $\delta > 0$.

Theorem 4. *If the initial assignment is not necessary a NE, and the modified schedule is reached by better-RD, then the PoA is at most $m_0 + m'$.*

3   Machines' Removal

In this section we study the scenario in which the system's modification involves the removal of machines. Every job assigned to a removed machine must be reassigned. As a result, additional jobs might also be interested in migrating. Recall that M_0, M' denote the sets of initial and removed machines, respectively. Let $M_1 = M_0 \setminus M'$ denote the set of remaining machines. Let m_0, m', m_1 denote the corresponding numbers of machines, that is $m_1 = m_0 - m'$. Throughout this section we assume that the initial schedule, s_0, is a NE. The last result in this section, Theorem 9, considers the case in which s_0 is not a NE.

3.1   Equilibrium Existence and Computation

We prove the existence of a NE and analyze the convergence rate of several policies. When better-RD is applied, all jobs are activated in an arbitrary order. When activated, each job migrates if it is on M' or if it can improve its cost. For every job j, if $s_0(j) \in M'$, j must be activated at least once, move to a machine in M_1, and be extended. Clearly, jobs must not migrate into machines in M'.

Theorem 5. *Better-RD leads to a NE assignment for every instance of the load rebalancing game with removed machines and uniform extension penalty.*

Two-phase max-length best-RD: In the 1st phase all the jobs assigned to machines in M' are activated. In the 2nd phase all the jobs (now assigned to M_1) are activated in a non-increasing order of processing time p_j without taking into account the extension penalty. In both phases, jobs perform their best move.

Unlike the 'adding machines' scenario, when machines are removed, a single phase of max-length best-RD might not end up in a NE. However, linear time convergence to a NE is guaranteed by the above *two-phase max-length best-RD*.

Theorem 6. Two-phase max-length best-RD *leads to a pure NE schedule.*

3.2 Equilibrium Inefficiency

In this section we analyze the PoS and the PoA with various initial states and convergence algorithms. We show that the results differ from the classical load balancing game as well as from the machines' addition scenario.

We note that by the discussion in Section 2.2, the PoS of the selfish load rebalancing game with removed machines and any job extension penalty is 1. Also (see details in [4]), the PoA is unbounded if the NE is not reached by performing beneficial migrations. On the other hand, by assuming the NE is reached by better-RD, we can bound the PoA. The following is tight for every $m_1 \le m'$, such that $m_1 | m'$.

Theorem 7. *The PoA assuming that s_0 is a NE and the modified NE is reached by* better-RD *is m_1.*

While the PoA for arbitrary better-RD is m_1, a smaller bound can be shown if the NE is reached by *two-phase* better-RD. In the first phase, all the jobs that are assigned to machines in M' are activated, each performing its best move to a machine in M_1. In the second phase, all the jobs (now assigned to M_1) are activated, possibly several times, in an arbitrary order.

As this is a specific application of better-RD, convergence to a NE is guaranteed. The use of two-phase better-RD can reduce significantly the inefficiency of a NE. The following analysis is tight even for *two-phase best-RD*.

Theorem 8. *The PoA assuming that s_0 is a NE and the modified NE is reached by* two-phase better-RD *is $2 - \frac{1}{m_1}$.*

Finally, we bound the PoA assuming the initial schedule is not a NE. The upper bound proof is similar to the proof of Theorem 4. The lower bound follows from Theorem 7.

Theorem 9. *If the initial assignment is not necessary a NE, and the modified schedule is reached by better-RD, then the PoA is at most m_1 and this is tight.*

4 Analysis of Coordinated Deviations

In this section we assume that agents can coordinate their strategies and perform a coordinated deviation. Recall that a set of players $\Gamma \subseteq N$ forms a *coalition* if there exists a move where each job $j \in \Gamma$ strictly reduces its cost. A schedule s is a *strong equilibrium* (SE) if there is no coalition $\Gamma \subseteq N$ that has a beneficial move from s.

It is not difficult to see (using the arguments used for the classic load balancing game [2]) that for any instance of the dynamic game, a SE exists. In particular, an assignment in which the vector of loads is lexicographically minimal is a SE. On the other hand, as we show, finding a SE schedule, or deciding whether a NE schedule s is a SE is NP-hard. Moreover, given a set of jobs, it is NP-hard to determine whether this set has a beneficial coordinated deviation.

Theorem 10. *Let s be a NE schedule in a system after a modification took place. For any $\delta \ge 0$, it is NP-hard to determine whether s is a SE.*

4.1   Equilibrium Inefficiency

We present tight bounds for the *strong price of anarchy*. By the discussion in Section 2.2, the *strong price of stability* is 1.

Theorem 11. *The SPoA of load rebalancing games with uniform extension penalty with added or removed machines is at most 3.*

We show that the above analysis is tight even when the initial schedule is a SE in the cases of adding and removing machines. For simplicity, the instance below is described for specific values of m_0 and m'. It can be generalized by scaling and/or adding dummy jobs.

Theorem 12. *For any $\rho > 0$, there exist instances with added machines and with removed machines for which $SPoA \geq 3 - \rho$.*

Proof. We show the construction for added machines, see [4] for removed machines. Given ρ, let $\varepsilon < 1$ be a small constant and let B be an integer such that $\rho \geq \frac{4\varepsilon}{B+\varepsilon}$. Fix $\delta = B - \varepsilon$. The initial schedule, on $m_0 = 3$ machines, is given in Fig. 2(a). Note that each machine accommodates one long job of length B and one tiny job of length ε (job-lengthes' indices in the figure denote $s_0(j)$ - to help follow the migrations). Since the load is perfectly balanced, s_0 is a strong equilibrium. Assume that $m' = 2$ machines are added. Consider the schedule s given in Fig. 2(b). We have $L_{max}(s) = 2B + \delta = 3B - \varepsilon$.

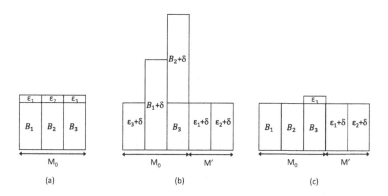

Fig. 2. An instance achieving $SPOA = 3$. (a) the initial assignment, (b) a possible SE, and (c) the best SE.

We show that s is a SE by showing that no job can be part of a coalition. Note that the current cost of each ε-job is $\varepsilon + \delta = B$, therefore, no job will join an ε-job on its current machine. Moreover, an ε-job will participate in a coalition only if it returns to its original machine alone. Therefore, after any coalitional move, three different machines will be dedicated to the ε-jobs. Since there are three B_i jobs and two machines without ε-jobs, there is a machine with two B_i

jobs on it. At least one of which is extended. Thus, in any coalitional move, one machine has load $2B + \delta$ which is not beneficial for the jobs assigned to it. Thus, no coalition exists. An optimal schedule for the modified instance is given in Fig. 2(c). $L_{max}(OPT) = B+\varepsilon$. Therefore, $SPoA \geq \frac{3B-\varepsilon}{B+\varepsilon} = \frac{3B+3\varepsilon}{B+\varepsilon} - \frac{4\varepsilon}{B+\varepsilon} \geq 3-\rho$.

$\square$

It is possible to provide a tighter analysis of the strong price of anarchy, by bounding this value as a function of the ratio between δ and OPT. The proof of the following theorem is similar to the proof of Theorem 11. The bound is tight even if the initial schedule is a SE, and the final SE is reached by a sequence of coalitional improvement steps (see [4]).

Theorem 13. *The SPoA of load rebalancing games with uniform extension penalty is at most* $2 + \frac{\delta}{OPT}$.

5   Summary and Future Work

We considered a dynamic variant of the classic load balancing game, in which machines are added or removed and job migrations are associated with job's extension.

To the best of our knowledge, these are the first results considering games with migration costs. We provided answers to the basic questions arising in this model. Specifically, we explored the existence and calculation of Nash equilibrium and strong equilibrium and provided tight bounds for NE and SE inefficiency - in general and under various dynamics. Our results show that the existence of migration penalty might lead to poor stable schedules, however, if the modification is a result of a sequence of improvement steps or, better, if the sequence of improvement steps can be supervised in some way (by forcing the jobs play in a specific order, or select their best response) then the modified schedule approximates well an optimal one, with approximation ratio similar to the classic load balancing game. Thus, while migration costs discourage changes and increase the stability of any given configuration, it is still guaranteed that any stable configuration that is reached by natural dynamics has a reasonable social cost.

Possible directions for future work include the study of heterogenous systems, in particular unrelated machines, or non-uniform extension penalty. That is, for each i, i', j we are given an extension parameter $\delta_{i,i',j}$ such that job j is extended by $\delta_{i,i',j}$ if it migrates from machine i to machine i'. Another interesting variant is proportional extension, i.e., a migration of job j extends its processing time from p_j to $p_j(1 + \delta)$. Studying modifications that include both addition and removal of machines is another open direction.

Our analysis of coordinated deviation show that the $SPoA$ heavily depends on the value of δ. Possible future work is to analyze instances in which δ is bounded by the instance parameters, e.g., when $\delta \leq p_{min}$. Analyzing equilibrium inefficiency with respect to the objective of minimizing the total players' cost is another challenge.

Finally, in our model a migration of job j affects all the jobs assigned to j's target machine. Another possible game can be defined by assuming *individual* penalties. Specifically, migrations are associated with a cost, but this cost is covered by the job and does not affect other jobs. The cost of a job j assigned to machine i is L_i if $i = s_0(j)$ and $L_i + \delta$ otherwise, where the load is the total processing time of jobs assigned to machine i.

References

1. Anshelevich, E., Dasgupta, A., Kleinberg, J.M., Tardos, É., Wexler, T., Roughgarden, T.: The price of stability for network design with fair cost allocation. In: FOCS (2004)
2. Andelman, N., Feldman, M., Mansour, Y.: Strong Price of Anarchy. In: SODA (2007)
3. Azar, Y., Jain, K., Mirrokni, V.S.: (Almost) optimal coordination mechanisms for unrelated machine scheduling. In: SODA (2008)
4. Belikovetsky, S., Tamir, T.: Load Rebalancing Games in Dynamic Systems with Migration Costs. Full version, `http://www.faculty.idc.ac.il/tami/Papers/Rebalance.pdf`
5. Caragiannis, I., Flammini, M., Kaklamanis, C., Kanellopoulos, P., Moscardelli, L.: Tight Bounds for Selfish and Greedy Load Balancing. Algorithmica 61(3), 606–637 (2011)
6. Czumaj, A., Vöcking, B.: Tight bounds for worst-case equilibria. ACM Transactions on Algorithms 3(1) (2007)
7. Even-Dar, E., Kesselman, A., Mansour, Y.: Convergence time to Nash equilibria. In: Baeten, J.C.M., Lenstra, J.K., Parrow, J., Woeginger, G.J. (eds.) ICALP 2003. LNCS, vol. 2719, pp. 502–513. Springer, Heidelberg (2003)
8. Feldman, M., Tamir, T.: Conflicting congestion effects in resource allocation games. Journal of Operation Research 60(3), 529–540 (2012)
9. Fiat, A., Kaplan, H., Levy, M., Olonetsky, S.: Strong Price of Anarchy for Machine Load Balancing. In: Arge, L., Cachin, C., Jurdziński, T., Tarlecki, A. (eds.) ICALP 2007. LNCS, vol. 4596, pp. 583–594. Springer, Heidelberg (2007)
10. Finn, G., Horowitz, E.: A linear time approximation algorithm for multiprocessor scheduling. BIT 19(3), 312–320 (1979)
11. Fotakis, D., Kontogiannis, S., Koutsoupias, E., Mavronicolas, M., Spirakis, P.: The structure and complexity of Nash equilibria for a selfish routing game. In: Widmayer, P., Triguero, F., Morales, R., Hennessy, M., Eidenbenz, S., Conejo, R. (eds.) ICALP 2002. LNCS, vol. 2380, pp. 123–134. Springer, Heidelberg (2002)
12. Graham, R.L.: Bounds on Multiprocessing Timing Anomalies. SIAM J. Appl. Math. 17, 263–269 (1969)
13. Hochbaum, D.S., Shmoys, D.B.: Using dual approximation algorithms for scheduling problems: Practical and theoretical results. Journal of the ACM 34(1), 144–162 (1987)
14. Koutsoupias, E., Papadimitriou, C.: Worst-case Equilibria. Computer Science Review 3(2), 65–69 (1999)
15. Papadimitriou, C.: Algorithms, Games, and the Internet. In: STOC (2001)
16. Paes Leme, R., Syrgkanis, V., Tardos, É.: The curse of simultaneity. In: ITCS (2012)
17. Vöcking, B.: In: Nisan, N., Roughgarden, T., Tardos, E., Vazirani, V. (eds.) Algorithmic Game Theory: Selfish Load Balancing, ch. 20. Cambridge University Press (2007)

Stochastic Congestion Games with Risk-Averse Players*

Haris Angelidakis[1], Dimitris Fotakis[2], and Thanasis Lianeas[2]

[1] Toyota Technological Institute at Chicago, USA
hangel@ttic.edu
[2] Electrical and Computer Engineering, National Technical University of Athens, Greece
fotakis@cs.ntua.gr, tlianeas@mail.ntua.gr

Abstract. Congestion games ignore the stochastic nature of resource delays and the risk-averse attitude of the players to uncertainty. To take these aspects into account, we introduce two variants of atomic congestion games, one with *stochastic players*, where each player assigns load to her strategy independently with a given probability, and another with *stochastic edges*, where the latency functions are random. In both variants, the players are risk-averse, and their individual cost is a player-specific quantile of their delay distribution. We focus on parallel-link networks and investigate how the main properties of such games depend on the risk attitude and on the participation probabilities of the players. In a nutshell, we prove that stochastic congestion games on parallel-links admit an efficiently computable pure Nash equilibrium if the players have either the same risk attitude or the same participation probabilities, and also admit a potential function if the players have the same risk attitude. On the negative side, we present examples of stochastic games with players of different risk attitudes that do not admit a potential function. As for the inefficiency of equilibria, for parallel-link networks with linear delays, we prove that the Price of Anarchy is $\Theta(n)$, where n is the number of stochastic players, and may be unbounded, in case of stochastic edges.

1 Introduction

Congestion games provide an elegant and useful model of selfish resource allocation in large-scale networks. In an (atomic) *congestion game*, a finite set of players, each with an unsplittable unit of load, compete over a finite set of resources (or edges). All players using an edge experience a latency given by a non-negative and non-decreasing function of the edge's load (or congestion). Each player selects a path between her origin and destination, trying to minimize her *individual cost*, that is, the sum of the latencies on the edges in the chosen path. A natural solution concept is that of a *pure Nash equilibrium* (PNE), a configuration where no player can decrease her individual cost by unilaterally changing her path.

In a seminal work, Rosenthal [18] proved that the PNE of congestion games correspond to the local optima of a natural potential function, and thus every congestion

---

* This research was supported by the project Algorithmic Game Theory, co-financed by the European Union (European Social Fund - ESF) and Greek national funds, through the Operational Program "Education and Lifelong Learning" of the National Strategic Reference Framework (NSRF) - Research Funding Program: THALES, investing in knowledge society through the European Social Fund.

B. Vöcking (Ed.): SAGT 2013, LNCS 8146, pp. 86–97, 2013.

game admits a PNE. Following [18], the properties of congestion games and several variants of them have been extensively studied. The prevailing questions in recent work have to do with whether congestion games and some natural generalizations of them admit an (approximate) PNE and/or an (approximate) potential function (see e.g., [13], [11], [12] and [5]), with bounding the convergence time to a PNE if the players act selfishly (see e.g., [1], [7], and [10]), and with quantifying the inefficiency of PNE due to the players' selfish behavior (see e.g., [4], [8], [2], and [6]). Notably, a significant part of recent work concerns the properties of congestion games and their generalizations on parallel-link networks (see e.g., [13], [6], and [10], and the references therein).

However, most research work on congestion games essentially ignores the stochastic nature of edge delays and assumes that players have precise knowledge of the (deterministic) edge latencies. On the contrary, in real life situations, players cannot accurately predict the actual edge delays, not only because they cannot know the exact congestion of every edge, but also due to (a priori unknown) external events (e.g., some construction work, a minor accident, a link failure) that may affect the edge latencies and introduce uncertainty. It is therefore natural to assume that the players decide on their strategies based only on estimations of their actual delay and, most importantly, that they are fully aware of the uncertainty and of the potential inaccuracy of their estimations. So, to secure themselves from the event of an increased delay, players select their paths taking uncertainty into account (e.g., people either take a safe route or plan for a larger than usual delay when they head to an important meeting).

Such considerations give rise to congestion games with stochastic delays and risk-averse players, where instead of the path that minimizes her expected delay, each player selects a path that guarantees her a reasonably low actual delay with reasonably high confidence. Here, the actual delay of each player can be modeled by a random variable. Then, a common assumption is that players seek to minimize either a convex combination of the expectation and the variance of their delay, or a player-specific quantile of the delay distribution (see also [19], [9] about the cost functions of risk-averse players, and [17] about possible ways of risk quantification in optimization under uncertainty).

Previous Work. Following the research direction above, Ordóñez and Stier-Moses [15] considered *nonatomic* congestion games and suggested that each path should be penalized by an additive term that increases with the risk-aversion of the players and with the maximum deviation from the expected delay of the path (however, this term does not depend on the actual load of the edges). For each path, the additive term can be chosen either as a δ-fraction of (resp. a δ-quantile of a random variable depending on) the maximum deviation from the expected delay of the path, or simply, as the sum of the δ-fractions of the maximum deviation from the expected delay of each edge in the path, where δ quantifies the risk-aversion of the players. Under some general assumptions, [15] proves that an equilibrium exists and is essentially unique in all the cases above.

Subsequently, Nikolova and Stier-Moses [14] suggested a model of stochastic selfish routing with risk-averse players, where each player selects a path that minimizes the expected delay plus δ times the standard deviation of the delay, where δ quantifies the risk-aversion of the players. They considered nonatomic and atomic congestion games, mostly with homogeneous players, that share the same risk attitude, and distinguished between the case where the standard deviation of a path's delay is *exogenous*, i.e., it does not depend on the load of the edges in the path, and the case where it is *endogenous*, i.e., it is a function of the load. They proved that in the exogenous case, which is similar to the model of [15], stochastic routing games essentially retain the nice properties of standard congestion games: they admit a potential function and, in the nonatomic setting, a unique equilibrium, and the inefficiency of equilibria can be bounded as for standard congestion games. In the endogenous case, they proved that nonatomic stochastic routing games admit an equilibrium, which is not necessarily unique, but may not admit a cardinal potential. Moreover, atomic stochastic routing games may not admit a PNE even in simple extension-parallel networks with 2 players and linear delays.

Contribution. Following this research agenda, we seek a better understanding of the properties of congestion games with stochastic delays and risk-averse players. We focus on atomic congestion games and introduce two variants of stochastic congestion games. We start from the observation that the variability of edge delays comes either from the variability of the traffic demand, and the subsequent variability of the edge loads, or from the variability of the edge performance level. Decoupling them, we introduce two variants, namely *Congestion Games with Stochastic Players* and *Congestion Games with Stochastic Edges*, each capturing one of the two sources of uncertainty above.

Congestion Games with Stochastic Players aim to model the variability of the traffic demand. Specifically, each player i participates in the game independently with probability p_i. As a result, the total network load, the edge loads, and the edge and the path latencies are all random variables. On the other hand, Congestion Games with Stochastic Edges aim to model variability in the network operation. Now, each edge e may operate either at the "standard" mode, where its latency is given by a function $f_e(x)$, or at the "faulty" mode (e.g., after a minor accident or a link failure), where its latency is given by $g_e(x)$, with $g_e(x) \geq f_e(x)$. Each edge e switches to the "faulty" mode independently with a given probability p_e. Hence, the network load and the edge loads are now deterministic, but the edge and the path latencies are random variables. In both variants, players are risk-averse to the stochastic delays. Specifically, each player i has a (possibly different) desired *confidence level* δ_i, and her cost on a path q is the δ_i-quantile (a.k.a. value-at-risk) of the delay distribution of q. In words, the individual cost of player i is the minimum delay she can achieve along q with probability at least δ_i.

At the conceptual level, the model of Congestion Games with Stochastic Players is similar to the model with endogenous standard deviations of [14]. In fact, using Chernoff bounds, one can show that for linear latency functions, if the expected edge loads are not too small, our δ_i-quantile individual cost can be approximated by the individual cost used in [14]. However, we also consider stochastic demands, a direction suggested in [14, Sec. 7] to enrich the model, and players that are heterogeneous with respect to risk attitude. As for Congestion Games with Stochastic Edges, the model is conceptually similar to the model with exogenous standard deviations of [14].

In the technical part of the paper, we restrict ourselves to parallel-link networks with symmetric player strategies, and investigate how the properties of stochastic congestion games depend on the players' participation probabilities and confidence levels. We first observe that such games admit a potential function and an efficiently computable PNE, if the players are homogeneous, namely if they have the same confidence level δ and, in case of stochastic players, the same participation probability p (Theorems 1 and 7).

We also show that if the players have different confidence levels (and the same participation probability, if they are stochastic), stochastic congestion games belong to the class of player-specific congestion games [13], and thus admit a PNE computable in polynomial time (Corollaries 1 and 2). On the negative side, we prove that such games may not admit a potential function (Theorems 2 and 8). For Congestion Games with Stochastic Players that have the same confidence level and different participation probabilities, we show that they admit a lexicographic potential (Theorem 4), and thus a PNE, which can be computed by a simple greedy best response algorithm (Theorem 3). As for the inefficiency of PNE, in the case of linear latency functions, we prove that the Price of Anarchy (PoA) is $\Theta(n)$, if we have n stochastic players, and (Theorems 5 and 6), and may be unbounded, in the case of stochastic edges (Theorem 9).

Other Related Work. There is a significant volume of work on theoretical and practical aspects of transportation networks with uncertain delays, which however focuses on nonatomic games and adopts notions of individual cost and viewpoints quite different from ours (see e.g., the discussion in [14]). Motivated by applications with only partial knowledge of the number of players participating in the game, Ashlagi, Monderer, and Tennenholtz [3] considered congestion games on parallel links with stochastic players. However, the players in [3] are risk-neutral, since their individual cost is the expected delay of the chosen link. They proved that a generalization of the fully mixed equilibrium remains a mixed Nash equilibrium in this setting. Very recently, Piliouras, Nikolova, and Shamma [16] considered atomic congestion games with risk-averse players and delays determined by a randomized scheduler of the players on each edge. They obtained tight bounds on the PoA of such games with linear latencies under various notions of risk-averse individual cost. Interestingly, they proved that the PoA can be unbounded for the individual cost of [14].

2   Notation and Preliminaries

In this section, we introduce the notation and the basic terminology of standard congestion games. For a random variable X, we let $\mathbb{E}[X]$ denote the *expectation* and $\mathbb{V}\mathrm{ar}[X]$ denote the *variance* of X. For an event E, we let $\Pr[E]$ denote its probability. For a vector $x = (x_1, \ldots, x_n)$, we let $x_{-i} \equiv (x_1, \ldots, x_{i-1}, x_{i+1}, \ldots, x_n)$ and $(x_{-i}, x_i') \equiv (x_1, \ldots, x_{i-1}, x_i', x_{i+1}, \ldots, x_n)$.

Congestion Games. A *congestion game* is a tuple $\mathcal{G}(N, E, (\Sigma_i)_{i \in N}, (d_e)_{e \in E})$, where N is the set of players, E is the set of resources, $\Sigma_i \subseteq 2^E \setminus \{\emptyset\}$ is the strategy space of each player i, and $d_e : \mathbb{N} \mapsto \mathbb{R}_{\geq 0}$ is a non-negative and non-decreasing latency function associated with each resource e. A congestion game is *symmetric* if all players share the same strategy space. In what follows, we let n denote the number of players and m denote the number of resources.

A *configuration* is a vector $\sigma = (\sigma_1, \ldots, \sigma_n)$ consisting of a strategy $\sigma_i \in \Sigma_i$ for each player i. For every resource e, we let $\sigma_e = |\{i \in N : e \in \sigma_i\}|$ denote the congestion induced on e by σ. The individual cost of player i in the configuration σ is $c_i(\sigma) = \sum_{e \in \sigma_i} d_e(\sigma_e)$. A configuration σ is a *pure Nash equilibrium* (PNE) if no player can improve her individual cost by unilaterally changing her strategy. Formally, σ is a PNE if for every player i and every strategy $s_i \in \Sigma_i$, $c_i(\sigma) \leq c_i(\sigma_{-i}, s_i)$.

Next, we focus on symmetric congestion games on parallel-link networks, where the strategies are singletons and there is a strategy for every resource. Thus, we use the terms "resource" and "edge", and "strategy" and "path" interchangeably.

Social Cost. To quantify the inefficiency of PNE, configurations are usually evaluated using the *total cost* of the players. In standard congestion games, the total cost of a configuration σ, denoted $C(\sigma)$, is $C(\sigma) = \sum_{i \in N} c_i(\sigma)$. The *optimal configuration*, usually denoted o, minimizes the total cost among all possible configurations.

Price of Anarchy. The (pure) *Price of Anarchy* (PoA) of a congestion game $\mathcal{G}$ is the maximum ratio $C(\sigma)/C(o)$ over all PNE σ of $\mathcal{G}$. The PoA of a class of games is defined as the maximum PoA among all games in the class.

3   Congestion Games with Stochastic Players

3.1   The Model

In *Congestion Games with Stochastic Players*, each player i is described by a tuple (p_i, δ_i), where $p_i \in [0, 1]$ is the probability that player i participates in the game, by assigning a unit of load to her strategy, and $\delta_i \in \left[\frac{1}{2}, 1\right]$ is the *confidence level* (or risk-aversion) of player i. Essentially, each player i is associated with a Bernoulli random variable X_i that is 1 with probability p_i, and 0 with probability $1 - p_i$. Then, the load of each edge e in a configuration σ is the random variable $N_e(\sigma) = \sum_{i:e \in \sigma_i} X_i$, and the cost of a strategy q in σ is the random variable $D_q(\sigma) = \sum_{e \in q} d_e(N_e(\sigma))$.

Given that player i participates in the game, the delay of player i in σ is given by the random variable:

$$D_i(\sigma) = \sum_{e \in \sigma_i} d_e \left(1 + \sum_{j \neq i: e \in \sigma_j} X_j \right).$$

We note that conditional on $X_i = 1$, $D_i(\sigma) = D_{\sigma_i}(\sigma)$, i.e., the delay of i in σ is equal to the cost of her strategy in σ, conditional that i participates in the game.

The (risk-averse) individual cost $c_i(\sigma)$ perceived by player i in σ is the δ_i-quantile (or value-at-risk) of $D_i(\sigma)$. Formally, $c_i(\sigma) = \min\{t : \Pr[D_i(\sigma) \leq t] \geq \delta_i\}$. We note that for parallel-link networks, the (risk-averse) individual cost of the players can be computed efficiently. PNE are defined as before, but with respect to the risk-averse individual cost of the players.

Depending on whether players have the same participation probabilities p_i and/or the same confidence levels δ_i, we distinguish between four classes of Congestion Games with Stochastic Players:

- *homogeneous*, where all players have the same participation probability p and confidence level δ.
- *p-homogeneous*, where all players have the same participation probability p, but may have different confidence levels.
- *δ-homogeneous*, where all players have the same confidence level δ, but may have different participation probabilities.
- *heterogeneous*, where both the participation probabilities and the confidence levels may be different.

3.2 Stochastic Players on Parallel Links: Existence and Computation of PNE

In the following, we restrict ourselves to parallel-link networks, and investigate the existence and the efficient computation of PNE for the four cases considered above.

Homogeneous Stochastic Players. If the players are homogeneous, stochastic congestion games on parallel-links are equivalent to standard congestion games on parallel-links (but with possibly different latencies), because the (risk-averse) individual cost of each player in a configuration σ depends only on the link e and its congestion σ_e. The proof of the following employs Rosenthal's potential function and a Greedy Best Response dynamics that guarantee the existence and the efficient computation of a PNE.

Theorem 1. *Congestion Games with Homogeneous Stochastic Players on parallel-link networks admit an exact potential function. Moreover, a PNE can be computed in polynomial time.*

p-**Homogeneous Stochastic Players.** In this case, a stochastic game is equivalent to a congestion game on parallel links with player-specific costs [13], as the (risk-averse) individual cost of each player i in a configuration σ depends only on the link e, its congestion σ_e, and i's confidence level δ_i. Thus, we obtain that:

Corollary 1. *Congestion Games with p-Homogeneous Stochastic Players on parallel-link networks admit a PNE. Moreover, a PNE can be computed in polynomial time.*

Milchtaich [13] proved that parallel-link games with general player-specific costs may not admit a potential function. However, in our case the players' individual costs are correlated, as for any edge, there is a common distribution on which they depend. Nevertheless, we next show that parallel-link games with p-homogeneous stochastic players and linear latencies may not admit any (even generalized) potential function.

Theorem 2. *There are Congestion Games with p-Homogeneous Stochastic Players on parallel-link networks with linear delays that do not admit any potential function.*

Proof. It suffices to show that there is an infinite sequence of deviations in which each deviating player improves her cost. To this end, we adjust the example in [13, Section 5] to our setting. We recall that since players have the same participation probability p, the load on each edge e that player i considers is binomially distributed.

We let $p = 0.75$, and consider 3 parallel links, e_1, e_2, and e_3, 3 "special" players, that change their strategies and form a better response cycle, with $\delta_1 = 0.75$, $\delta_2 = 0.58$ and $\delta_3 = 0.6$, and $n_1 = 25$ additional players on e_1, $n_2 = 20$ additional players on e_2 and $n_3 = 9$ additional players on e_3. The latency functions of the 3 edges are $f_1(k) = 3k + 71$, $f_2(k) = 6k + 33$ and $f_3(k) = 15k + 1$.

We proceed to describe a better response cycle that consists of 6 different configurations $\sigma_1, \ldots, \sigma_6$. Each configuration is represented by a vector $[S_1, S_2, S_3]$, where S_i is the subset of the "special" players using edge e_i.

$$\sigma_1 = \Big[\{1,2\}, \{3\}, \emptyset\Big] \to \sigma_2 = \Big[\{1,2\}, \emptyset, \{3\}\Big] \to \sigma_3 = \Big[\{2\}, \emptyset, \{1,3\}\Big] \to$$
$$\sigma_4 = \Big[\emptyset, \{2\}, \{1,3\}\Big] \to \sigma_5 = \Big[\emptyset, \{2,3\}, \{1\}\Big] \to \sigma_6 = \Big[\{1\}, \{2,3\}, \emptyset\Big] \to \sigma_1$$

To verify that this is indeed a better response cycle, we give the vectors of the risk-averse individual cost of the "special" players in each configuration: $c(\sigma_1) = (137, 134, 135)$, $c(\sigma_2) = (137, 134, 121)$, $c(\sigma_3) = (136, 131, 136)$, $c(\sigma_4) = (136, 129, 136)$, $c(\sigma_5) = (136, 135, 135)$, and $c(\sigma_6) = (134, 135, 135)$. □

δ-Homogeneous Stochastic Players. In this case, players have the same confidence level δ, but their participation probabilities may be different. We next show how to efficiently compute a PNE in parallel-link networks by the p-Decreasing Greedy Best Response algorithm, or p-DGBR, in short, which proceeds as follows:

- Order the players in *non-increasing* order of their participation probabilities p_i.
- Assign the current player, in the previous order, to the edge corresponding to her best response strategy in the current configuration.
- Repeat until all players are added.

Theorem 3. *The p-DGBR algorithm computes, in $O(nm + n^2)$ time, a PNE for Congestion Games with δ-Homogeneous Stochastic Players on parallel-link networks with general latency functions.*

Proof. The proof is by induction on the number of players. We assume that we are at a PNE, and player i is assigned to edge e. Since players on other edges do not deviate, we only have to show that players on e do not deviate. Let k be any player already on e, which implies that $p_k \geq p_i$. It suffices to show that in the current configuration σ, with $\sigma_i = \sigma_k = e$, we have that $c_k(\sigma) \leq c_i(\sigma)$. This holds because $p_k \geq p_i$ and players i and k perceive the same cost on any other edge.

Formally, let us consider $c_k(\sigma)$ and $c_i(\sigma)$. We have that:

$$c_k(\sigma) = \min \left\{ t : \Pr\left[d_e\left(1 + X_i + \sum\nolimits_{j \neq i,k:\, \sigma_j = e} X_j\right) \leq t\right] \geq \delta \right\} \text{ and}$$

$$c_i(\sigma) = \min \left\{ t : \Pr\left[d_e\left(1 + X_k + \sum\nolimits_{j \neq i,k:\, \sigma_j = e} X_j\right) \leq t\right] \geq \delta \right\}.$$

Since $p_k \geq p_i$, for any $r \in \mathbb{N}$, we have that:

$$\Pr\left[X_k + \sum_{j \neq i,k:\, \sigma_j = e} X_j \leq r\right] = \Pr\left[\sum_{j \neq i,k:\, \sigma_j = e} X_j \leq r\right] - \Pr\left[\sum_{j \neq i,k:\, \sigma_j = e} X_j = r\right] p_k$$

$$\leq \Pr\left[\sum_{j \neq i,k:\, \sigma_j = e} X_j \leq r\right] - \Pr\left[\sum_{j \neq i,k:\, \sigma_j = e} X_j = r\right] p_i$$

$$= \Pr\left[X_i + \sum_{j \neq i,k:\, \sigma_j = e} X_j \leq r\right].$$

Thus, since the edge latency functions are non-decreasing, we obtain that:

$$\Pr\left[d_e\left(1 + X_k + \sum_{j \neq i,k:\, \sigma_j = e} X_j\right) \leq d_e(r+1)\right]$$

$$\leq \Pr\left[d_e\left(1 + X_i + \sum_{j \neq i,k:\, \sigma_j = e} X_j\right) \leq d_e(r+1)\right].$$

Therefore, $c_k(\sigma) \leq c_i(\sigma)$, as required. The total computation time is $O(nm + n^2)$, as at each step i, the computations for the newly inserted player take $O(m + i^2)$ time, and we can use memoization to avoid recalculations.   □

We next show that Congestion Games with δ-Homogeneous Stochastic Players admit a two-dimensional lexicographic potential function.

Theorem 4. *Congestion Games with δ-Homogeneous Stochastic Players on parallel-link networks admit a generalized potential function.*

Proof. We define, for each edge e and each configuration σ, a two-dimensional vector $v_{e,\sigma}$ and a total order on these vectors. Moreover, for each configuration σ, we define a vector $w_\sigma = (v_{e,\sigma})_{e \in E}$, where the vectors $v_{e,\sigma}$ appear in increasing lexicographic order. The crux of the proof is to show that for any improving deviation that changes the configuration from σ to σ', we have that $w_\sigma < w_{\sigma'}$. Thus, any decreasing function on the vectors w_σ can serve as a generalized potential function.

Formally, we let $c_e(\sigma) = \min\{t : \Pr[d_e(1 + N_e(\sigma)) \leq t] \geq \delta\}$ be the *outside δ-cost* of each edge e under σ, i.e. the cost that any player not in e perceives when she considers moving to e. By definition, we have that:

$$c_e(\sigma) = c_i(\sigma_{-i}, e) \quad \forall i : \sigma_i \neq e \tag{1}$$
$$c_e(\sigma) \geq c_i(\sigma) \qquad \forall i : \sigma_i = e \tag{2}$$

We let $v_{e,\sigma} = \big(c_e(\sigma), \sigma_e\big)$, and consider the lexicographic order on these pairs:

- $(x_1, y_1) < (x_2, y_2)$, if either $x_1 < x_2$ or $x_1 = x_2$ and $y_1 < y_2$.
- $(x_1, y_1) = (x_2, y_2)$, if $x_1 = x_2$ and $y_1 = y_2$.
- $(x_1, y_1) > (x_2, y_2)$, otherwise.

For any configuration σ, we let $w_\sigma = (v_{e,\sigma})_{e \in E}$ be the vector consisting of the pairs $v_{e,\sigma}$ in increasing lexicographic order. We next show that after any improving deviation, the new configuration σ' has $w_{\sigma'} > w_\sigma$.

Let us assume that player i performs an improving deviation from e to e', and let $\sigma = (\sigma_{-i}, e)$ be the initial configuration and $\sigma' = (\sigma_{-i}, e')$ be the final configuration. Since we consider an improving deviation of player i, $c_i(\sigma) > c_i(\sigma')$. Furthermore, by (1), $c_i(\sigma') = c_i(\sigma_{-i}, e') = c_{e'}(\sigma)$, and by (2), $c_e(\sigma) \geq c_i(\sigma)$. Thus, we obtain that $c_e(\sigma) > c_{e'}(\sigma)$, which implies that $v_{e',\sigma} < v_{e,\sigma}$. Hence, if we consider the coordinates of w_σ, we have that the pair $v_{e',\sigma}$ of e' appears before the pair $v_{e,\sigma}$ of e.

Since we consider a deviation from e to e', the only pairs affected are $v_{e,\sigma}$ and $v_{e',\sigma}$. Consequently, in order to show that $w_\sigma < w_{\sigma'}$, we need to show (i) that $v_{e',\sigma} < v_{e',\sigma'}$, and (ii) that $v_{e',\sigma} < v_{e,\sigma'}$. In words, we need to show that the pair of e' increases by i's move from e to e', and that the pair of e in σ', although it might decrease, still remains greater than the pair of e' in σ.

As for inequality (i), we observe that $\sigma_{e'} < \sigma'_{e'}$ and that $c_{e'}(\sigma) \leq c_{e'}(\sigma')$. Combining these inequalities, we conclude that $v_{e',\sigma} < v_{e',\sigma'}$.

To show inequality (ii), we combine (1) with the hypothesis that player i performs an improving deviation from e to e', and obtain that $c_i(\sigma) > c_i(\sigma_{-i}, e') = c_{e'}(\sigma)$. Also, considering the outside δ-cost of e in σ' and using that $\sigma'_i \neq e$, we obtain that

$c_e(\sigma') = c_i(\sigma'_{-i}, e) = c_i(\sigma)$, because σ'_{-i} and σ_{-i} are identical. Combining these, we conclude that $c_{e'}(\sigma) < c_e(\sigma')$, which immediately implies that $v_{e',\sigma} < v_{e,\sigma'}$.

We have thus established a correspondence between configurations σ and the vectors w_σ, and that for any improving deviation that changes the configuration from σ to σ', $w_\sigma < w_{\sigma'}$. Now, let us consider any strictly decreasing function Φ from the vectors w_σ to $\mathbb{R}$. Then, for any configuration σ, any edges e, e', and any player i, we have that

$$c_i(\sigma_{-i}, e) > c_i(\sigma_{-i}, e') \Rightarrow \Phi(\sigma_{-i}, e) > \Phi(\sigma_{-i}, e').$$

Consequently, Φ serves as generalized potential function for Congestion Games with δ-Homogeneous Stochastic Players. □

3.3 The Price of Anarchy for Stochastic Games with Affine Latencies

In Congestion Games with Stochastic Players, we let the *total cost* of a configuration σ be $C(\sigma) = \mathbb{E}\left[\sum_{i\in N} X_i D_i(\sigma)\right]$, which is a natural generalization of the total cost for standard congestion games. We let o denote an optimal configuration that minimizes the total cost. Then, as for standard congestion games, the Price of Anarchy (PoA) of a stochastic congestion game $\mathcal{G}$ is the maximum ratio $C(\sigma)/C(o)$ over all PNE σ of $\mathcal{G}$.

Next, we first convert the total cost $C(\sigma)$ to a more convenient form, and then present upper and lower bounds on the PoA of Stochastic Congestion Games with Stochastic Players and affine latency functions.

As observed in Section 3.1, if we condition on $X_i = 1$, i.e., that player i participates in the game, $D_i(\sigma) = D_{\sigma_i}(\sigma)$, and thus, $X_i D_i(\sigma) = X_i D_{\sigma_i}(\sigma)$. Therefore,

$$C(\sigma) = \mathbb{E}\left[\sum_{i\in N} X_i D_{\sigma_i}(\sigma)\right] = \sum_e \mathbb{E}\left[N_e(\sigma) d_e(N_e(\sigma))\right].$$

Hence, for affine latency functions $d_e(x) = a_e x + b_e$, we have that

$$C(\sigma) = \sum_e \mathbb{E}\left[N_e(\sigma)\big(a_e N_e(\sigma) + b_e\big)\right]$$
$$= \sum_e \left[a_e\big(\mathbb{E}[N_e(\sigma)]^2 + \mathbb{V}\mathrm{ar}[N_e(\sigma)]\big) + b_e \mathbb{E}[N_e(\sigma)]\right].$$

Theorem 5. *Congestion Games with n Stochastic Players on parallel-link networks with affine latency functions have* PoA $= O(n)$.

Proof. Let $d_e(x) = a_e x + b_e$ denote the affine latency of each edge e. We first observe that (i) since $\delta \geq 1/2$, the cost that a player i perceives on her edge e is at least as large as her expected delay on e due to the load caused by the other players on e, formally $c_i(\sigma) \geq a_e \mathbb{E}[\sum_{j\neq i:\, e=\sigma_j} X_j] + b_e$, and that (ii) at equilibrium, all players perceive a cost of at most $n(a + b)$, where $a + b = \min_e\{a_e + b_e\}$, since otherwise, some player would have an incentive to deviate to the edge with latency $ax + b$.

In what follows, we let f be any PNE, and let o be an optimal configuration. Based on the observations above, we next show that $C(f) \leq 3nC(o)$.

For convenience, we let $F_e = N_e(f)$ and $O_e = N_e(o)$. We have that:

$$C(f) = \sum_e \left(a_e \left(\mathbb{E}[F_e]^2 + \mathrm{Var}[F_e] \right) + b_e \, \mathbb{E}[F_e] \right)$$

$$= \sum_e \mathbb{E}[F_e] \left(a_e \, \mathbb{E}[F_e] + b_e + a_e \frac{\mathrm{Var}[F_e]}{\mathbb{E}[F_e]} \right)$$

$$\leq \sum_e \mathbb{E}[F_e] \left(c_{\max} + a_e + a_e \frac{\mathrm{Var}[F_e]}{\mathbb{E}[F_e]} \right) \leq 3 \sum_e \mathbb{E}[F_e] \, c_{\max} ,$$

where $c_{\max}$ denotes the largest cost of a player in f. The inequalities follow from observation (i) above, from $c_{\max} \geq a_e$ for all used edges e, and from $\mathrm{Var}[F_e] \leq \mathbb{E}[F_e]$.
Using observation (ii) above, with $a + b = \min_e\{a_e + b_e\}$, we obtain that:

$$C(f) \leq 3 \, c_{\max} \sum_e \mathbb{E}[F_e] \leq 3n(a+b) \sum_e \mathbb{E}[F_e] = 3n(a+b) \sum_{i \in N} p_i$$

$$= 3n(a+b) \sum_e \mathbb{E}[O_e] \leq 3n \sum_e \mathbb{E}[O_e](a_e + b_e)$$

$$\leq 3n \sum_e \mathbb{E}[O_e] \left(a_e \frac{\mathbb{E}[O_e]^2 + \mathrm{Var}[O_e]}{\mathbb{E}[O_e]} + b_e \right) = 3nC(o) ,$$

where the last inequality follows from $\mathbb{E}[O_e]^2 + \mathrm{Var}[O_e] = \mathbb{E}[O_e^2] \geq \mathbb{E}[O_e]$. $\qquad\square$

Theorem 6. *There are Congestion Games with n Homogeneous Stochastic Players on parallel-link networks with affine latency functions that have* $\mathrm{PoA} = \Omega(n)$.

Proof sketch. We consider a game with n stochastic players on $k + 1$ parallel edges. Edge e_1 has latency $d_1(x) = x$, and every other edge e_j has latency $d_j(x) = (n - k)x$, $j = 2, \ldots, k + 1$. The players have participation probability p and confidence level $\delta = 1$. The configuration where $n - k$ players use e_1 and each of the remaining k players uses a different edge e_j, $j = 2, \ldots, k + 1$, is a PNE. In the optimal configuration, all n players are assigned to e_1. Calculating the total cost of these configurations, and using $k = n/2$ and $p = 1/n$, we obtain that the PoA is roughly $n/8$. $\qquad\square$

4 Congestion Games with Stochastic Edges

The Model. In *Congestion Games with Stochastic Edges*, players are deterministic, i.e., they always participate in the game. As before, each player i has a confidence level $\delta_i \in [\frac{1}{2}, 1]$. On the other hand, edges have a stochastic behavior, in the sense that the latency function of each edge e is an independent random variable:

$$d_e(x) = \begin{cases} f_e(x) & \text{with probability } 1 - p_e \\ g_e(x) & \text{with probability } p_e. \end{cases}$$

The delay of edge e under congestion k is given by the random variable $X_e(k)$, which is equal to $f_e(k)$, with probability $1 - p_e$, and to $g_e(k)$, with probability p_e, and the delay of a player i in a configuration σ is given by the random variable $D_i(\sigma) = \sum_{e \in \sigma_i} X_e(\sigma_e)$. The risk-averse individual cost of player i in σ is $c_i(\sigma) = \min\{t : \Pr[D_i(\sigma) \leq t] \geq \delta_i\}$, and the total cost of σ is $C(\sigma) = \mathbb{E}\left[\sum_{i \in N} D_i(\sigma)\right]$.

For congestion Games with Stochastic Edges, we distinguish between the case of *homogeneous* players, where all players have the same confidence level δ, and the case of *heterogeneous* players, where each player i may have a different confidence level δ_i.

4.1   Stochastic Edges on Parallel Links: Existence and Computation of PNE

Next, we restrict ourselves to Congestion Games with Stochastic Edges on parallel-link networks, and investigate the existence and the efficient computation of PNE.

Homogeneous Players. If the players are homogeneous, any Congestion Game on stochastic parallel links can be transformed into a standard congestion game on parallel links, but possibly with different latency functions. This holds because the risk-averse individual cost of each player in a configuration σ depends only on the link e and its congestion σ_e. Based on this observation, we can show that:

Theorem 7. *Stochastic Congestion Games with Stochastic Edges and Homogeneous Players on parallel-link networks admit an exact potential function. Moreover, a PNE can be computed in $O(nm)$ time.*

Heterogeneous Players. In this case, a Congestion Game on stochastic parallel links is a congestion game on parallel links with player-specific costs [13]. This holds because the risk-averse individual cost of each player i in a configuration σ depends only on the link e, its congestion σ_e, and i's confidence level δ_i. Thus, we obtain that:

Corollary 2. *Congestion Games with Stochastic Edges and Heterogenous Players on parallel-link networks admit a PNE computable in polynomial time.*

Milchtaich [13] proved that parallel-link games with general player-specific costs may not admit a potential function. But here, as in Section 3.2, the players' individual costs on each edge are correlated with each other. Nevertheless, the following shows that Congestion Games with Stochastic Edges do not admit any (even generalized) potential function.

Theorem 8. *There are Congestion Games with Stochastic Edges and Heterogeneous Players on parallel-link networks with affine latency functions that do not admit any potential function.*

4.2   Price of Anarchy

The following shows that selfish risk-averse players on stochastic parallel links may cause an unbounded degradation in the network performance at equilibrium.

Theorem 9. *There are Congestion Games with Stochastic Edges and Homogeneous Players on parallel-link networks with affine latencies that have an unbounded PoA.*

Acknowledgements. We wish to thank Christos Tzamos for his help in the lexicographic ordering argument, used in the proof of Theorem 4.

References

1. Ackermann, H., Röglin, H., Vöcking, B.: On the Impact of Combinatorial Structre on Congestion Games. Journal of the Association for Computing Machinery 55(6), 25 (2008)
2. Aland, S., Dumrauf, D., Gairing, M., Monien, B., Schoppmann, F.: Exact Price of Anarchy for Polynomial Congestion Games. In: Durand, B., Thomas, W. (eds.) STACS 2006. LNCS, vol. 3884, pp. 218–229. Springer, Heidelberg (2006)
3. Ashlagi, I., Monderer, D., Tennenholtz, M.: Resource selection games with unknown number of players. In: Proc. of the 5th Conference on Autonomous Agents and Multiagent Systems (AAMAS 2006), pp. 819–825 (2006)
4. Awerbuch, B., Azar, Y., Epstein, A.: The Price of Routing Unsplittable Flow. In: Proc. of the 37th ACM Symposium on Theory of Computing (STOC 2005), pp. 57–66 (2005)
5. Caragiannis, I., Fanelli, A., Gravin, N., Skopalik, A.: Efficient computation of approximate pure Nash equilibria in congestion games. In: Proc. of the 52nd IEEE Symposium on Foundations of Computer Science (FOCS 2011), pp. 532–541 (2011)
6. Caragiannis, I., Flammini, M., Kaklamanis, C., Kanellopoulos, P., Moscardelli, L.: Tight Bounds for Selfish and Greedy Load Balancing. In: Bugliesi, M., Preneel, B., Sassone, V., Wegener, I. (eds.) ICALP 2006. LNCS, vol. 4051, pp. 311–322. Springer, Heidelberg (2006)
7. Chien, S., Sinclair, A.: Convergece to Approximate Nash Equilibria in Congestion Games. In: Proc. of the 18th ACM-SIAM Symposium on Discrete Algorithms (SODA 2007), pp. 169–178 (2007)
8. Christodoulou, G., Koutsoupias, E.: The Price of Anarchy of Finite Congestion Games. In: Proc. of the 37th ACM Symposium on Theory of Computing (STOC 2005), pp. 67–73 (2005)
9. Fiat, A., Papadimitriou, C.: When the players are not expectation maximizers. In: Kontogiannis, S., Koutsoupias, E., Spirakis, P.G. (eds.) SAGT 2010. LNCS, vol. 6386, pp. 1–14. Springer, Heidelberg (2010)
10. Fotakis, D.: Congestion games with linearly independent paths: Convergence time and price of anarchy. Theory of Computing Systems 47(1), 113–136 (2010)
11. Fotakis, D., Kontogiannis, S., Spirakis, P.: Selfish Unsplittable Flows. Theoretical Computer Science 348, 226–239 (2005)
12. Harks, T., Klimm, M.: On the existence of pure Nash equilibria in weighted congestion games. Mathematics of Operations Research 37(3), 419–436 (2012)
13. Milchtaich, I.: Congestion Games with Player-Specific Payoff Functions. Games and Economic Behavior 13, 111–124 (1996)
14. Nikolova, E., Stier-Moses, N.E.: Stochastic selfish routing. In: Persiano, G. (ed.) SAGT 2011. LNCS, vol. 6982, pp. 314–325. Springer, Heidelberg (2011)
15. Ordóñez, F., Stier-Moses, N.: Wardrop equilibria with risk-averse users. Transportation Science 44(1), 63–86 (2010)
16. Piliouras, G., Nikolova, E., Shamma, J.S.: Risk sensitivity of price of anarchy under uncertainty. In: Proc. of the 14th ACM Conference on Electronic Commerce (EC 2013), pp. 715–732 (2013)
17. Rockafellar, R.T.: Coherent Approaches to Risk in Optimization Under Uncertainty. In: Tutorials in Operations Research, pp. 38–61 (2007)
18. Rosenthal, R.W.: A Class of Games Possessing Pure-Strategy Nash Equilibria. International Journal of Game Theory 2, 65–67 (1973)
19. Tversky, A., Kahneman, D.: Prospect Theory: An analysis of decision under risk. Econometrica 47(2), 263–291 (1979)

Congestion Games with Player-Specific Costs Revisited[*]

Martin Gairing[1] and Max Klimm[2]

[1] Department of Computer Science, University of Liverpool, UK
[2] Institut für Mathematik, Technische Universität Berlin, Germany

Abstract. We study the existence of pure Nash equilibria in congestion games with player-specific costs. Specifically, we provide a thorough characterization of the maximal sets of cost functions that guarantee the existence of a pure Nash equilibrium.

For the case that the players are unweighted, we show that it is necessary and sufficient that for every resource and for every pair of players the corresponding cost functions are affine transformations of each other. For weighted players, we show that in addition one needs to require that all cost functions are affine or all cost functions are exponential.

Finally, we construct a four-player singleton weighted congestion game where the cost functions are identical among the resources and differ only by an additive constant among the players and show that it does not have a pure Nash equilibrium. This answers an open question by Mavronicolas et al. [15] who showed that such games with at most three players always have a pure Nash equilibrium.

1 Introduction

The theory of congestion games is an important topic in the operations research and algorithmic game theory literature that has driven the innovation in that field for many years. E.g., the central notions of the price of anarchy and the price of stability were first introduced and studied for special classes of congestion games; see Koutsoupias and Papadimitriou [13] and Anshelevich et al. [3].

In a congestion game, as introduced by Rosenthal [20], we are given a set of resources and each player selects a subset of them. The private cost of each player is the sum of the costs of the chosen resources which depends on the number of players using them. Congestion games appear in a variety of applications ranging from traffic and telecommunication networks to real-world and virtual market places. A fundamental problem in game theory is to characterize conditions that guaranty the existence of a pure Nash equilibrium, a state in which no player can improve by unilaterally changing her (pure, i.e., deterministic) strategy.

Rosenthal proved in a seminal paper [20] that every unweighted congestion game has a pure Nash equilibrium. In contrast to this remarkable positive result,

---

[*] Supported by EPSRC grant EP/J019399/1 and the German Research Foundation (DFG) under contract KL 2761/1-1.

B. Vöcking (Ed.): SAGT 2013, LNCS 8146, pp. 98–109, 2013.

it is well known that many natural generalizations of congestion games need not admit a pure Nash equilibrium. This behavior can be observed, e.g., for congestion games with integer-splittable demands (Rosenthal [21], Tran et al. [22]), for weighted congestion games (Fotakis et al. [6], Goemans et al. [9], Libman and Orda [14], Harks and Klimm [10]) and for congestion games with player-specific cost functions (Gairing et al. [7], Milchtaich [16,17]). For these generalizations of congestion games it is NP-hard to decide whether a given instance admits a pure Nash equilibrium, as shown by Dunkel and Schulz [5] for integer-splittable congestion games and weighted congestion games and Ackermann and Skopalik [2] for games with player-specific costs. Milchtaich [18] showed that every finite game is isomorphic to both a weighted congestion game and a congestion game with player-specific costs which makes both classes of games universal.

In light of these results, it is an important problem to find subclasses of these games that are on the one hand narrow enough to guarantee the existence of a pure Nash equilibrium and on the other hand rich enough to model many interesting interactions. Ackermann et al. [1] gave a characterization of the existence of equilibria in terms of the players' strategy space. They showed that games in which the strategy space of each player is the basis of a matroid always possess an equilibrium and that this is basically the maximal property of the strategy space that guarantees the existence of a pure Nash equilibrium in weighted congestion games. They also show that their characterization carries over to games with player-specific costs. For weighted congestion games also the impact of the cost functions on the existence of pure Nash equilibria is now relatively well understood. It is known that games with affine costs or exponential costs always possess a pure Nash equilibrium, and that these are basically the only sets of cost functions that one may allow to guarantee the existence of such an equilibrium point; see Fotakis et al. [6], Panagopoulou and Spirakis [19], and Harks and Klimm [10].

For congestion games with player-specific costs and arbitrary strategy spaces, much less is known. The only known existence result we are aware of is due to Mavronicolas et al. [15] who showed that every weighted congestion game with player-specific costs in which the costs are linear and differ by a player-specific additive constant only, always possess a pure Nash equilibrium.

Voorneveld et al. [23] showed that the class of games considered by Konishi et al. [12] is equivalent to the class of singleton unweighted congestion games with player-specific cost functions. Konishi et al. [12] even proved the existence of a strong equilibrium [4], a strengthening of the pure Nash equilibrium concept that is even robust against coordinated deviations of coalitions of players. Georgiou et al. [8] showed that singleton weighted congestion games with linear player-specific cost functions and three players always admit a pure Nash equilibrium. Also for the case of three players, Mavronicolas et al. [15] showed the existence of a pure Nash equilibrium if the cost functions are non-decreasing and differ only by a player-specific constant.

1.1 Our Results

As our main results, we give a complete characterization of the existence of pure Nash equilibria in congestion games with player-specific costs with unweighted and weighted players, respectively. To formally state our results, let N be a finite set of players and let R be a finite set of resources. We say that a collection $\mathcal{C} = (\mathcal{C}_i^r)_{i \in N, r \in R}$ of cost functions is *consistent* if every congestion game with player-specific costs, in which the cost function c_i^r of player i on resource r is an element of $\mathcal{C}_i^r$, possesses a pure Nash equilibrium. Clearly, a player i, with the property that $\bigcup_{r \in R} \mathcal{C}_i^r$ contains constant functions only, has no impact on the existence of pure Nash equilibria since we may let such player choose a best reply and remove her from the game. Such a player will be called *trivial* henceforth.

We first characterize the consistency of cost functions for games with unweighted players. We show that $\mathcal{C}$ is consistent if and only if for each two non-trivial players $i, j \in N$, there is a constant $a_{i,j} \in \mathbb{R}_{>0}$ such that for each $r \in R$ and each $c_i^r \in \mathcal{C}_i^r$ and $c_j^r \in \mathcal{C}_j^r$, there is a constant $b \in \mathbb{R}$ with $c_i^r(x) = a_{i,j} c_j^r(x) + b$ for all $x \in \mathbb{N}$.

Based on this result and the characterization of the consistency of cost functions for weighted congestion games obtained in [10], we also give a similar characterization for weighted congestion games with player-specific costs. Specifically, we show that a collection $\mathcal{C}$ of cost functions is consistent if and only if at least one of the following two cases holds: (i) For every player i and every resource r, there is a constant $a_i^r \in \mathbb{R}$ such that the set $\mathcal{C}_i^r$ contains only affine functions of type $c_i^r(x) = a_i^r x + b$ where $b \in \mathbb{R}$ is arbitrary, while the ratio a_i^r / a_j^r is independet of r for each two non-trivial players $i, j \in N$; or (ii) There is a constant $\phi \in \mathbb{R}$ and, for every player i and every resource r, a constant a_i^r such that the set $\mathcal{C}_i^r$ contains only exponential functions of type $c_i^r(x) = a_i^r \exp(\phi x) + b$, where $b \in \mathbb{R}$ is arbitrary, while the ratio a_i^r / a_j^r is independent of r for each two non-trivial players $i, j \in N$.

We complement our results constructing an instance of a *singleton* weighted congestion game with costs that differ only by player-specific constants that does not possess a pure Nash equilibrium. Interestingly, this game involves *four* players and thus contrasts a result of Mavronicolas et al. [15] who showed that weighted singleton games with player-specific constants and *three* players always possess a pure Nash equilibrium. To the best of our knowledge, this is the first time that the threshold between existence and non-existence of pure Nash equilibria for a class of games is between three players and four players.

2   Preliminaries

Let $N = \{1, \ldots, n\}$ be a non-empty and finite set of n players and let $R = \{1, \ldots, m\}$ be a non-empty and finite set of m resources. Each player is associated with a *demand* $d_i \in \mathbb{R}_{>0}$ and a set of strategies $S_i \subseteq 2^R$, where each strategy $s_i \in S_i$ is a non-empty subset of the resources. A tuple of n strategies $s = (s_1, \ldots, s_n)$, one for each player, is called a *strategy profile*. The set of all strategy profiles

$S = S_1 \times \cdots \times \ldots S_n$ is called the *strategy space*. The private cost of each player in a strategy profile is defined in terms of a set of *player-specific* cost functions on the resources. Specifically, we are given for each player i and each resource r a *cost function* $c_i^r : \mathbb{R}_{\geq 0} \to \mathbb{R}_{\geq 0}$. The private cost of each player i in strategy profile s is defined as $\pi_i(s) = \sum_{r \in s_i} c_i^r(x^r(s))$, where $x^r(s) = \sum_{j \in N : r \in s_j} d_j$ is the *aggregated demand* of resource r under strategy profile s. We call the tuple $G = (N, S, (\pi_i)_{i \in N})$ a *weighted congestion game with player-specific costs*. For the special case that $d_i = 1$ for all $i \in N$, we call the game *unweighted*, instead. We call the game simply a weighted (or unweighted) congestion game, if the players' cost functions do not differ, i.e., $c_i^r = c_j^r$ for all $r \in R$ and all $i, j \in N$. If all strategies are singletons, i.e., $|s_i| = 1$ for all $s_i \in S_i$ and all $i \in N$, we call the game a *singleton* game.

Let N and R be given and let $\mathcal{C} = (\mathcal{C}_i^r)_{i \in N, r \in R}$ be a collection of cost functions. We say that $(\mathcal{C}_i^r)_{i \in N, r \in R}$ is *consistent* if there is a pure Nash equilibrium in every congestion game with player-specific costs G that satisfies the constraint that $c_i^r \in \mathcal{C}_i^r$ for all $i \in N$ and $r \in R$. Note that we allow for arbitrarily many copies of a resource in G. Given $\mathcal{C}$, we call player i a *trivial player*, if c_i^r is constant for all $r \in R$.

3   Player-Specific Constants

We start with the positive part of our characterization, i.e., we show that congestion games in which the players' cost function of each resource differ by an (additive) player-specific constant only always have a pure Nash equilibrium. In fact we show the more general result that each such game is isomorphic to a congestion game (without player-specific constants).

Formally, for two strategic games $G = (N, S, \pi)$ and $G = (N, S', \pi')$, we say that G and G' are *isomorphic*, if for each i there is a bijection $B_i : S_i \to S_i'$ such that $\pi(s) = \pi'(B_1(s_1), \ldots, B_n(s_n))$.

To prove the following observation, we model the player-specific constants of each player i by introducing an additional resource that is exclusively used by player i. The proof is omitted due to space constraints.

Proposition 1. *Let G be an unweighted (respectively, weighted) congestion game with player-specific costs such that $c_i^r - c_j^r$ is constant for each resource r and each two players i and j. Then, G is isomorphic to an unweighted (respectively, weighted) congestion game.*

4   A Characterization for Unweighted Players

The technically more challenging part of our characterization is to prove that it is indeed necessary that the cost functions of the players differ by a player-specific constants only in order to guarantee the existence of a pure Nash equilibrium. Before we prove this result, we need the following technical lemma.

Lemma 1. *Let N and R be arbitrary and let $\mathcal{C} = (\mathcal{C}_i^r)_{i \in N, r \in R}$ be a collection of cost functions. If $\mathcal{C}$ is consistent for unweighted congestion games with player-specific costs, then for each two non-trivial players $i, j \in N$ the following two conditions are satisfied:*

1. $\{x \in \mathbb{N} : c_i^r(x + 1) = c_i^r(x)\} = \{x \in \mathbb{N} : c_j^r(x + 1) = c_j^r(x)\}$ *for all* $r \in R$, $c_i^r \in \mathcal{C}_i^r$ *and* $c_j^r \in \mathcal{C}_j^r$.

2. $\dfrac{c_i^r(x + 1) - c_i^r(x)}{c_j^r(x + 1) - c_j^r(x)} = \dfrac{c_i^t(y + 1) - c_i^t(y)}{c_j^t(y + 1) - c_j^t(y)}$
   for all $r, t \in R$, $c_k^l \in \mathcal{C}_k^l$, $k \in \{i, j\}$, $l \in \{r, t\}$ *and* $x, y \in \mathbb{N}$ *with* $c_j^r(x + 1) \neq c_j^r(x)$ *and* $c_j^t(y + 1) \neq c_j^t(y)$.

Proof (Sketch). Due to space constraints, we show the claimed results only under the additional assumption that all cost functions are non-decreasing. The general case can be proven with similar arguments.

We start proving the first part of the claim. For a contradiction, let us assume that there are $i, j \in N$, $r \in R$, $c_i^r \in \mathcal{C}_i^r$, $c_j^r \in \mathcal{C}_j^r$ and $x \in \mathbb{N}$ such that $c_i^r(x + 1) = c_i^r(x)$ and $c_j^r(x+1) \neq c_j^r(x)$. As player i is non-trivial, there are $t \in R$, $c_i^t \in \mathcal{C}_i^t$ and $y \in \mathbb{N}$ such that $c_i^t(y+1) \neq c_i^t(y)$. Using the additional assumption that the cost functions are non-decreasing, we obtain $c_j^r(x + 1) > c_j^r(x)$ and $c_i^t(y + 1) > c_i^t(y)$.

Let $k \in \mathbb{N}$ be such that $k \left(c_j^r(x+1) - c_j^r(x)\right) > c_i^t(y+1) - c_i^t(y)$. We introduce $2k$ copies $r_1, \ldots, r_k, r_1', \ldots, r_k'$ of resource r. On all those resources the players have cost functions c_i^r and c_j^r, respectively. Moreover there are two resources t and t' with cost functions $c_i^t = c_i^{t'}$ and $c_j^t = c_j^{t'}$, respectively .

Player i has two strategies. She chooses either $\{r_1 \ldots, r_k, t\}$ or $\{r_1', \ldots, r_k', t'\}$. Player j chooses either $\{r_1, \ldots, r_k, t'\}$ or $\{r_1', \ldots, r_k', t\}$. Furthermore, we introduce $x - 1$ additional players with a single strategy only that always choose $\{r_1, \ldots, r_k, r_1', \ldots, r_k'\}$ and $y - 1$ additional players that always choose $\{t, t'\}$.

We claim that the thus defined game does not have a pure Nash equilibrium. To see this claim, note that in any strategy profile the two players either share k resources of type r or one resource of type t. Now assume we are in a strategy profile in which the players share one resource of type t. Then, player i may deviate to her other strategy since sharing k resources of type r doesn't increase her costs as we have $c_i^r(x + 1) = c_i^r(x)$. On the other hand, her cost is strictly decreased since $c_i^t(y + 1) > c_i^t(y)$. For player j, however, the situation is exactly converse. She prefers not to share k resources of type r since $k \left(c_j^r(x+1) - c_j(x)\right) > c_j^t(y + 1) - c_j^t(y)$. This observation finishes the first part of the proof.

For the second part of the claim, let us assume for a contradiction, that there are $i, j \in N$, $r, t \in R$, $x, y \in \mathbb{N}$, and $c_k^l \in \mathcal{C}_k^l$ with $k \in \{i, j\}$, $l \in \{r, t\}$ such that $c_j^r(x + 1) \neq c_j^r(x)$, $c_j^t(y + 1) \neq c_j^t(y)$ and

$$\frac{c_i^r(x + 1) - c_i^r(x)}{c_j^r(x + 1) - c_j^r(x)} > \frac{c_i^t(y + 1) - c_i^t(y)}{c_j^t(y + 1) - c_j^t(y)}. \tag{1}$$

Using the additional assumption that all cost functions are non-decreasing, we obtain that the denominators $c_j^r(x + 1) - c_j^r(x)$ and $c_j^t(y + 1) - c_j^t(y)$ are strictly positive and together with the first part of the statement of the lemma, this implies that the nominators $c_i^r(x+1) - c_i^r(x)$ and $c_i^t(y+1) - c_i^t(y)$ are strictly positive as well. For $\alpha = c_j^t(y+1) - c_j^t(y)/(c_j^r(x+1) - c_j^r(x))$, we obtain

$$\alpha\big(c_i^r(x+1) - c_i^r(x)\big) > c_i^t(y+1) - c_i^t(y).$$

As this expression is continuous in α, we may find $\alpha' < \alpha$ with $\alpha' \in \mathbb{Q}$ such that we still have

$$\alpha'\big(c_i^r(x+1) - c_i^r(x)\big) > c_i^t(y+1) - c_i^t(y).$$

On the other hand for player j, we derive

$$\alpha'\big(c_j^r(x+1) - c_j^r(x)\big) < \alpha\big(c_j^r(x+1) - c_j^r(x)\big) = c_j^t(y+1) - c_j^t(y).$$

Writing $\alpha' = k/l$ for some $k, l \in \mathbb{N}$, we obtain the following inequalities:

$$k\big(c_i^r(x+1) - c_i^r(x)\big) > l\big(c_i^t(y+1) - c_i^t(y)\big), \tag{2a}$$
$$k\big(c_j^r(x+1) - c_j^r(x)\big) < l\big(c_j^t(y+1) - c_j^t(y)\big). \tag{2b}$$

Next, we will use these inequalities to construct a congestion game with player-specific costs that does not have a pure Nash equilibrium.

The game has $2k$ resources $r_1, \ldots, r_k, r_1', \ldots, r_k'$ with cost functions c_i^r respectively c_j^r and $2l$ resources $t_1, \ldots, t_l, t_1', \ldots, t_l'$ with cost function c_i^t respectively c_j^t. Player i has two strategies, she chooses either $\{r_1, \ldots, r_k, t_1, \ldots, t_l\}$ or $\{r_1', \ldots, r_k', t_1', \ldots, t_l'\}$. Player j has two strategies as well and chooses either $\{r_1, \ldots, r_k, t_1', \ldots, t_l'\}$ or $\{r_1', \ldots, r_k', t_1, \ldots, t_l\}$. Furthermore, there are $x-1$ players with the single strategy $\{r_1, \ldots, r_k, r_1', \ldots, r_k'\}$ and $y-1$ players with the single strategy $\{t_1, \ldots, t_l, t_1', \ldots, t_l'\}$.

We claim that the thus constructed game does not have a pure Nash equilibrium. To see this note that for the strategy profile $s_i = \{r_1, \ldots, r_k, t_1, \ldots, t_l\}$ and $s_j = \{r_1, \ldots, r_k, t_1', \ldots, t_l'\}$, the two players i and j share k resources of type r together. In that case, player i improves switching to her alternative strategy $s_i' = \{r_1', \ldots, r_k', t_1', \ldots, t_l'\}$ as

$$\pi_i(s_i', s_j, \ldots) - \pi_i(s_i, s_j, \ldots) = k\,c_i^r(x) + l\,c_i^t(y+1) - k\,c_i^r(x+1) - l\,c_i^t(y)$$
$$= -k\big(c_i^r(x+1) - c_i^r(x)\big) + l\big(c_i^t(y+1) - c_i^t(y)\big),$$

which is negative using (2a). This strategy profile, in turn, is not a pure Nash equilibrium, since player j may deviate profitably to $s_j' = \{r_1', \ldots, r_k', t_1, \ldots, t_l\}$ as

$$\pi_j(s_i', s_j', \ldots) - \pi_i(s_i', s_j, \ldots) = k\,c_j^r(x+1) + l\,c_j^t(y) - k\,c_j^r(x) - l\,c_j^t(y+1)$$
$$= k\big(c_j^r(x+1) - c_j^r(x)\big) - l\big(c_j^t(y+1) - c_j^t(y)\big),$$

which is negative using (2b). By symmetry of the strategy space of the game, the other two strategy profiles are also not a pure Nash equilibrium. □

We are now ready to prove our main result.

Theorem 1. *For a collection $\mathcal{C} = (\mathcal{C}_i^r)_{i \in N, r \in R}$ of cost functions the following are equivalent:*

1. *$\mathcal{C}$ is consistent for unweighted congestion games with player-specific costs.*
2. *For each two non-trivial players i, j, there is a constant $a_{i,j} \in \mathbb{R}_{>0}$ such that for each $r \in R$ and each $c_i^r \in \mathcal{C}_i^r$ and $c_j^r \in \mathcal{C}_j^r$, there is a constant $b \in \mathbb{R}$ with $c_i^r(x) = a_{i,j}\, c_j^r(x) + b$ for all $x \in \mathbb{N}$.*
   (Note that b may depend on c_i^r and c_j^r while $a_{i,j}$ is equal for all $r \in R$, $c_i^r \in \mathcal{C}_i^r$ and $c_j^r \in \mathcal{C}_j^r$.)

Proof. $2. \Rightarrow 1.$: Let G be a congestion game with player-specific costs as required in the statement, i.e., for each two non-trivial players i, j there is $a_{i,j} \in \mathbb{R}_{>0}$ and $b_{i,j}^r \in \mathbb{R}$ such that $c_i^r(x) = a_{i,j} c_j^r(x) + b_{i,j}^r$ for all $x \in \mathbb{N}$. It is a useful observation that the existence of pure Nash equilibria is invariant under player-specific scaling of the private cost functions. We consider a normalized congestion game $\tilde{G}$ with player-specific costs $\tilde{c}_i^r$, which are defined as

$$\tilde{c}_i^r(x) = \frac{c_i^r(x)}{a_{1,i}} = c_1^r(x) + \frac{b_{1,i}^r}{a_{1,i}}$$

for all $x \in \mathbb{N}$ and $i \in N \setminus \{1\}$. Clearly, $\tilde{G}$ has the same set of pure Nash equilibria as G. Moreover, the set of pure Nash equilibria of $\tilde{G}$ is non-empty, as every game with player-specific constants is isomorphic to an unweighted congestion game (Proposition 1).

$1. \Rightarrow 2.$: Lemma 1 implies that for two non-trivial players $i, j \in N$, each resource $r \in R$, and each $c_i^r \in \mathcal{C}_i^r$ and $c_j^r \in \mathcal{C}_j^r$ the sets $\{x \in \mathbb{N} : c_i^r(x+1) = c_i(x)\}$ and $\{x \in \mathbb{N} : c_j^r(x+1) = c_j(x)\}$ coincide and that $\frac{c_i^r(x+1) - c_i^r(x)}{c_j^r(x+1) - c_j^r(x)}$ is constant for all resources $r \in R$ and all cost functions $c_i^r \in \mathcal{C}_i^r, c_j^r \in \mathcal{C}_j^r$ and $x \in \mathbb{N}$ for which this ratio is defined. Let us call this constant $a_{i,j}$. This implies that

$$c_i^r(x+1) - c_i^r(x) = a_{i,j}\big(c_j^r(x+1) - c_j^r(x)\big).$$

for all $r \in R$, $c_i^r \in \mathcal{C}_i^r$, $c_j^r \in \mathcal{C}_j^r$ and $x \in \mathbb{N}$. Using telescoping sums, we obtain

$$c_i^r(x) - c_i^r(1) = a_{i,j}\big(c_j^r(x) - c_j^r(1)\big),$$

or equivalently

$$c_i^r(x) = a_{i,j} c_j^r(x) + \big(c_i^r(1) - a_{i,j} c_j^r(1)\big),$$

for all $r \in R$, $c_i^r \in \mathcal{C}_i^r$, $c_j^r \in \mathcal{C}_j^r$ and $x \in \mathbb{N}$. Setting $b = c_i^r(1) - a_{i,j} c_j^r(1)$, the claimed result follows. $\square$

5   A Characterization for Weighted Players

Combining the results obtained in the last section and the characterization of consistency for weighted congestion games obtained in [10], we can also give a complete characterization of consistency for games with weighted players.

Theorem 2. *For a collection $\mathcal{C} = (\mathcal{C}_i^r)_{i \in N, r \in R}$ of continuous utility functions the following are equivalent:*

1. *$\mathcal{C}$ is consistent for weighted congestion games with player-specific costs.*
2. *One of the following two conditions is satisfied:*
   (a) *For every non-trivial player i and every resource r, there is a constant $a_i^r \in \mathbb{R}$ such that the set $\mathcal{C}_i^r$ contains only affine functions of type $c_i^r(x) = a_i^r x + b$ where $b \in \mathbb{R}$ is arbitrary, while the ratio a_i^r / a_j^r is independent of r for each two non-trivial players $i, j \in N$.*
   (b) *There is a constant $\phi \in \mathbb{R}$ and, for every non-trivial player i and every resource r, a constant $a_i^r \in \mathbb{R}$ such that the set $\mathcal{C}_i^r$ contains only exponential functions of type $c_i^r(x) = a_i^r \exp(\phi x) + b$, where $b \in \mathbb{R}$ is arbitrary, while the ratio a_i^r / a_j^r is independent of r for each two non-trivial players $i, j \in N$.*

Proof. 1. $\Rightarrow$ 2.: Since the set of unweighted congestion games is a subset of the set of weighted congestion games our characterization of consistency for unweighted games obtained in Theorem 1 implies that for each two non-trivial players i, j, there is a constant $a_{i,j} \in \mathbb{R}_{>0}$, such that for all $c_i^r \in \mathcal{C}_i^r$ and $c_j^r \in \mathcal{C}_j^r$ we have $c_i^r(x) = a_{i,j} \, c_j^r(x) + b$ for all $x \in \mathbb{N}$ and some $b \in \mathbb{R}$. Regarding games in which the demand of each players is equal to an arbitrary but fixed $\epsilon > 0$, we obtain along the same lines that this statement holds for all $x \in \mathbb{R}_{\geq 0}$ which are an integer multiple of ϵ. Letting ϵ go to zero and using the continuity of all cost functions in $\mathcal{C}$, we conclude that $c_i^r(x) = a_{i,j} \, c_j^r(x) + b$ for all $x \in \mathbb{R}_{\geq 0}$ and some $b \in \mathbb{R}$. As we already argued in the proof of Theorem 1, it is without loss of generality to assume that $a_{i,j} = 1$ for all $i, j \in N$, i.e., $c_i^r(x) = c_j^r(x) + b$ for all $x \in \mathbb{R}_{\geq 0}$ and some $b \in \mathbb{R}$.

Weighted congestion games (without a player-specific additive constant) are guaranteed to have a pure Nash equilibrium if and only if one of the following two cases holds: (i) the set of cost functions contains only affine functions $c = a\,x + b$; or the set of cost functions contains only exponential functions $c(x) = a \exp(\phi x) + b$, where ϕ is equal for all cost functions [10]. For the proof of this result, one considers three-player games in which two players have two strategies each, and one player has a single strategy only. The two strategies of the first two players have the property that they contain only resources with cost functions of at most two types, and each of the types occurs with the same cardinality, i.e., there are two cost functions c and c' and two integers a, a' such that each strategy of each player consists of exactly a resources of type c and a' resources of type c'. Now imagine that the cost functions c and c', in fact, differ by a player-specific additive constant. Adding these player-specific additive constants, however, shifts the private cost of each player for each of her strategies by a constant value and, thus, does not affect the existence of a pure Nash equilibrium. This observation establishes that the characterization for weighted congestion games obtained in [10] translates to weighted congestion games with player-specific additive constants, which completes the proof of the claim.

2. $\Rightarrow$ 1: Let G be a game as required in (a) or (b). We will transform G into an equivalent game (with the same set of pure Nash equilibria) for which we then show the existence of a pure Nash equilibrium.

In the first step, we scale the private cost functions of the players such that $a_i^r = a_j^r$ for each two non-trivial players $i, j \in N$ and obtain a weighted congestion game with player-specific costs G'. By Proposition 1, there is a weighted congestion game G'' that is isomorphic to G. Furthermore, if we started from a game G as required in (a) all cost functions in G'' are affine. Weighted congestion games with affine costs admit a potential function and, hence, a pure Nash equilibrium; see [6,11]. If, on the other hand, we started from a game G as required in (b), the game G'' has the property that the (player-independent) cost function of each resource is of type $c^r(x) = a \exp(\phi x) + b$, where ϕ is a common constant of all cost functions. Under this assumption, a weighted potential function exists [11], implying the existence of a pure Nash equilibrium. $\square$

6   Singleton Games

In this section, we consider singleton congestion games with player-specific costs. Milchtaich [16] proved that a pure Nash equilibrium always exists, if the players are unweighted and the player-specific cost functions are non-decreasing. He also provided a counterexample of a three-player game with weighted players that does not have a pure Nash equilibrium. On the positive side, Mavronicolas et al. [15] showed that each *three-player* game in which the cost functions are non-decreasing and differ by an additive constant only, have a pure Nash equilibrium. It has been open whether such a positive result holds for an arbitrary number of players. As the main result of this section, we answer this question to the negative, i.e., we give a counterexample of a *four-player* singleton weighted congestion game with non-decreasing and concave costs that differ by player-specific constants only but does not have a pure Nash equilibrium.

Proposition 2. *There is a singleton weighted congestion game, in which the cost functions are non-decreasing and concave and differ by player-specific additive constants only, that does not have a pure Nash equilibrium.*

Proof. There are four players $N = \{1, 2, 3, 4\}$ with demands $d_i = i$ for all $i \in N$. Further, we are given four resources $\{t, u, v, w\}$. The players' strategy sets are given as $S_1 = \{\{t\}, \{u\}\}$, $S_2 = \{\{u\}, \{v\}\}$, $S_3 = \{\{t\}, \{v\}\}$, $S_4 = \{\{v\}, \{w\}\}$. We first define player-independent cost functions c^t, c^u, c^v, c^w as

$$c^t(x) = \min\{6x, 24\} \qquad\qquad c^u(x) = \min\{20x, 40\}$$
$$c^v(x) = \min\{2x, 14\} \qquad\qquad c^w(x) = 0$$

for all $x \in \mathbb{R}_{\geq 0}$. For $i \in N$ and $r \in s_i \in S_i$, we obtain the player-specific cost functions c_i^r by adding a player-specific constant b_i^r to the cost function c^r. The player-specific constants are given as

$$b_1^t = 15, \qquad b_2^u = 0, \qquad b_3^t = 0, \qquad b_4^v = 0,$$
$$b_1^u = 0, \qquad b_2^v = 29, \qquad b_3^v = 9, \qquad b_4^w = 13.$$

We proceed to show that the thus defined congestion game does not have a pure Nash equilibrium. For the proof, we distinguish between the set of players $N_v(s)$ that uses the critical resource v.

We first note that the cost functions of player 4 are designed so as to ensure that she uses resource v if and only if the load on v is smaller or equal to 6. This rules out the possibilities $\{2\}, \{3, 4\}, \{2, 3, 4\}$ for $N_v(s)$ as in these cases player 4 would always prefer to switch. Next, note that the cost functions of player 2 are such that she uses v if and only if the load on v is smaller or equal 5 which rules out the possibilities $\{3\}, \{2, 4\}, \{2, 3, 4\}$ for $N_v(s)$ as player 2 would prefer to switch in these cases. This leaves use with the following two cases that can occur in equilibrium, which we will consider separately.

First case: $N_v(s) = 4$. This implies that $s_2 = u$ and $s_3 = t$. If $s_1 = t$ as well, the load on t is 4 and thus player 3 would be better of switching to v where the cost for her is at most 23. If, on the other hand, $s_1 = u$, then she would improve switching to t where the cost for her is at most 39.

Second case: $N_v(s) = \{2, 3\}$. Note that this implies that player 1 is on u, as she prefers u over t when both resources are not used by other players. From the strategy profile (u, v, v, w), however, player 3 improves switching to t where the cost for her equals 18 which is strictly less than the 19 cost units she experiences on v. $\qquad \square$

We can slightly strengthen the negative result showing that even for *identical* cost functions in the presence of player-specific additive constants a pure Nash equilibrium need not exist.

Corollary 1. *There is a singleton weighted congestion game with player-specific constants and identical cost functions that does not have a pure Nash equilibrium.*

Proof. As shown in Proposition 2, there is a weighted singleton congestion game with player-specific additive constants that does not admit a pure Nash equilibrium. We proceed to show how to transform G into an equivalent game $\tilde{G}$ that has the claimed properties and does not admit a pure Nash equilibrium as well. To this end, let N denote the set of players and R the set of resources of G. For ease of exposition, we assume that $R = \{0, \ldots, m - 1\}$ for some $m \in N$ and that $c^r(0) = 0$ for all $r \in R$. Let $D = \sum_{i \in N} d_i$ and $M = \max_{r \in R} c^r(D)$. We introduce $m - 1$ additional players $i_1, \ldots, i_{m-1}$ with demand $d_{i_j} = j \cdot D$ and a single strategy $S_{i_j} = \{j\}, j \in \{1, \ldots, m - 1\}$. The cost function $\tilde{c}$ of all resources in $\tilde{G}$ is defined as

$$\tilde{c}(x) = \begin{cases} c^0(x), & \text{if } x \in [0, D], \\ c^1(x - D) + M, & \text{if } x \in (D, 2D], \\ \quad \vdots \\ c^r(x - 2D) + r \cdot M, & \text{if } x \in \big(r \cdot D, (r+1)D\big], \\ \quad \vdots \\ c^{m-1}\big(x - (m-1)D\big) + (m-1)M, & \text{if } x \in \big((m-1)D, m \cdot D\big]. \end{cases}$$

Finally, we redefine the player-specific constants as $\tilde{b}_i^r = b_i^r - r \cdot M$.

Next, for every strategy profile s of G we associate the strategy profile $\tilde{s} = (s_1, \ldots, s_n, s_{i_1}, \ldots, s_{i_{m-1}})$ of $\tilde{G}$ in which the additional players use their unique strategy. Using the particular definitions of D, M and $\tilde{c}$, it is easy to see, that the private costs of each player $i \in N$ in s and $\tilde{s}$ coincide. Using that the additional players have a single strategy only, we derive that $\tilde{G}$ does not have a pure Nash equilibrium. $\qquad\square$

References

1. Ackermann, H., Röglin, H., Vöcking, B.: Pure Nash equilibria in player-specific and weighted congestion games. In: Spirakis, P., Mavronicolas, M., Kontogiannis, S. (eds.) WINE 2006. LNCS, vol. 4286, pp. 50–61. Springer, Heidelberg (2006)
2. Ackermann, H., Skopalik, A.: On the complexity of pure Nash equilibria in player-specific network congestion games. In: Deng, X., Graham, F.C. (eds.) WINE 2007. LNCS, vol. 4858, pp. 419–430. Springer, Heidelberg (2007)
3. Anshelevich, E., Dasgupta, A., Kleinberg, J., Tardos, É., Wexler, T., Roughgarden, T.: The price of stability for network design with fair cost allocation. SIAM J. Comput. 38(4), 1602–1623 (2008)
4. Aumann, R.: Acceptable points in general cooperative n-person games. In: Luce, R.D., Tucker, A.W. (eds.) Contributions to the Theory of Games IV, pp. 287–324. Princeton University Press, Princeton (1959)
5. Dunkel, J., Schulz, A.: On the complexity of pure-strategy Nash equilibria in congestion and local-effect games. Math. Oper. Res. 33(4), 851–868 (2008)
6. Fotakis, D., Kontogiannis, S., Spirakis, P.: Selfish unsplittable flows. Theoret. Comput. Sci. 348(2-3), 226–239 (2005)
7. Gairing, M., Monien, B., Tiemann, K.: Routing (un-)splittable flow in games with player-specific linear latency functions. In: Bugliesi, M., Preneel, B., Sassone, V., Wegener, I. (eds.) ICALP 2006. Part I. LNCS, vol. 4051, pp. 501–512. Springer, Heidelberg (2006)
8. Georgiou, C., Pavlides, T., Philippou, A.: Selfish routing in the presence of network uncertainty. Parallel Process. Lett. 19(1), 141–157 (2009)
9. Goemans, M., Mirrokni, V., Vetta, A.: Sink equilibria and convergence. In: Proc. 46th Annual IEEE Sympos. Foundations Comput. Sci., pp. 142–154 (2005)
10. Harks, T., Klimm, M.: On the existence of pure Nash equilibria in weighted congestion games. Math. Oper. Res. 37(3), 419–436 (2012)
11. Harks, T., Klimm, M., Möhring, R.: Characterizing the existence of potential functions in weighted congestion games. Theory Comput. Syst. 49(1), 46–70 (2011)

12. Konishi, H., Le Breton, M., Weber, S.: Equilibria in a model with partial rivalry. J. Econom. Theory 72(1), 225–237 (1997)
13. Koutsoupias, E., Papadimitriou, C.: Worst-case equilibria. In: Meinel, C., Tison, S. (eds.) STACS 1999. LNCS, vol. 1563, pp. 404–413. Springer, Heidelberg (1999)
14. Libman, L., Orda, A.: Atomic resource sharing in noncooperative networks. Telecommun. Syst. 17(4), 385–409 (2001)
15. Mavronicolas, M., Milchtaich, I., Monien, B., Tiemann, K.: Congestion games with player-specific constants. In: Kučera, L., Kučera, A. (eds.) MFCS 2007. LNCS, vol. 4708, pp. 633–644. Springer, Heidelberg (2007)
16. Milchtaich, I.: Congestion games with player-specific payoff functions. Games Econom. Behav. 13(1), 111–124 (1996)
17. Milchtaich, I.: The equilibrium existence problem in finite network congestion games. In: Spirakis, P., Mavronicolas, M., Kontogiannis, S. (eds.) WINE 2006. LNCS, vol. 4286, pp. 87–98. Springer, Heidelberg (2006)
18. Milchtaich, I.: Representation of finite games as network congestion games. In: Proc. 5th Internat. Conf. on Network Games, Control and Optimization, pp. 1–5 (2011)
19. Panagopoulou, P., Spirakis, P.: Algorithms for pure Nash equilibria in weighted congestion games. ACM J. Exp. Algorithmics 11, 1–19 (2006)
20. Rosenthal, R.: A class of games possessing pure-strategy Nash equilibria. Internat. J. Game Theory 2(1), 65–67 (1973)
21. Rosenthal, R.: The network equilibrium problem in integers. Networks 3, 53–59 (1973)
22. Tran-Thanh, L., Polukarov, M., Chapman, A., Rogers, A., Jennings, N.R.: On the existence of pure strategy Nash equilibria in integer-splittable weighted congestion games. In: Persiano, G. (ed.) SAGT 2011. LNCS, vol. 6982, pp. 236–253. Springer, Heidelberg (2011)
23. Voorneveld, M., Borm, P., van Megen, F., Tijs, S., Facchini, G.: Congestion games and potentials reconsidered. Int. Game Theory Rev. 1(3-4), 283–299 (1999)

Using Reputation Instead of Tolls in Repeated Selfish Routing with Incomplete Information[*][**]

Kun Hu, Jingjing Huang, and George Karakostas

McMaster University,
Dept. of Computing and Software,
1280 Main St. West, Hamilton, Ontario L8S 4K1, Canada
{huangi25,huk6,karakos}@mcmaster.ca

Abstract. We study the application of reputation as an instigator of beneficial user behavior in selfish routing and when the network users rely on the network operator for information on the network traffic. Instead of the use of tolls or artificial delays, the network operator takes advantage of the users' insufficient information, in order to manipulate them through the information he himself provides. The issue that arises then is what can the operator's gain be, without compromising by too much the trust users put on the information provided, i.e., by maintaining a reputation for (at least some) trustworthiness. Our main contribution is the modeling of such a system as a repeated game of incomplete information in the case of single-commodity general networks. This allows us to apply known folk-like theorems to get bounds on the price of anarchy that are better in the worst-case (if that is possible at all) than the well-known price of anarchy bounds in selfish routing without information manipulation.

1   Introduction

It is well known [18,5] that the price of anarchy (as defined by [11]) of non-atomic selfish routing may be bounded from above (by, for example, 4/3 in case of linear latency functions), but, nevertheless, still away from the optimal 1 [16]. A way of 'forcing' the infinitesimal users to a traffic equilibrium with optimal social cost (total latency) is by imposing (monetary) tolls on the edges of the network; then tolls behave as a coordination mechanism, and the utility function for every user has the general form $u_P := l_P(f) + \tau_P$ for every path P, where f is the flow pattern, $l_P(f)$ is the actual path latency, and τ_P is the tolls paid on P, possibly weighted by a different factor by each user (heterogeneous users) or the same (homogeneous users). For homogeneous users it has been known for many years that *marginal tolls* achieve this goal. For heterogeneous users the existence of such optimal tolls (and their computation) were shown relatively recently [22],[9],[6].

The natural question that arises is whether tolls is the only mechanism employed by a network designer in order to achieve the same effect. One objection

---

[*]Research supported by an NSERC Discovery grant.

[**]A full version of the material in this extended abstract can be found in [8].

B. Vöcking (Ed.): SAGT 2013, LNCS 8146, pp. 110–121, 2013.
© Springer-Verlag Berlin Heidelberg 2013

to tolls, for example, is the form of the utility function: is it always acceptable to add delay times (latency) to money (tolls)? An obvious answer to such issues could be that the designer can indeed achieve the same results by implementing the tolls part as artificial delays, say, by decreasing the available bandwidth on the network edges. This approach has been taken in the design of *Coordination Mechanisms* [3], especially in the work of Christodoulou et al. [4] for networks of parallel links and linear latency functions. But, apart from the possible objections raised by the users of such an engineered network, the obvious result of such a decision is that these delays now become part of the social cost, which is defined as the total delay experienced in the network. As a result, the price of anarchy may be reduced (to 5/4 instead of 4/3 for linear latencies [10]) but it is not optimal anymore. It would be optimal, though, if, somehow, this artificial delay didn't count towards the actual delay. For example, suppose that the network operator is also providing the path delay data to the users; then he could take advantage of the users' incomplete information to *lie* about the edge delays by an amount equal to the optimal tolls. In this case, a new challenge arises that didn't exist in the usual (one-shot) selfish routing game: if the game is infinitely repeated, how much lying (if any at all) can be tolerated by the users without their rendering the information they get from the network operator as completely bogus? Can the network operator manipulate the users in order to achieve a price of anarchy that may not be optimal but is still better than the known upper bounds? These are the issues addressed by this work.

We model the repeated interaction between the network operator and the infinitesimal users as a *repeated game* between a *long-term* player (the network operator) with a long-term objective of improving the average price of anarchy in a single-commodity network with linear latency functions, and a sequence of *short-term* players (the aggregation of the infinitesimal users) with the short-term objective of minimizing the individual path latencies as dictated by Wardrop's principle. This game is an infinite repetition of an one-shot *stage game*, where the long-term player knows everything about the game (and the network), including the payoff function of the short-term players, while the short-term players not only aren't aware of the network operator's payoff, but they rely crucially on information about the network provided by that player. The latter can then take advantage of short-term players' *incomplete information* to manipulate the information he provides. The only problem is that the short-term players keep a record of what has happened in the previous rounds (all of them or a finite recent past, depending on whether we assume unbounded or bounded memory for the infinitesimal users respectively). This means that the network operator acquires a *reputation* with the users: (i) he may be a consistent player, i.e., even when he lies, his lies are the same, as happens, for example, when latency measurements of a computer network may be off their real values by the same constant, or (ii) he may be a truly untrustworthy source of information. This reputation is crucial, since it may lead the users to play something different than their usual best response, and therefore leading the price of anarchy to values that are higher than even the worst-case value achieved by a truthful network operator, thus negating the short-term gains that the latter achieved in the first few rounds.

Our main contribution is the modeling of the repeated game. We use a version of the well-known product-choice (P-C) game (see, e.g., [13]) to model the stage game: just as the P-C game is playing the product quality promised by a manufacturer against the price a customer is willing to pay, our game plays the quality of information supplied by the network provider against the trust the users put on that information. If the users are adamant about not using any corrupted data, then, obviously, the network provider must always provide the correct information to avoid further degradation, and the worst-case price of anarchy achieved is equal to the well-known bounds (e.g., 4/3 for linear latencies). The interesting case appears when the users are willing to somehow use that information even when they *know* that it may be corrupted, since, at the end of the day, this is the only data they get. Then we use known results in economics to get bounds on the price of anarchy achieved by the network operator; namely, we use known folk theorem-like results by Fudenberg and Levine [7] and Liu and Skrzypacz [12] to get bounds for the case of users with unlimited or limited memory respectively. It is very interesting that in the latter case [12] can also characterize exactly the moves of the players for every round at equilibrium. Our results work for a single origin-destination pair in a general topology network with linear latencies, and, under certain assumptions, with more general functions, both deterministic and stochastic. We believe that such *bounded-rationality* users better capture automated (i.e., algorithmic) players, and are more relevant in a computer science context; we see our work as only a first step towards applying well-known lessons learned by economists (see, e.g., [2] and [13]) to selfish routing problems.

2   Preliminaries

A directed network $G = (V, E)$, with parallel edges allowed, is given on which a set of *identical* users want to route each an infinitesimal amount of flow (traffic) from a specified origin to a destination node in G. Users are divided into k classes (commodities). The demand of class $i = 1, \ldots, k$, is $d_i > 0$ and the corresponding origin–destination pair is (s_i, t_i). A *feasible* vector x is a valid flow vector (defined on the path or edge space as appropriate) that satisfies the standard multicommodity flow conventions and routes demands d_i for every commodity i. Each edge e is assigned a latency function $l_e(f_e) \geq 0$ that gives the delay experienced by any user on e due to congestion caused by the total flow f_e that passes through e. For a path P, $l_P(f) = \sum_{e \in P} l_e(f_e)$. We define the *cost* of a flow f that satisfies all demands as the total latency experienced by all users, i.e., $C(f) := \sum_{e \in E} f_e l_e(f_e)$. In the standard selfish routing setting, the infinitesimal users try to minimize their travel time, resulting in a *traffic equilibrium* that obeys Wordrop's principle [21]: all used flow paths have the same latency, which is no greater than the latency of the unused paths. If f^*, f^{opt} are a traffic equilibrium flow of maximum total cost and the optimal (minimum total cost) flow respectively, then the *price of anarchy* ρ (PoA) for the network is defined [11] as $\rho := \frac{C(f^*)}{C(f^{opt})}$. Assuming that the latency functions are strictly increasing, then the edge flow pattern for traffic equilibria is unique (see, e.g., [1]).

It is well known that by imposing *marginal tolls* $\tau_e := f_e^{opt} \frac{\partial l_e}{\partial x}(f_e^{opt})$ on the network edges, i.e., if the latency functions are modified to be $l_e^{new}(f_e) := l_e(f_e) + f_e^{opt} \frac{\partial l_e}{\partial x}(f_e^{opt})$, the traffic equilibrium edge flows $f_e^*, \forall e \in E$ coincide with the optimal flow f^{opt}, and therefore $C(f^*) = C(f^{opt})$. In this work we deviate from the traditional view of tolls as monetary compensation (possibly returned to the society); we will try to achieve the same effect by manipulating the information the infinitesimal users (whose aggregation is Player 2 below) receive from the network operator (who is Player 1 below) about the flow in the network. Player 2 has some internal estimate about the actual flow, but the success of this deception cannot lie only on this internal uncertainty, since we are more ambitious than playing the routing game just once; it is *repeated* indefinitely with the *same* players. Therefore, the players in this *repeated game* know of the past history (and past deceptions) every time they play a new round of the routing game (the *stage game*). Nevertheless, we will show that, under certain assumptions, Player 1 can build up his *reputation* in the eyes of Player 2, so that the latter's (believed) best response increases the former's overall payoff. Unfortunately, our results currently hold only for a single origin-destination pair (commodity) (s, t); this case already covers some non-trivial applications, such as the scheduling of jobs arriving at a single queue to different servers, but we leave the multicommodity case to future work.

3   The Stage Game

In what follows, the players try to minimize their cost, but since payoffs are understood to be maximized, we will set the payoffs to be the negative of cost functions.

The stage game played in every round is played by two players, Player 1 and Player 2, and is a version of the classic *product-choice game* (cf. [13]). The pure strategies space is the continuum $[0, 1]$, i.e., the two players pick *simultaneously* numbers x, y respectively in that range. Intuitively, Player 1's x indicates how much truthful that player is willing to be towards Player 2 (e.g., $x = 1$ means no deception whatsoever, and $x = 0$ means Player 1 is as deceitful as possible); Player 2's y indicates how trustful this player is of Player 1 (e.g., $y = 1$ means that Player 2 completely trusts Player 1's transmitted information, and $y = 0$ means that Player 2 completely mistrusts Player 1). Actions x, y control the *extra* flow f^{extra} (beyond the known to both players flow f of total demand d) that Player 2 *perceives* as being injected into the selfish routing game. In this work we study a specific simple tactic of deception for Player 1:

Definition 1. *The* SCALE *tactic by Player 1, is the announcement of extra flow* $f^{extra} = (1 - x)f^{opt}$.

Note that, for simplicity, we assume that the maximum possible extra flow Player 1 can announce is d.

Let $f^{opt}(x,y)$, $f^*(x,y)$ be the optimal and equilibrium (actual) flows routed in the network.[1] The payoff functions of the two players are as follows:

Player 1's Payoff: It is the negation of the PoA $\rho_{x,y}(G, l, d)$ of the selfish routing game played on the network by the infinitesimal users after x and y have been chosen:

$$\Gamma_1(x,y) := -\frac{C(f^*(x,y))}{C(f^{opt}(x,y))} \left(= -\frac{\sum_{e \in E} f_e^* l_e(f_e^*)}{\sum_{e \in E} f_e^{opt} l_e(f_e^{opt})} \right). \tag{1}$$

Player 2's Payoff: Recall that Player 2 is a fictitious player that is the aggregation of homogeneous infinitesimal users of the network. Before defining her payoff, we define the *perceived latency* $\hat{l}(f)$ of the users, as follows:

$$\hat{l}_P(f) := l_P(f + (1 - x)y f^{opt}) + (1 - y)m, \quad \forall P \in \mathcal{P}. \tag{2}$$

The perceived latency is different to the actual latency $l(f)$ in two important aspects: (i) The perceived total flow is comprised of the normal flow f and the extra flow $(1 - x)y f^{opt}$, which is the extra flow announced by Player 1, but weighted by Player 2's trust y. (ii) There is an additive internal estimate $m \geq 0$, by the infinitesimal users, of how much bigger the latency of every path is due to extra flow; in essence, Player 2 pits her own extra latency estimate m against Player 1's claimed extra flow, weighing the former by $(1 - y)$ and the latter by y.[2] The payoff for Player 2 is the (common) path latency of the used paths at equilibrium, when the path latency is the perceived latency. I.e., if f^* is the traffic equilibrium flow with perceived latencies and extra flow $(1-x)y f^{opt}$, then

$$\Gamma_2(x,y) := -L^*(x,y) \tag{3}$$

where $L^*(x,y) = l_P(f^* + (1 - x)y f^{opt}) + (1 - y)m$, $\forall P \in \mathcal{P}$ s.t. $f_P^* > 0$ is the common latency on the paths used by f^*. Note that after the extra flow $(1 - x)y f^{opt}$ has been announced, the only variable for the selfish routing game is normal flow f. Since the infinitesimal users know everything Player 1 knows about the network (including Player 1's claim to extra flow for every edge $(1 - x)f^{opt}$) *except* the fact that there isn't really any extra flow at all, Player 2 can always calculate Γ_2.

We emphasize that when the two players play their simultaneous strategies (x,y), the resulting selfish routing game will be played with edge latencies $\hat{l}$. Afterwards the *actual* latency for each infinitesimal user is revealed (since the infinitesimal user actually travelled the chosen route), but by then it is too late for Player 2 to use this information in order to determine y; the stage game has already been played.

---

[1]We will drop the parameters x, y from the notation when their presence is clear from the context.

[2]The fact that m is the same for *all* paths seems too restrictive, but, in view of Wardrop's principle used to define Player 2's payoff below, it is actually as general as the single commodity setting we study here.

3.1 Stackelberg Strategy[3]

If the two players play actions $(x, y) = (0, 0)$, then the stage game becomes the classic selfish routing game with just an additional path latency m. It is well-known [16] that, in this case, there are networks for which the PoA is the worst possible. The question we are trying to answer here is whether Player 1 can be guaranteed a PoA *strictly* better than the worst case, *independently* of the network topology, *and* when the selfish routing is done *repeatedly*. We address the last issue in the next section. Here we study the *Stackelberg strategy* of Player 1, i.e., the strategy that ensures the biggest payoff for Player 1, provided Player 2 chooses a best response.

Definition 2 (Stackelberg strategy [20]). *Let* $y^*(x)$ *be the best response[4] of Player 2 to Player 1's playing* x. *Player 1's* Stackelberg strategy x_s *is*

$$x_s := \arg\max_{x \in [0,1]} \Gamma_1(x, y^*(x))$$

It is important to notice that $(x_s, y^*(x_s))$ does not have to be a Nash equilibrium, so it doesn't need to be the final outcome of the game. E.g., if $(0, 0)$ is the only equilibrium of the stage game, then the worst-case PoA will be the only outcome. In fact, in our results we don't even require the existence of a Stackelberg strategy; the next section shows that Player 1 can drive the Nash equilibrium (extended to the definition of repeated games) arbitrarily close to the Stackelberg payoff (if it exists) or at least come up with a strategy that guarantees strictly better payoff than the payoff at $(0, 0)$, under certain assumptions. Still, it may be possible that Γ_1^s is equal to the worst-case $\Gamma_1(0, 0)$, and in this case nothing can be done. We show that this is not the case for non-trivial latency functions (e.g. linear).

3.2 Linear Latencies

For linear latency functions $l_e(f_e) = a_e f_e + b_e, a_e, b_e \geq 0, \forall e \in E$, it is well known [18] that the worst-case $\Gamma_1(0, 0)$ is $-4/3$. We show the following

Lemma 1. *For any* $m > 0$, $\Gamma_1(x_s, y^*(x_s)) > -\frac{4}{3}$.

The proof of Lemma 1 is left for the full version. It implies that, as long as the infinitesimal users have *any* inclination ($m > 0$) to believe that there may be extra flow in the system, the Stackelberg payoff for Player 1 is guaranteed to be better than the worst-case PoA bound.

---

[3]What follows should not be confused with *Stackelberg routing* (e.g., [17]), where there is a central coordinator that controls a fraction of the actual flow. Here there is no such coordinator.

[4]If the set $B(x)$ of Player 2's best responses to x is not a singleton, we assume that Player 2 picks the best response that is the worst possible for Player 1.

3.3   General Latencies

Before we tackle the general latency functions case, we recall a couple of well-known definitions.

Definition 3 ([5]). *If $\mathcal{L}$ is a family of latency functions, we define*

$$\beta(l) := \sup_{0 < y < x} \frac{y[l(x) - l(y)]}{xl(x)}, \ \forall l \in \mathcal{L}, \ and \ \beta(\mathcal{L}) := \sup_{l \in \mathcal{L}} \beta(l).$$

For simplicity, we will use $\beta := \beta(\mathcal{L})$ below.

We will also use the notion of *Jacobian similarity* as used in [15]. Namely, if $\nabla l(f) = \left[\frac{\partial l_e}{\partial f_{e'}}\right]_{(e,e') \in E^2}$ is the Jacobian matrix of function $l(f)$, then there exists a constant J satisfying

$$\frac{1}{J} \mathbf{w}^T \nabla l(f) \mathbf{w} \leq \mathbf{w}^T \nabla l(f') \mathbf{w} \leq J \mathbf{w}^T \nabla l(f) \mathbf{w} \tag{4}$$

for all feasible flows f, f', and for all $\mathbf{w} \in \mathbb{R}^{|E|}$. The smallest J satisfying the property is referred to as the *Jacobian similarity factor*.

In the case of general latency functions, the worst-case PoA upper bound is $\frac{1}{1-\beta}$ [5]. We are able to guarantee a Stackelberg payoff that is greater than this bound, in case the following assumptions hold:

Assumption 1. *Functions $l_e(x)$ are convex and non-decreasing continuous function of x, with the first and second derivative existing everywhere.*

Assumption 1 is not very restrictive in practice, since it captures the fact that the latency deterioration rate increases as the congestion on an edge increases. But the next two assumptions are quite technical, and are due to our proof methods; we leave lifting them as an open problem.

Assumption 2. *We assume that $\beta(\mathcal{L}) < \frac{1}{2}$ (i.e., $\mathcal{L}$ is a family of not too "non-linear" functions).*

Assumption 3. *The Jacobian similarity property holds for the instance (G, l, d), and the Jacobian similarity factor J satisfies $J < \frac{1}{1-\beta}$.*

Note that linear functions satisfy all three assumptions.

Lemma 2. *When $m > 0$, and under Assumptions 1-3, $\Gamma_1^s > -\frac{1}{1-\beta}$.*

The proof of Lemma 2 is left for the full version.

In what follows we denote by X, Y (both equal to $[0, 1]$) the sets of pure strategies for Players 1 and 2 in the stage game, and by Σ_1, Σ_2 the sets of mixed strategies for Players 1 and 2 (note that the two sets are the same, i.e., the set of distributions over $[0, 1]$).

4   The Repeated Game

If the stage game is played repeatedly without a memory of the past history to influence the players' decision, then there is no reason for them to deviate from playing a stage game Nash equilibrium; if this equilibrium happens to be $(x, y) = (0, 0)$ every time, then it is impossible for Player 1 to induce Player 2 into deviating from playing $y = 0$. It is exactly the fact that the players have a record of the past history of the game that allows Player 1 to achieve a PoA strictly better than the worst-case $\Gamma_1(0, 0)$, by exploiting a *reputation* that he can built in his interaction with Player 2. We formulate this new setting using the standard notions of repeated games, as they are used in game theory and economics.

A *repeated game* between two players 1 and 2 is an infinite repetition of the playing of a game (called the *stage game*) in rounds or times $t = 0, 1, 2, \ldots, \infty$. In our case the stage game is the one defined in Section 3. Player 1 is a *long-run* player, i.e., his total payoff is a summation of his stage payoff over all periods discounted by a *discount factor* $\delta \in [0, 1)$, which is

$$(1 - \delta) \sum_{t=0}^{\infty} \delta^t g_1^t(x^t, y^t)$$

(the factor $(1 - \delta)$ in front is a normalization factor that brings the repeated game payoff to the same units as the stage payoff). The closer δ is to 1, the more equivalent (in terms of importance) stage payoffs in the distant future are to the ones closer to the present. In our case, the network operator Player 1 is almost equally interested to the payoffs of all periods, i.e., $\delta \to 1$. On the other hand, Player 2 acts as a *short-run* player in every period, since in each period she acts to maximize myopically that period's payoff.

Of central importance in order to escape the stage game Nash equilibrium is the notion of *history* $h^t = \{(x_0, y_0), (x_1, y_1), \ldots, (x_{t-1}, y_{t-1})\}$, defined for every time length t as the sequence of pure strategies played by the two players in the first t periods or $h^0 = \emptyset$ at the beginning of the game. Each player always records all his past actions (has *perfect recall*), but we will later distinguish between a Player 2 with unlimited memory who has a perfect record of Player 1's actions, and a Player 2 that has a limited memory and can only record the last K actions of Player 1. Let $\mathcal{H}^t = (X \times Y)^t$ be the set of all possible histories of length $t \geq 0$ ($\mathcal{H}^0 = \emptyset$), and $\mathcal{H} = \cup_{t=0}^{\infty} \mathcal{H}^t$ the set of all possible histories. Then the *behavioral strategy* of (long-run) Player 1 is defined as $\sigma_1 : \mathcal{H} \to \Sigma_1$. Things are a little bit more complicated for Player 2, since she acts as a short-run player in every period. She can be replaced by an infinite sequence of players $i_0, i_1, i_2, \ldots$, each with a behavioral strategy of $\sigma_2^{i_t} : \mathcal{H}^t \to \Sigma_2$ and payoff Γ_2; each such player enters the game in only one specific round, but has available the whole history available to Player 2 in that round. A *Nash equilibrium* then is defined in the usual way, as a behavioral strategy profile $\sigma = (\sigma_1, \sigma_2^{i_0}, \sigma_2^{i_1}, \ldots)$ with the property that no deviation by any player will improve his payoff if the other players' strategies remain the same.

In order to exploit reputation phenomena in repeated games, we define two types for Player 1's strategy profile:

- **committed type** ω_c: If Player 1 is of this type, he always plays $c \in (0, 1]$, independently of the history of the repeated game. The strategy c will be chosen to be the Stackelberg strategy s.
- **rational type** ω_0: Player 1 is not restricted in playing any strategy in every round (he is opportunistic), and the payoff for the moves of this type of Player 1 is given by $\Gamma_1(x, y)$ defined above.

Player 2's *perception* of the type of Player 1 is captured by Player 2 assigning a probability (initial belief) μ^* to Player 1 being of commitment type ω_c (and, hence, probability $1 - \mu^*$ of being of rational type ω_0).

Let $\underline{V}_1(\delta, \mu^*)$ be the least payoff achievable by Player 1 in the repeated game with discount factor δ and prior belief μ^* for the type of Player 1 held by Player 2. If the latency functions l are continuous, and since it is well-known that the equilibrium flow f^{eq} is also continuous on (x, y) as the solution of a parametric mathematical program with a closed and bounded feasibility region, the following holds:

Lemma 3. *If latency functions l are continuous, then functions Γ_1, Γ_2 are continuous on (x, y).*

Then Theorem 4 in [7] (folk theorem) holds in our case:

Theorem 1 ([7]). *If $0 < \mu^* < 1$, then for all $\varepsilon > 0$ there exists a $\underline{\delta} < 1$ such that for all $\delta \in (\underline{\delta}, 1)$*

$$\underline{V}_1(\delta, \mu^*) \geq (1 - \varepsilon)\Gamma_1^s - \varepsilon\Gamma_1^{min}.$$

where $\Gamma_1^{min} \geq -\frac{1}{1-\beta}$ is the minimum possible payoff for Player 1.

This version of the folk theorem implies that Player 1 can almost achieve Γ_1^s when $\delta \to 1$. We also emphasize that the theorem provides an improvement on the *worst-case* behavior of PoA over *all possible* instances, but it may be the case that for a particular instance, this worst case never happens. The study of particular instances, other than worst case ones, (e.g., networks of parallel links), is not the subject of this work.

4.1 Weak Payoffs

A stronger version of Theorem 1 can be shown, in case Player 1 compromises over his payoff function in the following way: Although the payoff function $\Gamma_1(x, y)$ captures exactly the PoA, the fact that we are studying only worst-case instances allows Player 1 to relax his payoff function to be *directly* the upper bound (calculated in the full version of the paper) rather, than the actual PoA,

$$\bar{\Gamma}_1(x, y) = -\frac{1 + (J - 1)(1 - x)y}{1 - \beta + \beta(1 - x)y}, \tag{5}$$

(which becomes $\frac{4}{3+y(1-x)}$ in the case of linear latency functions). We continue to assume Assumptions 1-3 apply. Then the following holds:

Fact 1.

1. *(myopic incentive of Player 1)* $\bar{\Gamma}_1(x, y)$ *is strictly decreasing in x if $y > 0$, and constant if $y = 0$.*
2. *(Player 1 wants to be trusted)* $\bar{\Gamma}_1(x, y)$ *is strictly increasing in y, unless $x = 1$, in which case it is constant.*
3. *(sub-modularity of Player 1)* $\bar{\Gamma}_1(x, y) - \bar{\Gamma}_1(x', y)$ *is strictly increasing in y, for any $x < x'$.*

In addition, one can show that

Fact 2. *(valuable reputation for Player 1) If $m > \frac{\beta}{1-\beta} \frac{S(f^{opt})}{d}$, then $\bar{\Gamma}_1^s > -\frac{1}{1-\beta}$.*

Facts 1 and 2 will help us to use a more powerful result by Liu and Skrzypacz [12] in case Player 2 is of *bounded rationality* in the sense that Player 2's record keeping is limited (e.g., by memory limitations) to recording only the K most recent actions of Player 1, for some parameter K (Player 2 has still perfect recall of her actions in all past history). Unlike the folk theorem of [7], this limitation allows [12] to describe exactly the equilibrium strategies for the two players, and prove a payoff bound for Player 1's payoff similar to the bound in Theorem 1 *at any point of the game* (and not just at the beginning as the bound in Theorem 1 does). This is important for the study of games that have already been played for a number of periods which we don't know (or don't care about), and we want to evaluate the quality of Player 1's payoff at the moment we start our observation.

Let $P(t), \mu(\omega|h)$ be Player 2's prior belief of whether the current round is t (i.e., she doesn't keep track of time, so she must have a prior belief on which is the current round), and her posterior belief over Player 1's type being ω given a history h (truncated to the most recent K rounds for Player 1's actions). Note that if h contains an action $x \neq c$, then $\mu(\omega_c|h) = 1 - \mu(\omega_0|h) = 0$. In this case, the notion of equilibrium used is that of *stationary Perfect Bayesian Equilibrium (PBE)* that is more sophisticated than the simple Nash equilibrium considered above since it takes into account Player 2's beliefs, when the latter are updated using Bayes' rule[5]. To simplify their analysis, [12] assume the following

Assumption 4. *For any (mixed) action x (ν) by Player 1, Player 2 has a unique pure best response $y^*(x)$ ($y^*(\nu)$), and $y^*(\nu)$ increases if ν increases in the first-order stochastic dominance sense.*

Then Theorem 3 in [12] holds in our case:

Theorem 2 ([12]). *Assume that Assumptions 1-4 hold and $m > \frac{\beta}{1-\beta} \frac{S(f^{opt})}{d}$. Then for any $\varepsilon > 0$, $\mu^* \in (0, 1)$, there exists integer $K(\varepsilon, \mu^*)$ independent of the equilibrium and δ, such that if record keeping length $K > K(\varepsilon, \mu^*)$, we have*

$$\underline{V}_1(\delta, \mu^*) \geq \delta^K \bar{\Gamma}_1^s - (1 - \delta^K) \frac{1}{1 - \beta} - \varepsilon$$

which converges to $\bar{\Gamma}_1^s - \varepsilon$ as δ goes to 1.

---

[5]See [12] for a formal definition.

In fact, the theorem in [12] gives also a description of the strategies the players play at every round in order for Player 1 to achieve the payoff bound; these strategies are pure for Player 2, and mixed with a support of 2 for Player 1. Note that for this stronger result, it's not enough for the infinitesimal users to have *any* inclination ($m > 0$) to believe that there may be extra flow in the system, but they must have *significant* inclination ($m > \frac{\beta}{1-\beta}\frac{S(f^{opt})}{d}$).

5   Discussion and Conclusions

In the previous, we assumed that the perceived latency of the users is always deterministic, since even their internal estimate for delay due to extra flow m is fixed. We can generalize this framework to the *stochastic* case. i.e., the case where the users are uncertain for the exact latency of a path, and, therefore, their perceived latency contains a random component. Hence the perceived latency becomes $\hat{l}_P(f) := l_P(f + (1-x)yf^{opt}) + (1-y)\varepsilon_P, \quad \forall P \in \mathcal{P}$, where ε_P is a random variable. The details are left out of this extended abstract.

Our main goal was to make a first step towards modeling incentives for selfish routing that are based on reputation built by repeated rounds of the basic selfish routing game. Bounded rationality plays a very important role in proving a uniform payoff bound in [12] that goes beyond the folk theorem of [7]. As this is mainly a result of properties (1) and (2) in Fact 1, and Assumption 4 is introduced for technical reasons, an immediate open problem is to get rid of the latter; this can be done either for general functions $\Gamma_2(x, y)$, or by pinpointing further the exact payoff considerations for Player 2. Actually, there are three main modeling challenges that can lead to (i) better bounds and (ii) better characterization of equilibria actions by the players (the two are, in fact, interconnected):

- Different issues of bounded rationality will lead to different repeated games; we only give an example where bounded rationality means memory limitations.
- Different models of incomplete information arise with different signaling protocols between the players; the model depends on the particular application (e.g., signals announcing the waiting-time for different bank tellers).
- Related to the previous item, different specific applications imply different payoff functions for the players; we specified Player 2's payoff exactly for a specific perceived latency model, but such a specification really depends on the application and the nature of information available to her. We leave the study of other models and/or the removal of the assumptions made above as an open problem.
- Unfortunately we don't currently know how to tackle the multicommodity case of our model; this extension would generalize nicely our results, since we already have a general network topology.

References

1. Aashtiani, H.Z., Magnanti, T.L.: Equilibria on a congested transportation network. SIAM Journal of Algebraic and Discrete Methods 2, 213–226 (1981)

2. Aumann, R.J., Maschler, M.B.: Repeated Games with Incomplete Information. MIT Press (1995)
3. Christodoulou, G., Koutsoupias, E., Nanavati, A.: Coordination mechanisms. Theor. Comput. Sci. 410(36), 3327–3336 (2009)
4. Christodoulou, G., Mehlhorn, K., Pyrga, E.: Improving the Price of Anarchy for Selfish Routing via Coordination Mechanisms. In: Demetrescu, C., Halldórsson, M.M. (eds.) ESA 2011. LNCS, vol. 6942, pp. 119–130. Springer, Heidelberg (2011)
5. Correa, J.R., Schulz, A.S., Stier Moses, N.E.: Selfish routing in capacitated networks. Mathematics of Operations Research 29(4), 961–976 (2004)
6. Fleischer, L., Jain, K., Mahdian, M.: Tolls for heterogeneous selfish users in multicommodity networks and generalized congestion games. In: Proceedings of the 45th Annual IEEE Symposium on Foundations of Computer Science, pp. 277–285 (2004)
7. Fundenberg, D., Levine, D.K.: Reputation and equilibrium selection in games with a patient player. Econometrica 57(4), 759–778 (1989)
8. Hu, K.: Using Reputation in Repeated Selfish Routing with Incomplete Information. Open Access Dissertations and Theses, paper 7845, McMaster University (2013), http://digitalcommons.mcmaster.ca/opendissertations/7845
9. Karakostas, G., Kolliopoulos, S.G.: Edge pricing of multicommodity networks for heterogoneous selfish users. In: Proceedings of the 45th Annual IEEE Symposium on Foundations of Computer Science, pp. 268–276 (2004)
10. Karakostas, G., Kolliopoulos, S.G.: The efficiency of optimal taxes. In: López-Ortiz, A., Hamel, A.M. (eds.) CAAN 2004. LNCS, vol. 3405, pp. 3–12. Springer, Heidelberg (2005)
11. Koutsoupias, E., Papadimitriou, C.: Worst-case equilibria. In: Proceedings of the 16th Annual Symposium on Theoretical Aspects of Computer Science, pp. 404–413 (1999)
12. Liu, Q., Skrzypacz, A.: Limited Records and Reputation. Stanford GSB Research Paper No. 2030, Rock Center for Corporate Governance Working Paper No. 54 (2009), http://gsbapps.stanford.edu/researchpapers/library/RP2030.pdf
13. Mailath, G.J., Samuelson, L.: Repeated Games and Reputations, Oxford (2006)
14. Mcleod, R.M.: Mean value theorems for vector valued functions. Proceedings in Edinburgh Math. and Soc. Ser., pp. 197–209 (1964)
15. Perakis, G.: The Price of Anarchy Under Nonlinear and Asymmetric Costs. Mathematics of Operations Research 32(3), 614–628 (2007)
16. Roughgarden, T.: The price of anarchy is independent of the network topology. Journal of Computer and System Sciences 67, 341–364 (2003)
17. Roughgarden, T.: Stackelberg scheduling strategies. SIAM Journal on Computing 33, 332–350 (2004)
18. Roughgarden, T., Tardos, É.: How bad is selfish routing? Journal of the ACM 49, 236–259 (2002)
19. Sheffi, Y.: Urban Transportation Networks. Prentice-Hall (1985)
20. von Stackelberg, H.: The theory of market economy. Oxford University Press, Oxford (1952)
21. Wardrop, J.G.: Some theoretical aspects of road traffic research. Proc. Inst. Civil Engineers, Part II 1, 325–378 (1952)
22. Yang, H., Huang, H.-J.: The multi-class, multi-criteria traffic network equilibrium and systems optimum problem. Transportation Research B 38, 1–15 (2004)

Anti-coordination Games and Stable Graph Colorings

Jeremy Kun, Brian Powers, and Lev Reyzin

Department of Mathematics, Statistics, and Computer Science
University of Illinois at Chicago
{jkun2,bpower6,lreyzin}@math.uic.edu

Abstract. Motivated by understanding non-strict and strict pure strategy equilibria in network anti-coordination games, we define notions of stable and, respectively, strictly stable colorings in graphs. We characterize the cases when such colorings exist and when the decision problem is NP-hard. These correspond to finding pure strategy equilibria in the anti-coordination games, whose price of anarchy we also analyze. We further consider the directed case, a generalization that captures both coordination and anti-coordination. We prove the decision problem for non-strict equilibria in directed graphs is NP-hard. Our notions also have multiple connections to other combinatorial questions, and our results resolve some open problems in these areas, most notably the complexity of the strictly unfriendly partition problem.

1 Introduction

Anti-coordination games form some of the basic payoff structures in game theory. Such games are ubiquitous; miners deciding which land to drill for resources, company employees trying to learn diverse skills, and airplanes selecting flight paths all need to mutually anti-coordinate their strategies in order to maximize their profits or even avoid catastrophe.

Two-player anti-coordination is simple and well understood. In its barest form, the players have two actions, and payoffs are symmetric for the players, paying off 1 if the players choose different actions and 0 otherwise. This game has two strict pure-strategy equilibria, paying off 1 to each player, as well as a non-strict mixed-strategy equilibrium paying off an expected $1/2$ to each player.

In the real world, however, coordination and anti-coordination games are more complex than the simple two-player game. People, companies, and even countries play such multi-party games simultaneously with one another. One straightforward way to model this is with a graph, whose vertices correspond to agents and whose edges capture their pairwise interactions. A vertex then chooses one of k strategies, trying to anti-coordinate with all its neighbors simultaneously. The payoff of a vertex is the sum of the payoffs of its games with its neighbors – namely the number of neighbors with which it has successfully anti-coordinated. It is easy to see that this model naturally captures many applications. For example countries may choose commodities to produce, and their value will depend on how many trading partners do not produce that commodity.

B. Vöcking (Ed.): SAGT 2013, LNCS 8146, pp. 122–133, 2013.
© Springer-Verlag Berlin Heidelberg 2013

In this paper we focus on finding pure strategies in equilibrium, as well as their associated social welfare and price of anarchy, concepts we shall presently define. We look at both strict and non-strict pure strategy equilibria, as well as games on directed and undirected graphs. Directed graphs characterize the case where only one of the vertices is trying to anti-coordinate with another. The directed case turns out to not only generalize the symmetric undirected case, but also captures coordination in addition to anti-coordination.

These problems also have nice interpretations as certain natural graph coloring and partition problems, variants of which have been extensively studied. For instance, a pure strategy equilibrium in an undirected graph corresponds to what we call a stable k-coloring of the graph, in which no vertex can have fewer neighbors of any color different than its own. For $k = 2$ colors this is equivalent to the well-studied *unfriendly partition* or *co-satisfactory partition* problem. The strict equilibrium version of this problem (which corresponds to what we call a strictly stable k-coloring) generalizes the *strictly unfriendly partition problem.* We establish both the NP-hardness of the decision problem for strictly unfriendly partitions and NP-hardness for higher k.

1.1   Previous Work

In an early work on what can be seen as a coloring game, Naor and Stockmeyer [19] define a *weak k-coloring* of a graph to be one in which each vertex has a differently colored neighbor. They give a locally distributed algorithm that, under certain conditions, weakly 2-colors a graph in constant time.

Then, in an influential experimental study of anti-coordination in networks, Kearns et al. [15] propose a true graph coloring game, in which each participant controlled the color of a vertex, with the goal of coloring a graph in a distributed fashion. The players receive a reward only when a proper coloring of the graph is found. The theoretical properties of this game are further studied by Chaudhuri et al. [7] who prove that in a graph of maximum degree d, if players have $d + 2$ colors available they will w.h.p. converge to a proper coloring rapidly using a greedy local algorithm. Our work is also largely motivated by the work of Kearns et al., but for a somewhat relaxed version of proper coloring.

Bramoullé et al. [3] also study a general anti-coordination game played on networks. In their formulation, vertices can choose to form links, and the payoffs of two anti-coordinated strategies may not be identical. They go on to characterize the strict equilibria of such games, as well as the effect of network structure on the behavior of individual agents. We, on the other hand, consider an arbitrary number of strategies but with a simpler payoff structure.

The game we study is related to the MAX-k-CUT game, in which each player (vertex) chooses its place in a partition so as to maximize the number of neighbors in other partitions. Hoefer [14], Monnot & Gourvès [13], research Nash equlibria and coalitions in this context. Our Propositions 1 and 2 generalize known facts proved there, and we include them for completeness.

This paper also has a strong relationship to *unfriendly partitions* in graph theory. An unfriendly partition of a graph is one in which each vertex has at least as many neighbors in other partitions as in its own. This topic has been extensively studied, especially in the combinatorics community [1,4,8,23]. While locally finite graphs admit 2-unfriendly partitions, uncountable graphs may not [23].

Friendly (the natural counterpart) and unfriendly partitions are also studied under the names *max satisfactory* and *min co-satisfactory partitions* by Bazgan et al. [2], who focus on partitions of size greater than 2. They characterize the complexity of determining whether a graph has a k-friendly partition and asked about characterizing k-unfriendly partitions for $k > 2$. Our notion of stable colorings captures unfriendly partitions, and we also solve the $k > 2$ case.

A natural strengthening of the notion above yields *strictly unfriendly partitions*, defined by Shafique and Dutton [22]. A strictly unfriendly partition requires each vertex to have strictly more neighbors outside its partition than inside it. Shafique and Dutton characterize a weaker notion, called *alliance-free partition*, but leave characterizing strictly unfriendly partitions open. Our notion of strictly stable coloring captures strictly unfriendly partitions, giving some of the first results on this problem. Cao and Yang [5] also study a related problem originating from sociology, called the *matching pennies game*, where some vertices try to coordinate and others try to anti-coordinate. They prove that deciding whether such a game has a pure strategy equilibrium is NP-Hard. Our work on the directed case generalizes their notion (which they suggested for future work). Among our results we give a simpler proof of their hardness result for $k = 2$ and also tackle $k > 2$, settling one of their open questions.

There are a few related games on graphs that involve coloring, but they instead focus on finding good proper colorings. In [20] Panagopoulou and Spirakis define a coloring game in which the payoff for a vertex is either zero if it shares a color with a neighbor, and otherwise the number of vertices in the graph with which it shares a color. They prove pure Nash equilibria always exist and can be efficiently computed, and provide nice bounds on the number of colors used. Chatzigiannakis, et al. [6] extend this line of work by analyzing distributed algorithms for this game, and Escoffier, et al. [10] improve their bounds.

1.2   Results

We provide proofs of the following, the last two being our main results.

1. *For all $k \geq 2$, every undirected graph has a stable k-coloring, and such a coloring can be found in polynomial time.*
   Our notion of stable k-colorings is a strengthening of the notion of k-unfriendly partitions of Bazgan et al. [2], solving their open problem number 15.
2. *For undirected graphs, the price of anarchy for stable k-colorings is bounded by $\frac{k}{k-1}$, and this bound is tight.*
3. *In undirected graphs, for all $k \geq 2$, determining whether a graph has a strictly stable k-coloring is NP-hard.*
   For $k = 2$, this notion is equivalent to the notion that is defined by Shafique and Dutton [22], but left unsolved.
4. *For all $k \geq 2$, determining whether a directed graph has even a non-strictly stable k-coloring is NP-hard.*

Because directed graphs also capture coordination, this solves two open problems of Cao and Yang [5], namely generalizing the coin matching game to more than two strategies and considering the directed case.

2   Preliminaries

For an unweighted undirected graph $G = (V, E)$, let $C = \{f | f : V \to \{1, \ldots, k\}\}$. We call a function $c \in C$ a **coloring**.

We study the following anti-coordination game played on a graph $G = (V, E)$. In the game, all vertices simultaneously choose a color, which induces a coloring $c \in C$ of the graph. In a given coloring c, an agent v's **payoff**, $\mu_c(v)$, is the number of neighbors choosing colors different from v's, namely

$$\mu_c(v) := \sum_{\{v,w\} \in E} 1_{\{c(v) \neq c(w)\}}.$$

Note that in this game higher degree vertices have higher potential payoffs.

We also have a natural generalization to directed graphs. That is, if $G = (V, E)$ is a directed graph and c is a coloring of V, we can define the payoff $\mu_c(v)$ of a vertex $v \in V$ analogously as the sum over outgoing edges:

$$\mu_c(v) := \sum_{(v,w) \in E} 1_{\{c(v) \neq c(w)\}}$$

Here a directed edge from v to w is interpreted as "v cares about w." We can then define the social welfare and price of anarchy for directed graphs identically using this payoff function.

Given a graph G, we define the **social welfare** of a coloring c to be

$$W(G, c) := \sum_{v \in V} \mu_c(v).$$

We say a coloring c is **stable**, or in equilibrium, if no vertex can improve its payoff by changing its color from $c(v)$ to another color. We define Q to be the set of stable colorings.

We call a coloring function c **strictly stable**, or in strict equilibrium, if every vertex would decrease its payoff by changing its color from $c(v)$ to another color. If a coloring function is stable and at least one vertex can change its color without decreasing its payoff, then the coloring is **non-strict**.

We define the **price of anarchy** for a graph G to be

$$\text{PoA}(G) := \frac{\max_{c' \in C} W(G, c')}{\min_{c \in Q} W(G, c)}.$$

This concept was originally introduced by Koutsoupias and Papadimitriou in [16], where they consider the ratio of social payoffs in the best and worst-case Nash equilibria. Much work has since focused on the price of anarchy, e.g. [11,21].

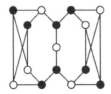 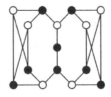

Fig. 1. The strictly stable 2-coloring on the left attains a social welfare of 40 while the non-strictly stable coloring on the right attains 42, the maximum for this graph

Mixed and Pure Strategies. It is natural to consider both pure and mixed strategies for the players in our network anti-coordination game. A pure strategy solution does not in general exist for every 2 player game, while a mixed strategy solution will. However, in this coloring game not only will a pure strategy solution always exist, but for any mixed strategy solution there is a pure strategy equilibrium solution which achieves a social welfare at least as good, and where each player's payoff is identical with its expected payoff under the mixed strategy.

Strict and Non-strict Stability It is worthwhile to note that a strictly stable coloring c need not provide the maximum social welfare. In fact, it is not difficult to construct a graph for which a strictly stable coloring exists yet the maximum social welfare is achieved by a non-strictly stable coloring, as shown in Figure 1.

3   Stable Colorings

First we consider the problem of finding stable colorings in graphs. For the case $k = 2$, this is equivalent to the solved unfriendly partition problem. For this case our algorithm is equivalent to the well-studied local algorithm for MAX-CUT [9,18]. Our argument is a variant of a standard approximation algorithm for MAX-CUT, generalized to work with partitions of size $k \geq 2$.

Proposition 1. *For all $k \geq 2$, every finite graph $G = (V, E)$ admits a stable k-coloring. Moreover, a stable k-coloring can be found in polynomial time.*

Proof. Given a coloring c of a graph, define $\Phi(c)$ to be the number of properly-colored edges. It is clear that this function is bounded and that social welfare is $2\Phi(c)$. Moreover, the change in a vertex's utility by switching colors is exactly the change in Φ, realizing this as an exact potential game [17]. In a given coloring, we call a vertex v *unhappy* if v has more neighbors of its color than of some other color. We now run the following process: while any unhappy vertex exists, change its color to the color

$$c'(u) = \underset{m \in \{1, \dots, k\}}{\operatorname{argmin}} \sum_{v \in N(u)} \mathbf{1}_{\{c(v) = m\}}. \tag{1}$$

As we only modify the colors of unhappy vertices, such an amendment to a coloring increases the value of Φ by at least 1. After at most $|E|$ such modifications, no vertex will be unhappy, which by definition means the coloring is stable. $\square$

We note that because, in the case of $k = 2$, maximizing the social welfare of a stable coloring is equivalent to finding the MAX-CUT of the same graph, which is known to be NP-hard [12], we cannot hope to find a global optimum for the potential function. However, we can ask about the price of anarchy, for which we obtain a tight bound. The following result also appears, using a different construction, in [14], but we include it herein for completeness.

Proposition 2. *The price of anarchy of the k-coloring anti-coordination game is at most $\frac{k}{k-1}$, and this bound is tight.*

Proof. By the pigeonhole principle, each vertex can always achieve a $\frac{k-1}{k}$ fraction of its maximum payoff by choosing its color according to Equation 1. Hence, if some vertex does not achieve this payoff then the coloring is not stable. This implies that the price of anarchy is at most $\frac{k}{k-1}$.

To see that this bound is tight take two copies of K_k on vertices $v_1, \ldots, v_k$ and $v_{k+1}, \ldots, v_{2k}$ respectively. Add an edge joining v_i with v_{i+k} for $i \in \{1, \ldots, k\}$. If each vertex v_i and v_{i+k} is given color i this gives a stable k-coloring of the graph, as each vertex has one neighbor of each of the k colors attaining the social welfare lower bound of $2(\frac{k-1}{k})|E|$. If, however, the vertices v_{i+k} take color $i + 1$ for $i \in \{1, \ldots, k - 1\}$ and v_{2k} takes color 1, the graph achieves the maximum social welfare of $2|E|$. This is illustrated for $k = 5$ in Figure 2. $\square$

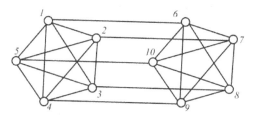

Fig. 2. A graph achieving PoA of $\frac{5}{4}$, for k=5

4   Strictly Stable Colorings

In this section we show that the problem of finding a strictly stable equilibrium with any fixed number $k \geq 2$ of colors is NP-complete. We give NP-hardness reductions first for $k \geq 3$ and then for $k = 2$. The $k = 2$ case is equivalent to the strictly unfriendly 2-partition problem [22], whose complexity we settle.

Theorem 1. *For all $k \geq 2$, determining whether a graph has a strictly stable k-coloring is NP-complete.*

Proof. This problem is clearly in NP. We now analyze the hardness in two cases.
1) $k \geq 3$: For this case we reduce from classical k-coloring. Given a graph G, we produce a graph G' as follows.

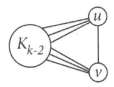

Fig. 3. The gadget added for each edge in G

Start with $G' = G$, and then for each edge $e = \{u, v\}$ in G add a copy H_e of K_{k-2} to G' and enough edges s.t. the induced subgraph of G' on $V(H_e) \cup \{u, v\}$ is the complete graph on k vertices. Figure 3 illustrates this construction.

Now supposing that G is k-colorable, we construct a strictly stable equilibrium in G' as follows. Fix any proper k-coloring φ of G. Color each vertex in G' which came from G (which is not in any H_e) using φ. For each edge $e = (u, v)$ we can trivially assign the remaining $k - 2$ colors among the vertices of H_e to put the corresponding copy of K_k in a strict equilibrium. Doing this for every such edge results in a strictly stable coloring. Indeed, this is a proper k-coloring of G' in which every vertex is adjacent to vertices of all other $k - 1$ colors.

Conversely, suppose G' has a strictly stable equilibrium with k colors. Then no edge e originally coming from G can be monochromatic. If it were, then there would be $k - 1$ remaining colors to assign among the remaining $k - 2$ vertices of H_e. No matter the choice, some color is unused and any vertex of H_e could change its color without penalty, contradicting that G' is in a strict equilibrium.

The only issue is if G originally has an isolated vertex. In this case, G' would have an isolated vertex, and hence will not have a strict equilibrium because the isolated vertex may switch colors arbitrarily without decreasing its payoff. In this case, augment the reduction to attach a copy of K_{k-1} to the isolated vertex, and the proof remains the same.

2) $k = 2$: We reduce from 3-SAT. Let $\varphi = C_1 \wedge \cdots \wedge C_k$ be a boolean formula in 3-CNF form. We construct a graph G by piecing together gadgets as follows.

For each clause C_i construct an isomorphic copy of the graph shown in Figure 4. We call this the *clause gadget* for C_i. In Figure 4, we label certain vertices to show how the construction corresponds to a clause. We call the two vertices labeled by the same literal in a clause gadget a *literal gadget*. In particular, Figure 4 would correspond to the clause $(x \vee y \vee \bar{z})$, and a literal assumes a value of true when the literal gadget is monochromatic. Later in the proof we will force literals to be consistent across all clause gadgets, but presently we focus on the following key property of a clause gadget.

Lemma 1. *Any strictly stable 2-coloring of a clause gadget has a monochromatic literal gadget. Moreover, any coloring of the literal gadgets which includes a monochromatic literal extends to a strictly stable coloring of the clause gadget (excluding the literal gadgets).*

Proof. The parenthetical note will be resolved later by the high-degree of the vertices in the literal gadgets. Up to symmetries of the clause gadget (as a graph)

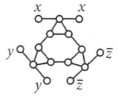

Fig. 4. The clause gadget for $(x \vee y \vee \bar{z})$. Each literal corresponds to a pair of vertices, and a literal being satisfied corresponds to both vertices having the same color.

and up to swapping colors, the proof of Lemma 1 is illustrated in Figure 5. The first five graphs show the cases where one or more literal gadgets are monochromatic, and the sixth shows how no strict equilibrium can exist otherwise. Using the labels in Figure 5, whatever the choice of color for the vertex v_1, its two uncolored neighbors must have the same color (or else v_1 is not in strict equilibrium). Call this color a. For v_2, v_3, use the same argument and call the corresponding colors b, c, respectively. Since there are only two colors, one pair of a, b, c must agree. WLOG suppose $a = b$. But then the two vertices labeled by a and b which are adjacent are not in strict equilibrium.   □

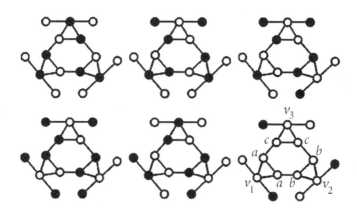

Fig. 5. The first five figures show that a coloring with a monochromatic literal gadget can be extended to a strict equilibrium. The sixth (bottom right) shows that no strict equilibrium can exist if all the literals are not monochromatic.

Using Lemma 1, we complete the proof of the theorem. We must enforce that any two identical literal gadgets in different clause gadgets agree (they are both monochromatic or both not monochromatic), and that any negated literals disagree. We introduce two more simple gadgets for each purpose.

The first is for literals which must agree across two clause gadgets, and we call this the *literal persistence gadget*. It is shown in Figure 6. The choice of colors for the literals on one side determines the choice of colors on the other, provided the coloring is strictly stable. In particular, this follows from the central connecting vertex having degree 2. A nearly identical argument applies to the second gadget, which forces negated literals to assume opposite truth values. We call this the *literal negation gadget*, and it is shown in Figure 6. We do not connect all matching literals pairwise by such gadgets but rather choose one reference literal x' per variable and connect all literals for $x, \bar{x}$ to x' by the needed gadget.

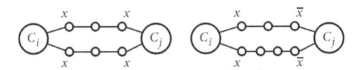

Fig. 6. The literal persistence gadget (left) and literal negation gadget (right) connecting two clause gadgets C_i and C_j. The vertices labeled x on the left are part of the clause gadget for C_i, and the vertices labeled x on the right are in the gadget for C_j.

The reduction is proved in a straightforward way. If φ is satisfiable, then monochromatically color all satisfied literal gadgets in G. We can extend this to a stable 2-coloring: all connection gadgets and unsatisfied literal gadgets are forced, and by Lemma 1 each clause gadget can be extended to an equilibrium. By attaching two additional single-degree vertices to each vertex in a literal gadget, we can ensure that the literal gadgets themselves are in strict equilibrium and this does not affect any of the forcing arguments in the rest of the construction.

Conversely, if G has a strictly stable 2-coloring, then each clause gadget has a monochromatic literal gadget which gives a satisfying assignment of φ. All of the gadgets have a constant number of vertices so the construction is polynomial in the size of φ. This completes the reduction and proves the theorem. □

5   Stable Colorings in Directed Graphs

In this section we turn to directed graphs. The directed case clearly generalizes the undirected as each undirected edge can be replaced by two directed edges. Moreover, directed graphs can capture coordination. For two colors, if vertex u wants to coordinate with vertex v, then instead of adding an edge (u, v) we can add a proxy vertex u' and edges (u, u') and (u', v). To be in equilibrium, the proxy has no choice but to disagree with v, and so u will be more inclined to agree with v. For k colors we can achieve the same effect by adding an undirected copy of K_{k-1}, appropriately orienting the edges, and adding edges $(u, x), (x, v)$ for each $x \in K_{k-1}$. Hence, this model is quite general.

Unlike in the undirected graph case, a vertex updating its color according to Equation 1 does not necessarily improve the overall social welfare. In fact, we cannot guarantee that a pure strategy equilibrium even exists – e.g. a directed 3-cycle has no stable 2-coloring, a fact that we will use in this section.

We now turn to the problem of determining if a directed graph has an equilibrium with k colors and prove it is NP-hard. Indeed, for strictly stable colorings the answer is immediate by reduction from the undirected case. Interestingly enough, it is also NP-hard for non-strict k-colorings for any $k \geq 2$.

Theorem 2. *For all $k \geq 2$, determining whether a directed graph has a stable k-coloring is NP-complete.*

Proof. This problem is clearly in NP. We again separate the hardness analysis into two parts: $k = 2$ and $k \geq 3$.

1) $k = 2$: We reduce from the balanced unfriendly partition problem. A balanced 2-partition of an undirected graph is called unfriendly if each vertex has at least as many neighbors outside its part as within. Bazgan et al. proved that the decision problem for balanced unfriendly partitions is NP-complete [2]. Given an undirected graph G as an instance of balanced unfriendly partition, we construct a directed graph G' as follows.

Start by giving G' the same vertex set as G, and replace each undirected edge of G with a pair of directed edges in G'. Add two vertices u, v to G', each with edges to the other and to all other vertices in G'. Add an additional vertex w with an edge (w, v), and connect one vertex of a directed 3-cycle to u and to w, as shown in Figure 7.

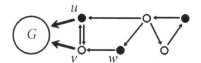

Fig. 7. The construction from balanced unfriendly partition to directed stable 2-coloring. Here u and v "stabilize" the 3-cycle. A bold arrow denotes a complete incidence from the source to the target.

An unbalanced unfriendly partition of G corresponds to a two-coloring of G in which the colors occur equally often. Partially coloring G' in this way, we can achieve stability by coloring u, v opposite colors, coloring w the same color as u, and using this to stabilize the 3-cycle, as shown in Figure 7. Conversely, suppose G does not have a balanced unfriendly partition and fix a stable 2-coloring of G'. WLOG suppose G has an even number of vertices and suppose color 1 occurs more often among the vertices coming from G. Then u, v must both have color 2, and hence w has color 1. Since u, w have different colors, the 3-cycle will not be stable. This completes the reduction.

2) $k \geq 3$: We reduce from the case of $k = 2$. The idea is to augment the construction G' above by disallowing all but two colors to be used in the G' part. We call the larger construction G''.

We start with $G'' = G'$ add two new vertices x, y to G'' which are adjacent to each other. In a stable coloring, x and y will necessarily have different colors (in our construction they will not be the tail of any other edges). We call these colors 1 and 2, and will force them to be used in coloring G'. Specifically, let n be the number of vertices of G', and construct n^3 copies of K_{k-2}. For each vertex v in any copy of K_{k-2}, add the edges $(v, x), (v, y)$. Finally, add all edges (a, b) where $a \in G'$ and b comes from a copy of K_{k-2}. Figure 8 shows this construction.

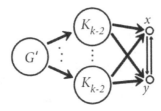

Fig. 8. Reducing k colors to two colors. A bold arrow indicates complete incidence from the source subgraph to the target subgraph.

Now in a stable coloring any vertex from a copy of K_{k-2} must use a different color than both x, y, and the vertex set of a copy of K_{k-2} must use all possible remaining $k-2$ colors. By being connected to n^3 copies of K_{k-2}, each $a \in G'$ will have exactly n^3 neighbors of each of the $k-2$ colors. Even if a were connected to all other vertices in G' and they all use color 1, it is still better to use color 1 than to use any of the colors in $\{3, \ldots, k\}$. The same holds for color 2, and hence we force the vertices of G' to use only colors 1 and 2.   □

6   Discussion and Open Problems

In this paper we defined new notions of graph coloring. Our results elucidated anti-coordination behavior, and solved some open problems in related areas.

Many interesting questions remain. For instance, one can consider alternative payoff functions. For players choosing colors i and j, the payoff $|i - j|$ is related to the *channel assignment problem* [24]. In the cases when the coloring problem is hard, as in our problem and the example above, we can find classes of graphs in which it is feasible, or study random graphs in which we conjecture colorings should be possible to find. Another variant is to study weighted graphs, perhaps with weights, as distances, satisfying a Euclidian metric.

Acknowledgements. We thank György Turán for helpful discussions.

References

1. Aharoni, R., Milner, E.C., Prikry, K.: Unfriendly partitions of a graph. J. Comb. Theory, Ser. B 50(1), 1–10 (1990)
2. Bazgan, C., Tuza, Z., Vanderpooten, D.: Satisfactory graph partition, variants, and generalizations. Eur. J. Oper. Res. 206(2), 271–280 (2010)

3. Bramoullé, Y., López-Pintado, D., Goyal, S., Vega-Redondo, F.: Network formation and anti-coordination games. Int. J. Game Theory 33(1), 1–19 (2004)
4. Bruhn, H., Diestel, R., Georgakopoulos, A., Sprüssel, P.: Every rayless graph has an unfriendly partition. Combinatorica 30(5), 521–532 (2010)
5. Cao, Z., Yang, X.: The fashion game: Matching pennies on social networks. In: SSRN (2012)
6. Chatzigiannakis, I., Koninis, C., Panagopoulou, P.N., Spirakis, P.G.: Distributed game-theoretic vertex coloring. In: Lu, C., Masuzawa, T., Mosbah, M. (eds.) OPODIS 2010. LNCS, vol. 6490, pp. 103–118. Springer, Heidelberg (2010)
7. Chaudhuri, K., Chung Graham, F., Jamall, M.S.: A network coloring game. In: Papadimitriou, C., Zhang, S. (eds.) WINE 2008. LNCS, vol. 5385, pp. 522–530. Springer, Heidelberg (2008)
8. Cowen, R., Emerson, W.: Proportional colorings of graphs (unpublished)
9. Elsässer, R., Tscheuschner, T.: Settling the complexity of local max-cut (Almost) completely. In: Aceto, L., Henzinger, M., Sgall, J. (eds.) ICALP 2011, Part I. LNCS, vol. 6755, pp. 171–182. Springer, Heidelberg (2011)
10. Escoffier, B., Gourvès, L., Monnot, J.: Strategic coloring of a graph. In: Calamoneri, T., Diaz, J. (eds.) CIAC 2010. LNCS, vol. 6078, pp. 155–166. Springer, Heidelberg (2010)
11. Fotakis, D., Kontogiannis, S., Koutsoupias, E., Mavronicolas, M., Spirakis, P.G.: The structure and complexity of nash equilibria for a selfish routing game. In: Widmayer, P., Triguero, F., Morales, R., Hennessy, M., Eidenbenz, S., Conejo, R. (eds.) ICALP 2002. LNCS, vol. 2380, pp. 123–134. Springer, Heidelberg (2002)
12. Garey, M.R., Johnson, D.S.: Computers and Intractability: A Guide to the Theory of NP-Completeness. W. H. Freeman & Co., New York (1979)
13. Gourvès, L., Monnot, J.: On strong equilibria in the max cut game. In: Leonardi, S. (ed.) WINE 2009. LNCS, vol. 5929, pp. 608–615. Springer, Heidelberg (2009)
14. Hoefer, M.: Cost sharing and clustering under distributed competition. PhD thesis, Universität Konstanz, Germany (2007)
15. Kearns, M., Suri, S., Montfort, N.: A behavioral study of the coloring problem on human subject networks. Science 313, 2006 (2006)
16. Koutsoupias, E., Papadimitriou, C.: Worst-case equilibria. In: Meinel, C., Tison, S. (eds.) STACS 1999. LNCS, vol. 1563, pp. 404–413. Springer, Heidelberg (1999)
17. Monderer, D., Shapley, L.S.: Potential games. Games and Economic Behavior 14(1), 124–143 (1996)
18. Monien, B., Tscheuschner, T.: On the power of nodes of degree four in the local max-cut problem. In: Calamoneri, T., Diaz, J. (eds.) CIAC 2010. LNCS, vol. 6078, pp. 264–275. Springer, Heidelberg (2010)
19. Naor, M., Stockmeyer, L.: What can be computed locally? In: STOC 1993, pp. 184–193. ACM (1993)
20. Panagopoulou, P.N., Spirakis, P.G.: A game theoretic approach for efficient graph coloring. In: Hong, S.-H., Nagamochi, H., Fukunaga, T. (eds.) ISAAC 2008. LNCS, vol. 5369, pp. 183–195. Springer, Heidelberg (2008)
21. Roughgarden, T., Tardos, É.: How bad is selfish routing? J. ACM 49(2), 236–259 (2002)
22. Shafique, K., Dutton, R.D.: Partitioning a graph into alliance free sets. Discrete Mathematics 309(10), 3102–3105 (2009)
23. Shelah, S., Milner, E.C.: Graphs with no unfriendly partitions. A tribute to Paul Erdös, pp. 373–384 (1990)
24. van den Heuvel, J., Leese, R.A., Shepherd, M.A.: Graph labeling and radio channel assignment. J. Graph Theory 29(4), 263–283 (1998)

Approximating Multi Commodity Network Design on Graphs of Bounded Pathwidth and Bounded Degree

Kord Eickmeyer* and Ken-ichi Kawarabayashi

National Institute of Informatics,
2-1-2 Hitotsubashi, Chiyoda-ku, Tokyo 101-8430, Japan
{eickmeye,k_keniti}@nii.ac.jp

Abstract. In the *Multicommodity Network Design* problem (MCND) we are given a digraph G together with latency functions on its edges and specified flow requests between certain pairs of vertices. A flow satisfying these requests is said to be at Nash equilibrium if every path which carries a positive amount of flow is a shortest path between its source and sink. The goal of MCND is to find a subgraph H of G such that the flow at Nash equilibrium in H is optimal. While this has been shown to be hard to approximate (with multiplicative error) for a fairly large class of graphs and latency functions, we present an algorithm which computes solutions with small additive error in polynomial time, assuming the graph G is of bounded degree and bounded path-width, and the latency functions are Lipschitz-continuous. Previous hardness results in particular apply to graphs of bounded degree and graphs of bounded path-width, so it is not possible to drop one of these assumptions.

1 Introduction

We model road networks by directed graphs whose edges are labelled with *latency functions*, i.e., functions which express the expected time it takes to traverse the edge depending on the amount of traffic taking it. Adding the assumption that each driver will take a route which, given the current traffic situation, has shortest travel time, one arrives at the model of *selfish routing*, which we review in Section 1.1. Surprisingly, simple examples show that, in this model, removing edges from the network may improve the perfomance of the network, in the sense that the travel time of all participants may be reduced. This phenomenon is called *Braess's paradox* after Dietrich Braess, who first described it in [1].

The obvious question of which edges should be removed to yield an optimal traffic situation is called Multicommodity Network Design Problem (MCND) and has been shown to be computationally hard to solve even approximately, see section 1.2 for details. Our main contribution is a polynomial time approximation algorithm on inputs in which

---

* This work was supported by a fellowship of the first author within the FIT-Programme of the German Academic Exchange Service (DAAD).

B. Vöcking (Ed.): SAGT 2013, LNCS 8146, pp. 134–145, 2013.

- the input graph is of bounded path-width and bounded degree and
- the time it takes to traverse an edges depends in a Lipschitz continuous way on the amount of traffic traversing that edge.

The algorithm returns a subgraph in which an ϵ-approximate Nash equilibrium exists which is at most by an additive term of ϵ worse than the best γ-Nash equilibrium in any other subgraph, for some γ depending on ϵ and which is smaller than ϵ, see Definition 6. Here, an ϵ-Nash equilibrium is a flow in which all traffic is routed along paths which are at most an additive term of ϵ worse than shortest paths. The proof is contained in Theorem 7 and Lemma 10.

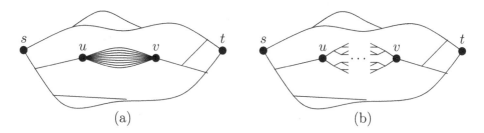

Fig. 1. (a) The reduction in [2] produces very simple graphs in which all of the input is encoded in a set of parallel edges between a certain pair of vertices (u, v). (b) By replacing the parallel edges with binary trees one obtains graphs of bounded degree.

Assuming the input graph to be simultaneously of bounded degree and bounded path-width is a strong restriction. However, previous hardness results (B2, B3 of Section 1.2) showed that MCND is hard even on very simple graphs (planar, acyclic) with just one source of complexity: There are pairs of nodes with many disjoint paths between them. If these paths are short (or even parallel edges), then there must be vertices of high degree (cf. Fig 1a). If we bound the maximum degree there may still be pairs of vertices with many paths between them, by replacing nodes of high degree with binary trees (cf. Fig 1b). We rule this out by bounding the path-width. Note that our approximation algorithm works with approximate Nash equilibria, so technically the hardness results mentioned here do not apply exactly. However, they can be adapted to include approximate Nash equilibria.

Previous attempts at obtaining approximation algorithms for the (single commodity) network design problem on restricted instances include Fotakis et al. [3]. Using a probabilistic argument, they show that *approximate* Nash equilibria can be found in a restricted search space, which yields a polynomial time algorithm on instances with only polynomially many paths from the source to the sink whose length is polylogarithmical in the number of edges of the graph.

1.1   Selfish Routing

We follow the definitions and notation of [4]. Let $G = (V, E)$ be a digraph, and let (s_i, t_i) be k pairs of designated vertices. Furthermore, for each edge $e \in E$

there is an associated *latency function* $l_e : \mathbb{R}_{\geq 0} \to \mathbb{R}_{\geq 0}$, and for each pair (s_i, t_i) we are given a flow request $r_i \in \mathbb{R}_{\geq 0}$. We assume all latency functions to be continuous and non-decreasing. We frequently denote the number of vertices by n and the number of edges by m. We use the notation $[l] := \{1, \ldots, l\}$ for every natural number l.

The intuition behind the problem is that we want to route r_i units of traffic from each of the source nodes s_i to the corresponding target node t_i. The traffic is supposed to consist of infinitesimally small pieces, so that it may be split arbitrarily among the possible paths connecting s_i to t_i in the graph. Formally, by a *u-v-path* P we mean a sequence $u = v_0, v_1, \ldots, v_n = v$ of pairwise distinct nodes $v_i \in V$ such that $(v_{i-1}, v_i) \in E$ for all $i \in [n]$. Denoting the set of all s_i-t_i-paths in G by $\mathcal{P}_i$ and the set of all paths between pairs of vertices in the problem by $\mathcal{P} = \cup_i \mathcal{P}_i$, a *flow f feasible for (G, r, l)* is any function $f : \mathcal{P} \to \mathbb{R}_{\geq 0}$ such that for each $i \in [k]$ the equation $\sum_{P \in \mathcal{P}_i} f_P = r_i$ is satisfied. We denote the set of all paths in the graph by $\mathcal{P}_{\text{all}} := \bigcup_{u,v \in V} \{P \mid P \text{ is a } u\text{-}v\text{-path}\}$. The *length* $|P|$ of a path $P \in \mathcal{P}_{\text{all}}$ is the number of edges in P.

The latency function l_e specifies the amount of time it takes to traverse the edge e, as a function of the amount of traffic being routed along this edge. More traffic might cause congestion and therefore increase this time. The flow function f can be seen as a way of assigning a path $P \in \mathcal{P}_i$ to each of the infinitesimal atoms (say, cars) composing the traffic from s_i to t_i.

For an edge $e \in E$ and a flow f, denote by f_e the total amount of traffic that is routed along e, i.e., $f_e = \sum_{P \in \mathcal{P}, e \in P} f_P$. The *latency $l_P(f)$ of a path* $P \in \mathcal{P}_{\text{all}}$ given a specific flow f is the sum of the latencies of all edges along this path, i.e., $l_P(f) = \sum_{e \in P} l_e(f_e)$.

The assumption that each atom behaves selfishly is captured in the notion of Nash equilibrium: A feasible flow f is said to be at *Nash equilibrium* if

$$l_P(f) \leq l_{P'}(f) \quad \text{for all } i \in [k] \text{ and all } P, P' \in \mathcal{P}_i \text{ with } f_P > 0.$$

Note that since the atoms are infinitesimally small, a single deviation will not change the latencies $l_e(f_e)$.[1]

1.2   Braess's Paradox

By the definition of Nash equilibrium, all s_i-t_i-paths $P \in \mathcal{P}_i$ actually carrying flow (i.e., $f_P > 0$) must have the same latency, which we denote by $L_i(f)$. It can be shown ([4, Cor. 2.6.2]) that for an instance (G, r, l), all Nash equilibria yield the same edge latencies, i.e., $l_e(f_e) = l_e(f'_e)$ for all edges $e \in E$ and flows f, f' both at Nash equilibrium. In particular the maximum latency

$$M(f) := \max_i L_i(f)$$

of a Nash equilibrium of an instance (G, r, l) is well-defined for each instance; we denote this by $M(G, r, l)$.

---

[1] Alternatively, taking some $\delta > 0$ as atomic amount of traffic and letting δ tend to zero yields the same notion of Nash equilibrium.

Braess's paradox amounts to the fact that there may be a subgraph $H \leq G$, obtained from G by removing edges, such that $M(H, r, l) < M(G, r, l)$. We define the *Braess ratio* of (G, r, l) as

$$B(G, r, l) := \max_{H \leq G} \frac{M(G, r, l)}{M(H, r, l)},$$

with the convention that $0/0 := 1$; if $M(H, r, l) = 0$ for some $H \leq G$ then also $M(G, r, l) = 0$. Braess's paradox has been studied intensively recently, and there are strong bounds on the Braess ratio:

(A1) In *single-commodity instances* (G, r, l), i.e., when $k = 1$, then $B(G, r, l) \leq \lfloor \frac{n}{2} \rfloor$, and this bound is optimal [5].[2] Moreover [2], if every matching of $V \setminus \{s, t\}$ using only edges in $G \setminus H$ has size at most c, then

$$L(G, r, l) \leq (c + 1)L(H, r, l).$$

In particular, removing c edges from G may only reduce $L(G, r, l)$ by a factor of $1/(c + 1)$.

(A2) In multi-commodity instances, there are (G, r, l) with $B(G, r, l) = 2^{\Omega(n)}$, and this ratio may be attained by removing a single edge from G (see [2]). The same paper contains an upper bound of $2^{O(\min\{m \log n, kn\})}$ on the Braess ratio in arbitrary networks.

(A3) In single-commodity instances with linear latency functions (i.e., $l_e(x) = a_e + b_e \cdot x$ for all $e \in E$), the Braess ratio is at most $4/3$. Again, this bound may actually be attained by removing a single edge (see [4]).

(A4) In [5], Roughgarden defines the *incline* $\Gamma(c)$ of a continuous monotonely increasing function $c : \mathbb{R}_{\geq 0} \to \mathbb{R}_{\geq 0}$ as $\Gamma(c) := \sup_{x > 0} \frac{x \cdot c(x)}{\int_0^x c(t)\, dt}$, with $0/0 := 1$. With this definition, he obtains the bound $B(G, r, l) \leq \gamma$ for all instances (G, r, l) with $\Gamma(l_e) \leq \gamma$ for all $e \in E$.

Finding a subgraph $H \leq G$ which minimises $M(H, r, l)$ is a natural algorithmic question. This problem, called *Multicommodity Network Design (MCND)* is hard to approximate in the following sense: For $\delta > 1$, we say that an algorithm is a δ-approximation algorithm for MCND if it computes, on input (G, r, l), a subgraph $H \leq G$ with

$$M(H, r, l) \leq \delta \cdot \min_{H' \leq G} M(H', r, l).$$

Assuming P $\neq$ NP,

(B1) there is no polynomial-time $(\frac{4}{3} - \epsilon)$-approximation algorithm for single-commodity network design with linear latency functions [5], for any $\epsilon > 0$,

(B2) there is no polynomial-time $(n/2 - \epsilon)$-approximation algorithm for single-commodity network design with arbitrary latency functions [5], for any $\epsilon > 0$,

---

[2] Note that, in single-commodity instances, $M(G, r, l) = L(G, r, l)$.

(B3) there is no polynomial-time $2^{o(n)}$-approximation algorithm for multi-commodity network design [2].

(B4) there is a constant $c > 0$ such that for all $\gamma \geq 1$, there is no $(c \cdot \gamma)$-approximation algorithm for network design for instances (G, r, l) with $\Gamma(l_e) \leq \gamma$ for all $e \in E$ [5].

Note that A1 and A3 imply that in the first two cases, the trivial algorithm (which always returns the whole graph G) is a best-possible approximation algorithm.

These results are proven by reducing NP-complete problems to the appropriate network design problems. For B2 and B3, the reductions produce acyclic planar graphs which contain pairs of vertices with many parallel edges (Fig. 1a). The latency functions used in these reductions are continuous approximations of step functions and therefore increase very steeply in very small regions.

Since step functions can not be approximated sufficiently well (for the purpose of the above reductions) by linear functions or functions with bounded incline, the proofs for B1 and B4 use an entirely different approach. Here, the reductions start from the problem 2-Directed Vertex Disjoint Paths (2DDP) of finding, in a directed graph with two designated pairs (s_1, t_1) and (s_2, t_2) of vertices, a pair of vertex-disjoint paths between them. This problem has been shown to be NP-hard for general graphs [6], but it is solvable in polynomial time on planar graphs [7]. On the other hand, by inspection we see that the reduction B1 actually uses only latency functions with slopes 0 and 1.

2   An Approximation Algorithm for MCND

In this section we will describe an approximation algorithm for MCND on a restricted set of instances. The restriction is two-fold: We impose restrictions on the graph G (see Definition 1 and section 3), and we bound the slope of the admissible latency functions.

In contrast to the inapproximability results mentioned in section 1, we obtain approximations up to an *additive*, rather than multiplicative, error. Because in all of the non-approximability results of section 1 there was no s_i-t_i-path of latency less than 1, these results also hold for additive errors (note that $c + \epsilon < c \cdot (1 + \epsilon)$ if $c \geq 1$).

We discretise the MCND problem in two ways:

1. We consider only flows which route integer multiples of γ along each edge, for some $\gamma > 0$ and
2. we approximate path latencies up to an additive error of γ.

With these discretisations, we are able to approximately solve MCND on instances for which the latency functions are Lipschitz continuous (see below) and for which the underlying graph is of bounded total degree and bounded pathwidth (ignoring the edge directions). For such graphs it is possible to compute the following kind of decomposition:

Definition 1. *Let* $G = (V, E)$ *be a directed graph. A* strong directed path-decomposition (sdpd) *is a sequence of subsets* $\emptyset = A_0, \ldots, A_m \subset V$ *(called* bags*) such that:*

(SDPD1) $|A_j \setminus A_{j-1}| \leq 1$ *for all* $j \in [m]$, *and if* $v \in A_j \setminus A_{j-1}$, *then all incoming edges to* v *come from vertices in* A_{j-1},

(SDPD2) for every $v \in V$ *there are* $1 \leq a \leq b \leq m$ *such that*

$$v \in A_j \quad \Leftrightarrow \quad a \leq j \leq b$$

for all $j = 0, \ldots, m$ *(in particular,* $\bigcup_j A_j = V$ *).*

For a vertex $v \in V$ *we denote by* $\iota(v) := \min\{j \mid v \in A_j\}$ *the index of the first bag which contains it. We call* m *the* length *and* $\max_i |A_i| - 1$ *the* width *of the decomposition. We define* $G_j := (V_j, E_j)$ *to be the subgraph of* G *induced on the set of vertices in the bags up to* A_j, *i.e.,*

$$V_j := \bigcup_{j' \leq j} A_{j'} \quad and \quad E_j := E \cap (V_j \times V_j).$$

We use the term *strong* to distinguish our definition from the definition of a directed path-decomposition given, e.g., by Barát in [8], where (SDPD1) is replaced by the weaker condition that for all edges uv there be indices $i \leq j$ with $u \in A_i$ and $v \in A_j$. In particular, any acyclic digraph has directed path-width 0, but the strong directed path-width may be arbitrarily high. Also, to simplify the presentation we require that $|A_j \setminus A_{j-1}|$ be at most 1, but it should be clear how to adjust the update step of Theorem 7 to the case $|A_j \setminus A_{j-1}| \geq 2$. In Section 3 we show how to find sdpds in polynomial time in graphs of bounded path-width and bounded degree.

The important feature of sdpds which we use in our algorithm is that

1. every bag A_j separates G_j from $\bar{G}_j := G \setminus G_j$ and
2. all edges between G_j and $\bar{G}_j$ are directed from the former to the latter.

Consequently, any path between vertices $u, v \in V_j$ stays entirely in G_j. Directed path-decompositions in the sense of [8] only have the second property, while undirected path-decompositions (see Definition 8) have only the first one.

In general, discretising the amount of traffic that may travel along each edge might drastically change the set of Nash equilibria. In order to prevent this, we use the following continuity condition:

Definition 2. *For* $\alpha > 0$ *we call an instance* (G, r, l) *of MCND* α-Lipschitz continuous *if all latency functions are Lipschitz continuous with Lipschitz constant* α, *i.e.,*

$$|l_e(x) - l_e(y)| \leq \alpha |x - y|$$

for all edges $e \in E$ *and* $x, y \in \mathbb{R}_{\geq 0}$.

Lipschitz continuity ensures that we may slightly change a flow without changing path latencies by too much:

Lemma 3. *Let $\epsilon, \alpha > 0$ and let (G, r, l) be an α-Lipschitz continuous instance of MCND. If f and f' are flows for which $|f_e - f'_e| < \epsilon$ for all edges e, and $P \in \mathcal{P}_{\text{all}}$ is a path, then*

$$|l_P(f) - l_P(f')| < |P|\,\alpha\epsilon$$

We omit the straightforward proof. Putting these two together we obtain the following lemma for discretised flows:

Lemma 4. *Let $\alpha > 0$ and let (G, r, l) be an α-Lipschitz continuous instance of MCND, and $A_0, \ldots, A_m$ is an sdpd of G. Let $\gamma > 0$ and assume that all r_i are integer multiples of γ. Then if $f : \mathcal{P} \to \mathbb{R}_{\geq 0}$ is a flow in (G, r, l), there is a flow f' such that*

- *f'_e is an integer multiple of γ for all edges $e \in E$ and*
- *$|f_e - f'_e| < \iota(v)\gamma$ for all $e \in E$ directed towards the node $v \in V$.*
- *$|l_P(f) - l_P(f')| < \alpha\gamma\,|P|\,\iota(v)$ for all simple paths $P \in \mathcal{P}_{\text{all}}$ ending in a vertex $v \in V$.*

Proof. We can obtain the flow f' by discretising along the decomposition, maintaining the conditions that the total flow which f' routes into vertices in A_j is an integer multiple of γ and does not differ from the flow which f routes into these vertices by more than $j\gamma$. The statement about path latencies is an easy consequence of Lemma 3. $\square$

We relax the notion of a Nash equilibrium as follows:

Definition 5. *Let (G, r, l) be an instance of selfish routing and $\epsilon > 0$. An ϵ-Nash equilibrium is a flow f for (G, r, l) such that for all i and all paths $P, P' \in \mathcal{P}_i$ we have*

$$l_P(f) \leq l_{P'}(f) + \epsilon \quad \text{if } f_P > 0.$$

Accordingly, we define

$$M_{\epsilon\text{-}Nash}(G, r, l) := \min\{M(f) \mid f \text{ is } \epsilon\text{-Nash equilibrium}\},$$

where $M(f)$ is the maximum latency of a path from some s_i to some t_i in the flow f.

Note that we use an additive error here, as opposed to a factor of $(1 + \epsilon)$ as in Roughgarden's definition of ϵ-approximate Nash equilibria [4, Sec. 4.2]. We can now make precise what we mean by "approximately solving MCND":

Definition 6. *Let (G, r, l) be an instance of MCND, and $\epsilon, \gamma > 0$. An (ϵ, γ)-approximate solution to MCND is a subgraph $H \subseteq G$ such that*

$$M_{\epsilon\text{-}Nash}(H, r, l) \leq M_{\gamma - Nash}(H', r, l) + \epsilon$$

for all subgraphs $H' \subseteq G$.

We will invoke this definition with γ much smaller than ϵ. With these definitions we are ready to state our main approximation result:

Theorem 7. *Let $\alpha, \epsilon > 0$ and $k, \rho \in \mathbb{N}$ be fixed. Assume we are given an α-Lipschitz continuous MCND instance (G, r, l) with k source-sink pairs $s_1, t_1, \ldots,$ s_k, t_k, together with a strong directed path-decomposition $A_0, \ldots, A_m \subset V$ of width ρ. Then it is possible to find an (ϵ, γ)-approximate solution to MCND in time polynomial in the size of G and $\max r_i$, for $\gamma := \frac{\epsilon}{1 + \alpha m^2}$.*

Proof. Let M be an upper bound on all flow requests r_i and on the latency of an s_i-t_i-path in a Nash equilibrium in the input graph G, and denote by k the number of source-sink pairs in (G, r, l). For a fixed Lipschitz constant α, this bound can be taken to be $\alpha |V| \max r_i$.

We discretise flows and approximate latencies with the γ stated in the theorem. For each A_j we compute a table T_j which is indexed by all tuples $(\sigma_j^{(1)}, \ldots, \sigma_j^{(k)}, \lambda_j^{(1)}, \ldots, \lambda_j^{(k)})$ of functions such that

- $\sigma_j^{(i)}$ is a function from A_j to $\mathbb{R}_{\geq 0} \cup \{\perp, \top\}$ such that if $\sigma_j^{(i)}(A_j) \subset \mathbb{R}_{\geq 0}$ then

$$\sum_{v \in A_j} \sigma_j^{(i)}(v) = r_i$$

  and $\sigma_j^{(i)}(v)$ is an integer multiple of γ for all $v \in A_j$. If $\sigma_j^{(i)}(v) \in \{\perp, \top\}$ for some $v \in A_j$ we demand $\sigma_j^{(i)}(w) = \sigma_j^{(i)}(v)$ for all $w \in A_j$.
- $\lambda_j^{(i)}$ is a function from A_j to $[0, M]$ such that every $\lambda_j^{(i)}(v)$ is an integer multiple of γ for all $v \in A_j$.

An index $(\sigma_j^{(1)}, \ldots, \sigma_j^{(k)}, \lambda_j^{(1)}, \ldots, \lambda_j^{(k)})$ is meant to represent a flow routing approximately $\sigma_j^{(i)}(v)$ of traffic from s_i to v, such that the common latency of all s_i-v-paths is roughly $\lambda_j^{(i)}(v)$. The special symbol $\perp$ signifies that $s_i \notin G_j$, while $\top$ signifies that $t_i \in G_j$. Note that the size of this index set is linear in k and polynomial of degree ρ in M for fixed values of ρ and ϵ (note that $m < M$). An entry in the table may be $\perp$ or a subset of the edges of G_j. We define

$$B_j := 1 + \alpha j^2$$

The table entries will satisfy the following conditions:

(a) If $T_j(\sigma_j^{(1)}, \ldots, \sigma_j^{(k)}, \lambda_j^{(1)}, \ldots, \lambda_j^{(k)})$ is a subset of E, then after removing these edges there is a flow f in G_j which routes an amount $\sigma_j^{(i)}(v)$ of traffic from source s_i to $v \in A_j$, such that for each path P from s_i to v with $f_P > 0$ the latency $l_P(f)$ is within $B_j \gamma$ of $\lambda_j^{(i)}(v)$, and such that the flow f is a $B_j \gamma$-Nash equilibrium. If $s_i \notin V_j$ then we demand $\sigma_j^{(i)} \equiv \perp$, and if $t_i \in G_j$ we demand $\sigma_j^{(i)} \equiv \top$.

(b) If $T_j(\sigma_j^{(1)}, \ldots, \sigma_j^{(k)}, \lambda_j^{(1)}, \ldots, \lambda_j^{(k)}) = \perp$, then no way of removing edges from G will yield the existence of a flow f routing $\sigma_j^{(i)}(v)$ traffic from source s_i to v in such a way that f is a γ-Nash equilibrium and such that the latency from source s_i to v under this flow is $\lambda_j^{(i)}(v)$.

We successively compute the entries of T_j as follows:

- For T_0 we set all entries to $\perp$ except for the one corresponding to $\sigma_0^{(i)} \equiv \perp$ and $\lambda_0^{(i)} \equiv 0$ for all $i \in [k]$, which we set to $\emptyset$.
- Let $A_j \setminus A_{j-1} = \{v\}$ for some node $v \notin \{s_1, \ldots, s_k, t_1, \ldots, t_k\}$, and let $U = \{u_1, \ldots, u_h\} \subseteq A_{j-1}$ be the starting points of all incoming edges to v. Let $\sigma_j^{(1)}, \ldots, \sigma_j^{(k)}, \lambda_j^{(1)}, \ldots, \lambda_j^{(k)}$ be an index into the table T_j. If $s_i \notin G_j$ we ignore commodity i in the following discussion and focus on indices (both into T_{j-1} and into T_j) with $\sigma_j^{(i)} \equiv \perp$ and $\lambda_j^{(i)} \equiv 0$. Similarly, if $t_i \in G_{j-1}$ we focus on indices with $\sigma_j^{(i)} \equiv \top$ and $\lambda_j^{(i)} \equiv 0$.

  To determine the entry $T_j(\sigma_j^{(1)}, \ldots, \sigma_j^{(k)}, \lambda_j^{(1)}, \ldots, \lambda_j^{(k)})$, we use the values in the table T_{j-1} to determine possible ways of routing traffic to the nodes in A_{j-1}, and try to extend these flows to a flow which is still an approximate Nash equilibrium and such that the new flow routes $\sigma_j^{(i)}$ of commodity i to the nodes in A_j, and such that the latency of travelling from source s_i to $v \in A_j$ is $\lambda_j^{(i)}(v)$, up to an additive error of $j\epsilon/m$. We need only change flows to A_{j-1} in such a way that we additionally route traffic from the nodes $u_1, \ldots, u_h \in A_{j-1}$ to v, as re-routing traffic between the nodes in A_{j-1} only results in flows which have already been considered when computing the table T_{j-1}.

  We are looking for an index $(\sigma_{j-1}^{(1)}, \ldots, \sigma_{j-1}^{(k)}, \lambda_{j-1}^{(1)}, \ldots, \lambda_{j-1}^{(k)})$ into the table T_{j-1}, a subset $S \subset \{u_1, \ldots, u_h\}$ of predecessors of v and real numbers $\phi_{i,w} \in \mathbb{R}_{\geq 0}$ for $w \in S$ and $i \in [k]$ such that:

  - There is a way of removing edges from G_{j-1} to yield the existence of an approximate Nash equilibrium up to A_{j-1}, i.e.,

    $$T_{j-1}(\sigma_{j-1}^{(1)}, \ldots, \sigma_{j-1}^{(k)}, \lambda_{j-1}^{(1)}, \ldots, \lambda_{j-1}^{(k)}) \neq \perp.$$

  - Routing $\phi_{i,w}$ of commodity i from w to v changes $\sigma_{j-1}^{(i)}$ into $\sigma_j^{(i)}$:

    $$\sum_{w \in S} \phi_{i,w} = \sigma_j^{(i)}(v) \quad \text{for all } i = 1, \ldots, k$$

    and for all $w \in S$ and $i = 1, \ldots, k$ we have

    $$\sigma_{j-1}^{(i)}(w) - \phi_{i,w} = \begin{cases} \sigma_j^{(i)}(w) & \text{if } w \in A_j \\ 0 & \text{otherwise} \end{cases}$$

    In particular, the $\phi_{i,w}$ are also integer multiples of ϵ/m.

  - If $\sigma_j^{(i)}(v) = 0$ then also $\lambda_j^{(i)}(v) = 0$. Otherwise, the latencies of s_i-v-paths are approximately given by $\lambda_j^{(i)}(v)$: If $\phi_{i,w} > 0$, then

    $$\left| \lambda_{j-1}^{(i)}(w) + l_{(wv)}\left(\sum_{i=1}^{k} \phi_{i,w} \right) - \lambda_j^{(i)}(v) \right| \leq (B_j - B_{j-1})\gamma$$

and the flow is still approximately at Nash equilibirium: For all $w \in S$ and $i \in [k]$ with $\sigma_{j-1}^{(i)}(w) > 0$,

$$\lambda_{j-1}^{(i)}(w) + l_{(wv)} \left(\sum_{i=1}^{k} \phi_{i,w} \right) \geq \lambda_j^{(i)}(v) - \frac{\epsilon}{2}.$$

- The approximate latencies for vertices in $A_j \cap A_{j-1}$ remain unchanged unless $\sigma_j^{(i)}(v) = 0$: $\lambda_j^{(i)}(u) = \lambda_{j-1}^{(i)}(u)$ for all $i \in [k]$ and $u \in A_j \cup A_{j-1}$. If there is such a combination, then we set

$$T_j(\sigma_j^{(1)}, \ldots, \sigma_j^{(k)}, \lambda_j^{(1)}, \ldots, \lambda_j^{(k)}) := T_{j-1}(\sigma_{j-1}^{(1)}, \ldots, \sigma_{j-1}^{(k)}, \lambda_{j-1}^{(1)}, \ldots, \lambda_{j-1}^{(k)})$$
$$\cup \{uv \mid u \in U \setminus S\}$$

- If $A_j \setminus A_{j-1} = \{s_i\}$ for a source node s_i we proceed essentially as in the previous step but demand that $\sigma_j^{(i)}(v) = r_i$ and $\lambda_j^{(i)} = 0$. The other latencies and flows of other commodities into s_i are handled as above. Note that this can be adjusted to the case where $s_i = s_{i'}$ for $i \neq i'$.
- Finally, if $A_j \setminus A_{j-1} = \{t_i\}$ for a source node t_i we demand that $\sigma_j^{(i)}(v) = \top$ treat this to mean that exactly an amount r_i of commodity enters t_i. The other latencies and flows of other commodities into t_i are handled as above.

That this way of filling the table will satisfy condition (a) is easily verified. We now turn to condition (b). Assume that for some bag A_j and some index $(\sigma, \lambda) = (\sigma_j^{(1)}, \ldots, \sigma_j^{(k)}, \lambda_j^{(1)}, \ldots, \lambda_j^{(k)})$ into T_j we have $T_j(\sigma, \lambda) = \bot$ but still there is a subset $F \subset E_j$ and a flow f the graph G_j with the edges in F removed such that f is a γ-Nash equilibrium which routes $\lambda_j^{(i)}(v)$ of commodity i into v.

Using Lemma 4, we discretise f to obtain a flow f' such that $|l_P(f) - l_P(f')| < \alpha j^2 \gamma$ for all paths $P \in \mathcal{P}_{\text{all}}$ with endpoints in A_j. Since we assumed f to be a γ-Nash equilibrium, for two s_i-u-paths P and P' with endpoint $u \in A_j$ we have

$$|l_P(f') - l_{P'}(f')| \leq |l_P(f') - l_P(f)| + |l_P(f) - l_{P'}(f)| + |l_{P'}(f) - l_{P'}(f')|$$
$$\leq \alpha\gamma j^2 + \gamma + \alpha\gamma j^2 \quad \leq \quad (2B_j - 1)\gamma$$

In particular, there is an integer multiple λ of γ such that all latencies $l_P(f')$ for s_i-u paths p are within distance B_j of λ. Following the computation path we see that the corresponding table entry in T_j can not be $\bot$.   □

3   Graphs of Bounded Path-Width

While the strong directed path-decompositions of Definition 1 are convenient for the purpose of our algorithm, they are non-standard and it is not clear what kinds of graphs allow for these decompositions and how they can be obtained. In this section we show that in particular graphs of bounded path-width and simultaneously bounded degree allow for such decompositions.

Path-width was defined by Robertson and Seymour in the first paper of their Graph Minors series [9]:

Definition 8. *Let $G = (V, E)$ be an undirected graph. A* path-decomposition *of G is a sequence $X_1, \ldots, X_s$ of subsets of V such that*

- *for every $e \in E$ there is an $i \in [s]$ such that both endpoints of e are in X_i and*
- *$X_i \cap X_k \subseteq X_j$ for all $1 \leq i \leq j \leq k \leq s$.*

The width *of the decomposition is $\max_i |X_i| - 1$. The* path-width *ρ of G is the minimum $\rho \in \mathbb{N}$ such that G has a path-decomposition of width ρ.*

We will need the following fact about graphs of bounded path-width:

Fact 9. *For every $\rho \in \mathbb{N}$ there is an $h \in \mathbb{N}$ such that no graph of path-width at most ρ has a minor isomorphic to the complete binary tree of height h.*

This follows easily from theorem (1.2) in [9] and the fact that every tree is a minor of a sufficiently large complete binary tree. As usual, a minor of a graph G is a graph obtained from a subgraph of G by contracting edges.

We are now ready to state the main result of this section:

Lemma 10. *Let $b \in \mathbb{N}$ and let $\mathcal{G}$ be a class of acyclic directed graphs such that for every $G \in \mathcal{G}$, the total degree (i.e., in-degree plus out-degree) of every vertex $v \in V(G)$ is at most b and the (undirected) path-width of G is at most b. Then there is a $\rho \in \mathbb{N}$ depending only on b such that, given a graph $G \in \mathcal{G}$, an sdpd of G of width at most ρ can be computed from G in linear time.*

Proof (Proof of Lemma 10). Using Fact 9, let $d_0 \in \mathbb{N}$ be such that none of the graphs in $\mathcal{G}$ contains a complete binary tree of depth d_0 as a minor (ignoring all edge directions).

Pick some $G \in \mathcal{G}$. We may assume that G has exactly one sink. Otherwise, let $X_1, \ldots, X_l$ be a path-decomposition of G, and let $v_1, \ldots, v_s$ be the sinks of G in the order in which they first appear in the path-decomposition, breaking ties arbitrarily. Adding a directed edge from v_i to v_{i+1} for $i \in [s-1]$ increases the path-width of G by at most 1, because we can obtain a path-decomposition for the new graph by adding v_i to all bags between the first entry of v_i and the first entry of v_{i+1}, increasing each bag-size by at most one. Furthermore, we only increase the degree of the graph by at most two. Acyclicity is also maintained, and by a result of Bodlaender and Kloks [10], for fixed path-width b a path-decomposition of width b can be computed in linear time.

To obtain an sdpd for G, we start from the sink and successively create new bags by taking all predecessors of all nodes in the current bag. If, at some point, the resulting bag has size exceeding b^{d_0}, then G has a minor isomorphic to a complete binary tree of depth d_0, a contradiction.   □

4   Conclusion

We complemented Roughgarden's [5] and Lin et al.'s [2] results on the hardness of approximation (up to a multiplicative error) of the multi-commodity network design problem by giving an approximation algorithm for this problem on a certain restricted class of inputs, namely graphs allowing for what we call a bounded path-decomposition with Lipschitz-continuous latency functions. For technical reasons, we have to work with approximate Nash equilibria, so our algorithm does not directly compare with previous hardness results.

For general latency functions, restrictions on the class of input graphs similar to ours seem to be necessary [2]. If the latency functions are polynomials of bounded degree, the proof technique used in [5] combined with Schrijver's algorithm for 2DDP on planar graphs [7] raises the question of whether efficient approximation algorithms exist for less severely restricted classes of input graphs such as planar graphs.

References

1. Braess, D.: Über ein Paradoxon aus der Verkehrsplanung. Mathematical Methods of Operations Research 12(1), 258–268 (1968)
2. Lin, H.C., Roughgarden, T., Tardos, É., Walkover, A.: Stronger bounds on Braess's paradox and the maximum latency of selfish routing. SIAM J. Discrete Math. 25(4), 1667–1686 (2011)
3. Fotakis, D., Kaporis, A.C., Spirakis, P.G.: Efficient methods for selfish network design. In: Albers, S., Marchetti-Spaccamela, A., Matias, Y., Nikoletseas, S., Thomas, W. (eds.) ICALP 2009, Part II. LNCS, vol. 5556, pp. 459–471. Springer, Heidelberg (2009)
4. Roughgarden, T.: Selfish Routing and the Price of Anarchy. MIT Press (2005)
5. Roughgarden, T.: On the severity of Braess's paradox: Designing networks for selfish users is hard. J. Comput. Syst. Sci. 72(5), 922–953 (2006)
6. Fortune, S., Hopcroft, J., Wyllie, J.: The directed subgraph homeomorphism problem. Theoretical Computer Science 10(2), 111–121 (1980)
7. Schrijver, A.: Finding k disjoint paths in a directed planar graph. SIAM Journal on Computing 23(4), 780–788 (1994)
8. Barát, J.: Directed path-width and monotonicity in digraph searching. Graphs and Combinatorics 22, 161–172 (2006)
9. Robertson, N., Seymour, P.D.: Graph minors I. Excluding a forest. Journal of Combinatorial Theory Series B (35), 39–61 (1983)
10. Bodlaender, H.L., Kloks, T.: Better algorithms for the pathwidth and treewidth of graphs. In: Leach Albert, J., Monien, B., Rodríguez-Artalejo, M. (eds.) ICALP 1991. LNCS, vol. 510, pp. 544–555. Springer, Heidelberg (1991)

Cooperative Equilibria in Iterated Social Dilemmas

Valerio Capraro[1], Matteo Venanzi[2], Maria Polukarov[2], and Nicholas R. Jennings[2]

[1] Mathematics, University of Southampton, United Kingdom
V.Capraro@soton.ac.uk
[2] Electronics and Computer Science, University of Southampton, United Kingdom
{mv1g10,mp3,nrj}@ecs.soton.ac.uk

Abstract. The implausibility of the extreme rationality assumptions of Nash equilibrium has been attested by numerous experimental studies with human players. In particular, the fundamental social dilemmas such as the Traveler's dilemma, the Prisoner's dilemma, and the Public Goods game demonstrate high rates of deviation from the unique Nash equilibrium, dependent on the game parameters or the environment in which the game is played. These results inspired several attempts to develop suitable solution concepts to more accurately explain human behaviour. In this line, the recently proposed notion of cooperative equilibrium [5, 6], based on the idea that players have a natural attitude to cooperation, has shown promising results for single-shot games. In this paper, we extend this approach to iterated settings. Specifically, we define the Iterated Cooperative Equilibrium (ICE) and show it makes statistically precise predictions of population average behaviour in the aforementioned domains. Importantly, the definition of ICE does not involve any free parameters, and so it is fully predictive.

1 Introduction

The standard assumption of economic models that players in strategic situations act perfectly rationally has been constantly rejected by numerous experiments over the years. These experiments, typically conducted on the fundamental social dilemmas such as the Prisoner's dilemma, the Traveler's dilemma, and the Public Goods game, have shown that cooperation between players (associated with the deviation from the unique, but inefficient, Nash equilibrium) is frequent, and appears to depend on both the game parameters and the environment in which the game is played. In particular, it has been observed that the rate of cooperation in the Traveler's dilemma depends on the bonus/penalty value, whenever the game is single-shot or iterated [7, 12]; the rate of cooperation in the Prisoner's dilemma depends on the payoff parameters or the way the players are matched to play together [11, 32]; and the rate of cooperation in the Public Goods game depends on the marginal return or on the frequency of interaction between free-riders and cooperators [13, 14, 17].

Considerable research efforts have been made in attempt to explain deviations from Nash equilibria. Some methods developed to this end are based on the idea that humans have bounded rationality and/or can make mistakes in computations[1] [4, 9, 20, 25]; others explain cooperation in terms of evolution [1, 3, 10, 19, 21–23, 29]. Finally,

---

[1] See [31] for a recent parallelism among these approaches.

B. Vöcking (Ed.): SAGT 2013, LNCS 8146, pp. 146–158, 2013.

much of work has been directed towards defining profoundly different solution concepts [24, 26], especially in the recent algorithmic game theory and artificial intelligence communities [2, 8, 15, 16, 18, 27, 30]. This interest is particularly motivated by the emerging applications of human-agent collectives, where artificial agents interact with humans. To build such systems effectively, it is highly important to understand and find accurate methods to predict human behaviour.

To this end, a new solution concept, termed *cooperative equilibrium*, has been recently proposed for one-shot games [5, 6]. This approach is inspired by the aforementioned experimental findings, which suggest that players are conditionally cooperative—that is, the same player may act more or less cooperatively in the same game scenario, depending on the actual payoffs. In other words, humans have an attitude to cooperation by nature: they do not act a priori as single players, but rather forecast how the game would have been played if they formed coalitions and then select actions according to their best forecast. It turns out, that direct implementation of this idea can predict human behaviour with impressively high precision, as demonstrated in [5, 6] on the aforementioned social dilemmas.

In this paper, we further explore this direction and extend the cooperative equilibrium approach to iterated settings. Specifically, we define the Iterated Cooperative Equilibrium (ICE), that combines this concept with some ideas developed in [7] for iterated games. Importantly, in contrast to other methods, ICE does not use any free parameters, and thus is fully predictive. We then evaluate our method on the iterated Traveler's dilemma, the Prisoner's dilemma, and the Public Goods game. To this end, we make use of the experimental data provided in [7], [32] and [14] for these three domains, respectively.[2] Our results confirm that the ICE makes accurate predictions of population average behaviour in social dilemmas. In particular, it clearly outperforms the Logit Learning Model (LLM) developed in [7] for the Traveler's dilemma.

The paper unfolds as follows. In Section 2 we define the social dilemmas in consideration. In Section 3 we formalise our approach. We then apply it to the iterative Traveler's dilemma in Section 4, to the Prisoner's dilemma in Section 5, and to the Public Goods game in Section 6. Section 7 concludes with directions for future work.

2 Preliminaries

We start with the definitions of the social dilemmas in consideration of this paper.

Prisoner's Dilemma (PD). Two players choose to either cooperate (C) or defect (D). If both players cooperate, each receives the monetary reward, R, for cooperating. If one player defects and the other cooperates, then the defector receives the temptation payoff, T, while the other receives the sucker payoff, S. If both players defect, they both receive the punishment payoff, P. Payoffs are subjected to the condition $T > R > P > S$.

Traveler's Dilemma (TD). Two travelers need to claim for a reimbursement between L and H monetary units for their (identical) luggage that has been lost by the same air company. To avoid high claims, the air company employs the following rule: the traveler who makes a lower claim, say m, gets a reimbursement of $m + b$ monetary units, and the other one gets a reimbursement of $m - b$ monetary units, for a fixed

---

[2] These were the only sources we could find that reported sufficient data for our purposes.

value of bonus/penalty, b. If both players claim the same amount, m, then they both get reimbursed by m monetary units.

Public Goods Game (PG). n players receive an initial endowment of $y > 0$ monetary units each and simultaneously choose an amount $0 \leq x_i \leq y$ to contribute to a public pool. The total amount in the pot is multiplied by α_0 and then divided equally among all group members. Thus, player i's utility is $u_i(x_1, \ldots, x_n) = y - x_i + \alpha(x_1 + \ldots + x_n)$, where $\alpha = \frac{\alpha_0}{n}$. The number α is termed the *constant marginal return* and assumed to belong to the interval $\left(\frac{1}{n}, 1\right)$.

3   Iterated Cooperative Equilibrium

We now introduce the concept of iterated cooperative equilibrium for the aforementioned social dilemmas.

Let $\mathcal{G} = (N, (S_i, u_i)_{i \in N})$ be a normal-form game with a set N of n players, and for all $i \in N$, a finite set of strategies S_i and a monetary payoff function $u_i : S \to \mathbb{R}$, where $S = \times_{j \in N} S_j$. As usual, we use $-i$ to denote the set $N \setminus \{i\}$ of all players but i. We denote by $\Delta(X)$ the set of probability distributions on a finite set X. Thus, $\Delta(S_i)$ defines the set of mixed strategies for player $i \in N$, and his expected payoff from a mixed strategy profile σ is given by $u_i(\sigma) = \sum_{s \in S} u_i(s) \sigma_1(s_1) \cdot \ldots \cdot \sigma_n(s_n)$.

The idea behind our approach is as follows. Suppose each agent i simply considers two possible scenarios: the fully selfish play p_s, where players take individual actions pursuing their private interests, and the fully cooperative play p_c, where players are assumed to pursue the collective interest. With each scenario p we associate a value $v_i(p)$, defined as an average $v_i(p) = e_i(p)\tau_i(p) + e_i(\overline{p})\tau_i(\overline{p})$, where, roughly speaking,

– $\tau_i(p)$ is the probability that all players follow scenario p, and $\tau_i(\overline{p}) = 1 - \tau_i(p)$ is the probability that (at least one of) the players $-i$ will deviate from p for the sake of their individual interests, knowing that player i follows scenario p. In particular, this implies that $\tau_i(\overline{p_s}) = 0$, since a Nash equilibrium cannot be improved by unilateral deviations;
– $e_i(p)$ is the payoff of i when scenario p is realised, and $e_i(\overline{p})$ is the infimum of gains player i achieves when other players deviate from p.

Then, the values $v_i(p)$ determine each player i's strategy as follows. Let $p_i^* \in \{p_s, p_c\}$ be the scenario that maximises the function v_i, and define the induced game $\mathcal{G}(p_i^*)$ to be the restriction of $\mathcal{G}$ where the set of allowed mixed strategy profiles is given by $\{\sigma | u_j(\sigma) \geq v_i(p_i^*), \forall j\}$. Note that $v_i(p_i^*)$ only reflects player i's beliefs, while in the induced game the strategies of all players are limited. That is, as is typical for human players, they extrapolate their own experience to others. Now, since this set of strategies is convex and compact, the induced game has Nash equilibria. The cooperative equilibrium is then given by a combination of strategies where each player i plays according to a Nash equilibrium of his induced game.

Formalising this idea is not completely trivial: while the payoffs e_i seem straightforward to define, the probabilities τ_i are much more delicate, since the event "players $-i$ deviate from scenario p_c" is not measurable in any universal sense. In iterated settings, we can approach this problem applying a sort of fictitious play. Specifically, we start with initial values $\tau_i(p_c) = \tau_i(\overline{p_c}) = \frac{1}{2}$, and then at each step we update these probabilities using observations made in previous rounds. To this end, we use the standard method for probabilistic modelling of binary random events based on the beta family of probability density functions [28]. If in the first round player i has observed cooperation, then $\tau_i(p_c)$ grows from $\frac{1}{2}$ to $\frac{2}{3}$, otherwise it drops from $\frac{1}{2}$ to $\frac{1}{3}$ and so forth: that is, if k is the number of cooperative plays observed in periods from 1 to $t-1$, then $\tau_i^{(t)}(p_c)$ is updated to $\frac{k+1}{t+1}$. We now define this procedure in detail.

Let $\mathcal{G} \in \{PD, TD, PG\}$. Then, $\mathcal{G}$ has a unique Nash equilibrium, $NE(\mathcal{G})$. Moreover, there is also a unique Pareto optimal strategy profile, $OPT(\mathcal{G})$. For each period $t \geq 1$, we set $v_i^{(t)}(p_s) = u_i(NE(\mathcal{G}))$ and $e_i^{(t)}(p_c) = u_i(OPT(\mathcal{G}))$. For other parameters, we consider the first and the later rounds separately.

Period 1. We define:

- $e_i^{(1)}(\overline{p_c}) = \inf\{u_i(\sigma)|\sigma_i = OPT(\mathcal{G})_i; \forall j \neq i, u_j(\sigma_j, OPT(\mathcal{G})_{-j}) \geq u_j(OPT(\mathcal{G}))\}$
  is the infimum payoff that player i obtains when he plays according to the Pareto optimum, while other players deviate from this profile if the corresponding *unilateral* deviation weakly improves the payoff to each deviator;
- $\tau_i^{(1)}(p_c) = \tau_i^{(1)}(\overline{p_c}) = \frac{1}{2}$;
- $v_i^{(1)}(p_c) = \tau_i^{(1)}(p_c)e_i^{(1)}(p_c) + \tau_i^{(1)}(\overline{p_c})e_i^{(1)}(\overline{p_c})$;
- $v_i^{(1)} = \max\{v_i^{(1)}(p_s), v_i^{(1)}(p_c)\}$;
- $\text{Ind}(\mathcal{G}, i, 1)$ is the restriction of game $\mathcal{G}$ where the set of allowed mixed strategy profiles is limited to $\left\{\sigma | u_j(\sigma) \geq v_i^{(1)}, \forall j\right\}$.

Period t. We update payoffs e_i and probabilities τ_i as follows.

- Let σ_{-i} be the *average* of strategies played by players $-i$ in periods from 1 to $t-1$. Then, $e_i^{(t)}(\overline{p_c}) = u_i(OPT(\mathcal{G})_i, \sigma_{-i})$;
- Let $\sigma_{-i}^{(s)}$ be the strategy played by players $-i$ in period $s < t$. We say that $\sigma_{-i}^{(s)}$ is a *cooperation* if there is a strategy $\sigma_i \neq (NE(\mathcal{G}))_i$ such that $(\sigma_i, \sigma_{-i}^{(s)})$ is allowed in $\text{Ind}(\mathcal{G}, i, s)$. Let k be the number of cooperations in periods from 1 to $t-1$. Then,

$$\tau_i^{(t)}(p_c) = \frac{k+1}{t+1} \quad \text{and} \quad \tau_i^{(t)}(\overline{p_c}) = 1 - \tau_i^{(t)}(p_c);$$

- $v_i^{(t)}(p_c)$, $v_i^{(t)}$ and $\text{Ind}(\mathcal{G}, i, t)$ are determined analogously to Period 1.

Given this, we can now make the following definition.

Definition 1. The *iterated cooperative equilibrium* (ICE) of game $\mathcal{G}$ in period t is a strategy profile σ where strategy σ_i for each player $i \in N$ corresponds to the strategy he plays in the Nash equilibrium of the induced game $\text{Ind}(\mathcal{G}, i, t)$.

Example 1. Consider the PD with payoffs $T = 20$, $R = 15$, $P = 5$ and $S = 0$. For a given player i, the selfish scenario has value $v_i^{(t)}(p_s) = 5$ for all periods $t \geq 1$, while the value of the cooperative scenario changes depending on what i has observed in previous iterations. At $t = 1$ it has value $v_i^{(1)}(p_c) = 15 \cdot \frac{1}{2} + 0 \cdot \frac{1}{2} = 7.5$. Thus, the

ICE at this step corresponds to the Nash equilibrium of the game where only strategies giving both players a payoff of at least 7.5 are allowed—i.e., $\sigma_i = 0.25C + 0.75D$, $\forall i$.

At $t = 2$ we have two cases: (i) If in the first period player i observes cooperation, then $\tau_i^{(2)}(p_c) = \frac{1}{3}$ and $e_i^{(2)}(p_c) = 15$. Hence, $v_i^{(2)}(p_c) = 15$, and the induced game allows only one strategy (C) for both players. So, in a cooperative equilibrium both players cooperate; (ii) Otherwise, if i observes defection, then $\tau_i^{(2)}(p_c) = \frac{2}{3}$ and $e_i^{(2)}(p_c) = 0$. Hence, $v_i^{(2)}(p_c) = 15 \cdot \frac{1}{3} + 0 \cdot \frac{2}{3} = 5 = v_i^{(2)}(p_s)$, and the induced game coincides with the original game. The cooperative equilibrium corresponds to the Nash equilibrium of the original game—both players defect.

4 Traveler's Dilemma

In this section, we demonstrate the predictive power of cooperative equilibrium on the iterated Traveler's dilemma. We make use of the experimental data provided by Capra-Goeree-Gomez-Holt in [7] for the setting with $L = 80$ and $H = 200$, and compare ICE predictions with the *logit learning model* (LLM), proposed in [7] to explain these data.

There are two main differences between the LLM and ICE we would like to stress:

- First, as have been previously mentioned, ICE does not use any free parameter, while the LLM involves two free parameters, a *learning* parameter and a *error* parameter. In other words, ICE is a *predictive* model, and the LMM is *descriptive*.
- Second, the models are different conceptually. ICE applies the idea that people have an *attitude to cooperation*: they do not act a priori as single players, but rather forecast how the game would be played if they formed coalitions, and then play according to their best forecast. In contrast, the LLM assumes *selfish*, *individual* decisions, and explains deviations from Nash equilibrium in terms of mistakes.

We now proceed to compare between the ICE and the LLM predictions, based on the experimental data collected in [7]. In this experiment, groups of 9, 10 and 12 subjects played a 10 rounds Traveler's dilemma with low ($b \in \{5, 10\}$), intermediate ($b \in \{20, 25\}$) or high ($b \in \{50, 80\}$) bonus/penalty values. After each round, the subjects' claims were casually matched to determine their payoffs. In this paper, we exclude the case with $b = 10$ since it involved an odd number of participants (9 players), and so at each turn one player remained unmatched and his payoff was not determined; we therefore cannot compute the ICE in this case. Following [7], the LLM predictions are calculated using the values $\rho = 0.75$ and $\mu = 10.9$ for the learning/error parameters.

Recall that the TD has a unique Nash equilibrium where each player chooses the minimal claim of $L = 80$, whichever is the value of bonus/penalty, b. The results in [7] show that in practice the players' behaviour is not independent of the value of b. Indeed, when the bonus/penalty is low, the players tend to make very high claims, especially in the last rounds; this to some extent is supported by the logit learning model proposed in [7]. However, as can be seen from Table 1 and Figure 1, for $b = 5$ the ICE predicted values fall much closer to the average observed claims than the LLM predictions.

Table 1. Observed and predicted claims in TD with low bonus/penalty of $b = 5$

t	Avg. obs. claim	ICE	LLM
1	180.08	195.00	167.75
2	180.00	182.06	175.09
3	185.30	185.77	179.53
4	191.34	188.15	181.88
5	194.98	190.03	183.81
6	196.62	191.35	185.14
7	196.86	192.70	186.32
8	196.68	193.34	186.82
9	195.48	194.05	187.02
10	194.34	194.03	186.80

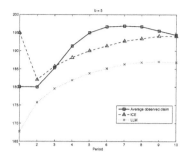

Fig. 1. ICE vs. LLM for $b = 5$. The solid line corresponds to the actual data. The ICE predictions are represented by the dashed line, and the LLM predicted values are depicted by the dotted line.

Table 2 and Figure 2 below present the data and predictions for the two cases with intermediate bonus/penalty values of $b = 20$ and $b = 25$.

Table 2. Observed and predicted claims in TD with intermediate bonus/penalty

	$b = 20$			$b = 25$		
t	Avg. obs. claim	ICE	LLM	Avg. obs. claim	ICE	LLM
1	131.20	180.00	152.64	139.96	175.00	146.60
2	127.20	134.53	151.32	137.59	134.68	145.77
3	128.35	135.57	150.63	159.90	137.94	146.73
4	108.70	133.02	148.74	154.27	146.38	150.66
5	103.30	125.69	144.38	146.49	150.19	151.17
6	117.30	120.66	142.55	161.44	148.51	147.84
7	105.80	119.37	145.71	151.88	150.65	150.60
8	117.30	117.60	146.60	139.12	150.99	149.47
9	119.20	117.73	146.82	132.09	147.04	142.74
10	119.20	117.66	149.14	143.04	143.62	135.32

For $b = 20$, ICE again clearly outperforms the LLM, as shown in Figure 2a. For $b = 25$, the two models show similar performance: ICE is closer to the actual average claim in periods 2, 5, 6, 7 and 10, while the LLM performs better in periods 1, 3, 4, 8, and 9 (see Figure 2b). Note that the observed data in this case is very noisy, with no clear tendency towards higher or lower claims across the rounds of the experiment.

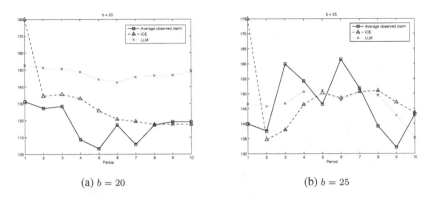

(a) $b = 20$ (b) $b = 25$

Fig. 2. ICE vs. LLM for $b \in \{20, 25\}$. Solid lines correspond to the actual data. The ICE predictions are represented by dashed lines, and the LLM predictions are depicted by dotted lines.

As the bonus/penalty values get higher, the players reduce their claims, and actually converge to the Nash equilibrium solution in the last rounds of the experiment for high b (see Table 3 and Figure 4). While both ICE and the LLM capture this tendency, yet again, the ICE predictions appear to be closer to the experimental data.

Table 3. Observed and predicted claims in TD with high bonus/penalty

t	$b = 50$			$b = 80$		
	Avg. obs. claim	ICE	LLM	Avg. obs. claim	ICE	LLM
1	155.86	150.00	117.17	120.07	120.00	98.04
2	125.37	122.15	130.95	112.18	103.33	103.38
3	125.77	121.66	121.63	106.16	101.66	103.26
4	109.13	119.06	117.15	88.75	93.55	92.43
5	89.47	114.75	106.95	85.00	91.66	92.23
6	102.26	106.46	95.13	84.91	85.71	88.44
7	101.68	100.67	101.74	82.41	83.33	85.79
8	84.38	96.99	108.54	81.58	82.96	83.77
9	82.00	91.43	105.42	80.00	80.00	83.35
10	88.27	88.27	100.63	80.00	80.00	83.34

In conclusion, the ICE model is much more accurate than the LLM in the prediction of population average behaviour in the TD. Next we show that it can be successfully applied to other relevant social dilemmas, such as in fact the PD and the PG game.

5   Prisoner's Dilemma

In this section, we test our method on the iterated Prisoner's dilemma, using the experimental data provided by Yang-Yue-Yu in [32]. Although it is a dominant strategy for each player to defect, irrespective of payoffs or any other factors, human behaviours observed in experimental studies show considerable rates of cooperation, which appear to depend on game parameters or the environment in which it is played. The study in [32] is particularly focused on the way the players are matched to play together. This feature is crucial since different matching rules entail different histories for a player, and hence, different beliefs regarding his opponents' play. These, in turn, ultimately reflect on the player's strategic decisions. Therefore, it is of great importance to provide prediction methods that would achieve robust performance in different environments. As we show, ICE can successfully tackle this challenge.

The experiment involved 70 subjects that played a 25 rounds Prisoner's dilemma with payoffs parameters $T = 12$, $R = 8$, $D = 3$ and $S = 1$, under different matching schemes. Specifically, it included the following treatments: (i) the *random matching* (RM) where subjects were randomly paired in each period; (ii) the *one-period correlated matching* (OP) where subjects who have selected identical strategies in a given round are randomly paired with one another in the next period; and (iii) *weighted-history correlated matching* (WH) where, after every round, subjects are matched with a player who has been choosing *similar* strategies in the previous five periods. In more detail, the history is weighted using Fibonacci numbers as follows. Each subject starts with a sorting score $T(t) = 0$, for all $t \leq 1$. At each round t, his score is updated to $T(t) = 5a(t-1) + 3a(t-2) + 2a(t-3) + 1a(t-4) + 1a(t-5)$, where $a(s)$ is 0 if he plays defection in period s, and 1 otherwise. In each period, subjects are paired in the order of their current scores.

Table 4 and Figure 3 summarise the data collected in this experiment, along with the corresponding values of iterated cooperative equilibrium. As these results demonstrate, ICE accurately predicts the players' behaviour in Prisoner's dilemma, especially for cases with correlated matching (see Figures 3b and 3c). In the case where the players were matched randomly (Figure 3a), the ICE predictions in the last rounds of the experiment appear slightly more pessimistic than the actual data, which is implied by relatively high rates of defection observed in the intermediate rounds.

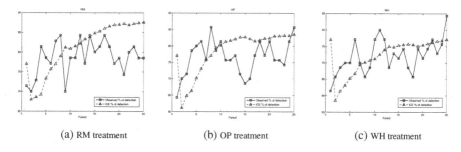

(a) RM treatment   (b) OP treatment   (c) WH treatment

Fig. 3. ICE in Prisoner's dilemma. Solid lines correspond to the actual data. The ICE predictions are represented by dashed lines.

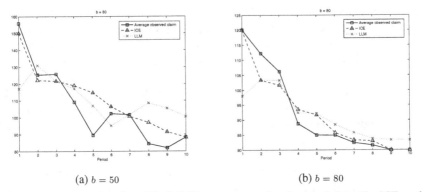

(a) $b = 50$ (b) $b = 80$

Fig. 4. ICE vs. LLM for $b \in \{50, 80\}$. Solid lines correspond to the actual data. The ICE predictions are represented by dashed lines, and the LLM predictions are depicted by dotted lines.

t	Avg. obs. contribution	s.t.d	ICE
1	41.00	18.92	0.00
2	29.36	18.11	18.50
3	31.89	17.89	15.86
4	27.80	20.55	17.37
5	16.97	15.24	11.09
6	10.50	9.80	7.24
7	10.33	8.10	5.16
8	7.91	5.45	4.42
9	6.39	9.54	1.56
10	4.39	7.41	1.77

Table 5. Observed and predicted contributions in Public Goods with marginal return of $\alpha = 0.3$.

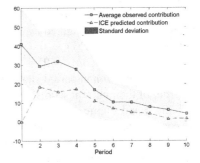

Fig. 5. ICE in Public Goods with $\alpha = 0.3$. The actual data are represented by the solid line. The shaded area shows the standard deviation. The dashed line corresponds to the ICE predictions.

6   Public Goods Game

In this section, we use ICE predictions to explain the experimental data on multi-round Public Goods game presented by Gunnthorsdottir-Houser-McCabe in [14].

The experiment consisted of three treatments with different constant marginal returns of $\alpha = 0.3$, $\alpha = 0.5$, and $\alpha = 0.75$. The first and the third treatment involved 36 subjects each, and 60 subjects participated in the second treatment. The subjects played a 10 rounds Public Goods game in groups of 4, to which they were randomly matched in each round. The average observed contributions and the corresponding ICE predictions for the first treatment with $\alpha = 0.3$ are given in Table 5 and Figure 5.

The data are very heterogenous – note the high rates of standard deviation. This is reflected on the fact that the ICE's predictions in this setting seem less accurate than in previously considered domains. Notice, however, that in all game rounds (except of the

Table 4. Observed and predicted percentages of defection in iterated PD

t	RM		OP		WH	
	Obs. % defect	ICE % defect	Obs. % defect	ICE % defect	Obs. % defect	ICE % defect
1	71.43	77.07	64.29	77.06	61.42	77.06
2	70.00	67.97	70.00	61.18	65.71	58.46
3	72.86	68.59	71.43	64.80	68.57	61.31
4	81.43	69.22	78.57	66.29	70.00	63.11
5	78.57	73.31	80.00	70.16	70.00	65.22
6	77.14	75.66	81.43	73.07	77.14	66.65
7	82.86	77.00	75.71	75.94	70.00	69.47
8	84.29	78.99	85.71	77.06	65.71	70.15
9	70.00	81.21	78.57	79.46	68.57	70.74
10	78.57	80.87	81.43	80.17	77.14	71.22
11	78.57	81.51	75.71	81.28	80.00	72.51
12	84.29	82.18	75.71	81.62	77.14	73.93
13	77.14	83.31	77.14	82.04	68.57	74.97
14	84.29	83.85	71.43	82.62	72.85	74.76
15	80.00	84.82	68.57	82.68	71.42	75.24
16	81.43	85.29	70.00	82.13	74.28	75.50
17	84.29	86.00	77.14	81.49	68.57	75.67
18	81.43	86.58	81.43	81.74	65.71	75.52
19	77.14	86.92	77.14	82.28	74.28	75.05
20	78.57	87.00	81.43	82.36	71.42	75.41
21	74.29	87.22	75.71	82.98	74.28	75.79
22	80.00	86.91	75.71	83.04	77.14	76.18
23	81.43	87.18	74.29	83.14	72.85	76.55
24	78.57	87.41	81.43	83.12	75.71	76.70
25	78.57	87.56	85.71	83.52	84.28	77.08

very first one where the players beliefs are yet completely fictitious), the ICE values fall within the standard deviation interval and their error decreases as the number of periods increases. Similar performance is also showed in treatments with higher marginal returns, presented in Table 6 and Figure 6.

7 Conclusions

In this paper, we introduced the Iterated Cooperative Equilibrium (ICE) which extends the approach of players' natural attitude to cooperation to games played in iterated fashion. In each round, the players forecast how the game would be played if they formed coalitions, and select their actions accordingly. The beliefs are initially defined through a sort of fictitious play, and then get updated at each step of the game, based on previous observations. We applied this concept to three fundamental social dilemmas: the Prisoner's dilemma, the Traveler's dilemma, and the Public Goods game. The novel and most important features of the ICE is that (1) it does not use any free parameters and

Table 6. Observed and predicted contributions in Public Goods with constant marginal returns of $\alpha = 0.5, 0.75$

t	$\alpha = 0.5$			$\alpha = 0.75$		
	Obs. contribution	s.t.d	ICE	Obs. contribution	s.t.c	ICE
1	55.48	19.77	25	65.00	17.47	43.75
2	58.88	20.69	74.31	62.08	17.67	84.01
3	55.83	22.09	69.57	71.11	13.53	79.29
4	49.03	22.06	64.62	67.78	14.76	78.97
5	42.16	21.67	59.67	67.02	15.64	78.07
6	44.16	21.29	54.75	63.02	14.88	77.18
7	42.33	19.84	53.10	57.16	19.47	75.48
8	35.38	22.17	50.86	54.02	21.46	73.40
9	31.60	22.72	48.94	54.52	18.26	71.56
10	31.10	17.93	45.53	57.78	24.12	69.81

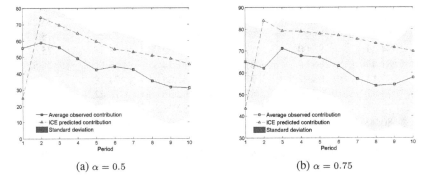

(a) $\alpha = 0.5$ (b) $\alpha = 0.75$

Fig. 6. ICE in Public Goods with $\alpha = 0.5, 0.75$. The actual data are represented by solid lines. Shaded areas show the standard deviation. Dashed lines correspond to the ICE predictions.

so it is completely predictive; (2) it makes statistically precise predictions of population average behaviour in the aforementioned domains.

This work opens a number of research directions, from the extension of the ICE to include other relevant game models to theoretical questions concerning, for instance, convergence of the iterative procedure. Regarding the latter point, one can easily see that the ICE can converge only to one of Rabin's fairness equilibria [24]: in the Traveler's dilemma, ICE can converge either to $(200, 200)$ or to $(80, 80)$; in the Prisoner's dilemma, ICE can converge either to (C, C) or to (D, D). But can actual human behaviour converge to a different strategy? The intuition suggests that the answer to this question is negative and that, in general, human behaviour may not converge at all. Indeed, if a player in the Traveler's dilemma would know that his opponent plays an intermediate strategy, say $s = 175$, then he would either reduce his claim to achieve a larger gain (which would finally lead the players to the Nash equilibrium), or rather decide to increase it to show his opponent that they both can gain more. This way of reasoning generates an oscillation, that is perfectly coherent with and reflected by ICE.

Finally, it would also be interesting to try and combine ICE with evolutionary models, in order to tackle the "cold start" effect—i.e., inaccurate predictions in early iterations. Indeed, ICE typically starts showing high performance only after a few rounds of iteration, since players have to form statistically robust beliefs. Now, in [19] the authors use an evolutionary model to explain the experimental data in the first two periods of iterated Traveler's dilemma presented in [7]. So, it is plausible that a clever combination of ICE with evolutionary models can fit the experimental data even better.

Acknowledgements. We would like to thank Anna Gunnthorsdottir for providing us with experimental data. We also gratefully acknowledge funding from the UK Research Council for project ORCHID (www.orchid.ac.uk), grant EP/I011587/1.

References

1. Axelrod, R.: The Evolution of Cooperation. Basic Books (1984)
2. Benisch, M., Davis, G.B., Sandholm, T.: Algorithms for rationalizability and CURB sets. In: AAAI 2006, pp. 598–604 (2006)
3. Boyd, R., Gintis, H., Bowles, S., Richerson, P.J.: The evolution of altruistic punishment. Proc. Nat. Acad. Sci. USA 100, 3531–3535 (2003)
4. Camerer, C., Ho, T., Chong, J.: A cognitive hierarchy model of games. Quaterly J. of Economics 119(3), 861–898 (2004)
5. Capraro, V.: A solution concept for games with altruism and cooperation, http://arxiv.org/pdf/1302.3988v2.pdf
6. Capraro, V.: A Model of Human Cooperation in Social Dilemmas. PLoS ONE (in press), Available at SSRN, http://ssrn.com/abstract=2246094
7. Capra, M., Goeree, J.K., Gomez, R., Holt, C.A.: Anomalous Behavior in a Traveler's Dilemma? American Economic Review 89(3), 678–690 (1999)
8. Conitzer, V., Sandholm, T.: A generalized strategy eliminability criterion and computational methods for applying it. In: AAAI 2005, pp. 483–488 (2005)
9. Costa-Gomes, M., Crawford, V., Broseta, B.: Cognition and behavior in normal form games: An experimental study. Econometrica 69(5), 1193–1235 (2001)
10. Fowler, J.H.: Altruistic punishment and the origin of cooperation. Proc. Nat. Acad. Sci. USA 102, 7047–7049 (2005)
11. Fudenberg, D., Rand, D.G., Dreber, A.: Slow to Anger and Fast to Forgive: Cooperation in an Uncertain World. American Economic Review 102, 720–749 (2012)
12. Goeree, J.K., Holt, C.A.: Ten Little Treasures of Game Theory and Ten Intuitive Contradictions. American Economic Review 91, 1402–1422 (2001)
13. Goeree, J.K., Holt, C.A., Laury, S.K.: Private Costs and Public Benefits: Unraveling the Effects of Altruism and Noisy Behavior. Journal of Public Economics 83(2), 255–276 (2002)
14. Gunnthorsdottir, A., Houser, D., McCabe, K.: Disposition, History and Contributions in Public Goods Experiments. Journal of Behavior and Organization 62, 304–315 (2007)
15. Halpern, J.Y., Pass, R.: Iterated Regret Minimization: a new solution concept. Games and Economic Behavior 74(1), 184–207 (2012)
16. Hyafil, N., Boutilier, C.: Regret minimizing equilibria and mechanisms for games with strict type uncertainty. In: UAI 2004, pp. 268–277 (2004)
17. Isaac, M.R., Walker, J.: Group size effects in public goods provision: The voluntary contribution mechanism. Quarterly Journal of Economics 103, 179–200 (1988)

18. Jamroga, W., Melissen, M.: Doubtful Deviations and Farsighted Play. In: Antunes, L., Pinto, H.S. (eds.) EPIA 2011. LNCS, vol. 7026, pp. 506–520. Springer, Heidelberg (2011)
19. Manapat, M.L., Rand, D.G., Pawlowitsch, C., Nowak, M.A.: Stochastic evolutionary dynamics resolve the Traveler's Dilemma. Journal of Theoretical Biology 303, 119–127 (2012)
20. McKelvey, R., Palfrey, T.: Quantal response equilibria for normal form games. Games and Economic Behavior 10(1), 6–38 (1995)
21. Nowak, M.A., Sigmund, K.: Evolution of indirect reciprocity. Nature 437, 1291–1298 (2005)
22. Nowak, M.A.: Five rules for the evolution of cooperation. Science 314, 1560–1563 (2006)
23. Panchanathan, K., Boyd, R.: Indirect reciprocity can stabilize cooperation without the second-order free rider problem. Nature 433, 499–502 (2004)
24. Rabin, M.: Incorporating Fairness into Game Theory and Economics. American Economic Review LXXXIII, 1281–1302 (1993)
25. Stahl, D., Wilson, P.: Experimental evidence on players' models of other players. J. Economic Behavior and Organization 25(3), 309–327 (1994)
26. Renou, L., Schlag, K.H.: Minimax regret and strategic uncertainty. Journal of Economic Theory 145, 264–286 (2009)
27. Tennenholtz, M.: Competitive safety analysis: Robust decision-making in multi-agent systems. JAIR 17, 363–378 (2002)
28. Teacy, W.T.L., Patel, J., Jennings, N.R., Luck, M.: Travos: Trust and reputation in the context of inaccurate information sources. Autonomous Agents and Multi-Agent Systems 12, 183–198 (2006)
29. Trivers, R.L.: The evolution of reciprocal altruism. Q. Rev. Biol. 46, 35–57 (1971)
30. Wellman, M., Greenwald, A., Stone, P.: Autonomous Bidding Agents: Strategies and Lessons from the Trading Agent Competition. MIT Press (2007)
31. Wright, J.R., Leyton-Brown, K.: Beyond Equilibrium: Predicting Human Behavior in Normal-Form Games. In: AAAI 2010, pp. 901–907 (2010)
32. Yang, C.L., Yue, C.D.J., Yu, I.T.: The rise of cooperation in correlated matching prisoners dilemma: An experiment. Experimental Economics 10, 3–20 (2007)

Symmetries of Quasi-Values

Ales A. Kubena[1] and Peter Franek[2]

[1] Institute of Information Theory and Automation of the ASCR,
Pod Vodarenskou vezi 4, 182 08, Prague, Czech Republic
kubena@utia.cas.cz
[2] Institute of Information Technologies, Czech Technical University, Thakurova 9,
Prague 160 00, Czech Republic
peter.franek@fit.cvut.cz

Abstract. According to Shapley's game-theoretical result, there exists
a unique game value of finite cooperative games that satisfies axioms
on additivity, efficiency, null-player property and symmetry. The original setting requires symmetry with respect to arbitrary permutations of
players. We analyze the consequences of weakening the symmetry axioms
and study quasi-values that are symmetric with respect to permutations
from a group $G \leq S_n$. We classify all the permutation groups G that
are large enough to assure a unique G-symmetric quasi-value, as well as
the structure and dimension of the space of all such quasi-values for a
general permutation group G.

We show how to construct G-symmetric quasi-values algorithmically
by averaging certain basic quasi-values (marginal operators).

1 Introduction

A cooperative game is an assignment of a real number to each subset of a given
set of players Ω. This illustrates an economic situation where a coalition profit depends on the involved players in a generally non-aditive way. Several approaches
deal with the question of redistributing the generated profit to the individual
players in a stable or in a "fair" way. The mathematical theory of cooperative
games was developed in forties by Neumann and Morgenstern [17]. Values of
games provide a tool for evaluating the contributions of the individual players
such that certain natural axioms are satisfied. The most famous value is the
Shapley value introduced in 1953 [22] that exists and is unique for all finite sets
of players.

There exist many axiomatic systems on game values such that the Shapley
value is their only solution: the original Shapley's axiomatics [22], Neyman's [18],
Young's [24], van den Brink's [3] and Kar's axiomatics [15]. One of its important
characteristics is the symmetry with respect to any permutation of players. This
means, roughly speaking, that the value of a player is calculated only from his
contributions to various coalitions and not from his identity. One may consider
this to represent the *equity* of players. However, this is probably not a realistic
assumption in many real-world situations where personal friendships and link-
age play a major role. Some examples of values with restricted symmetry were

B. Vöcking (Ed.): SAGT 2013, LNCS 8146, pp. 159–170, 2013.
© Springer-Verlag Berlin Heidelberg 2013

studied, such as the *Owen value* [20] or the *weighted Shapley value* in [14], and the formal concept of *quasi-value*, where one completely relaxes any symmetry requirement, was introduced by Gilboa and Monderer in 1991 [10]. It is known that for a particular player set, there exist infinitely many quasi-values.

In this work, we analyze one particular way of weakening the symmetry axiom. We suppose that a group G of permutations of Ω is given and define a G-*symmetric quasi-value* to be any quasi-value symmetric wrt. all permutations in G. Informally, the equity of players is restricted to a group of permutations of players, not necessarily to all permutations. The group expresses the measure of symmetry. If G is the full symmetry group, then the only G-symmetric quasi-value is the Shapley value; if G is the trivial group, then it carries no symmetry requirement and each quasi-value is G-symmetric. Our contribution is the classification of all permutation groups G of finite sets of players for which there exists a unique G-symmetric quasi-value. It turns out that while in the infinite setting for non-atomic games one may reduce the group of symmetries in a number of ways [16,19], in the finite setting, only few subgroups of the full permutation group assure uniqueness. Even if the group G acts transitively on Ω (i.e. for any two players a, b, there exists a permutation $\pi \in G$ such that $\pi(a) = b$), there may still exist many G-symmetric quasi-values different from the Shapley value. We also calculate the dimension of the space of all G-symmetric quasi-values for a general permutation group G.

In the second section, we give the formal definition of G-symmetric quasi-value and some necessary definitions from group theory, including our original definition of a supertransitive group action. In the third section, we show that the space of all G-symmetric quasi-values is an affine subspace of the vector space of all values, and derive a formula for its dimension. We further classify all permutation groups G such that there exists a unique G-symmetric quasivalue. In the fourth section, we give some examples of G-symmetric quasi-values and show how more examples can be constructed by averaging the marginal operators. The last section (Appendix) contains the proof of an auxiliary statement from group theory that we use in the proof of Theorem 2. We postpone this technical issue to the end in order to keep the rest of the text fluent.

2   Definitions and Notation

2.1   Cooperative Games

Let Ω be a set of players. In this paper, we always suppose that Ω is finite.

Definition 1. *A* cooperative game *is a function* $v : 2^\Omega \to \mathbb{R}$ *such that* $v(\emptyset) = 0$. *A cooperative game is* additive, *if for all* $T, R \in 2^\Omega$, $R \cap T = \emptyset$ *implies* $v(R \cup T) = v(R) + v(T)$. *We denote by* Γ *the set of all cooperative games and* Γ_1 *the set of all additive cooperative games. A* game value *is an operator* $\varphi : \Gamma \to \Gamma_1$. *For a game value* φ *and* $i \in \Omega$, *we define* $\varphi_i(v) := \varphi(v)(\{i\})$.

For each game v, $\varphi(v)$ is uniquelly determined by the numbers $\varphi_i(v)$. Shapley theorem [22] proves the existence and uniqueness of a game value φ assuming it satisfies the following four axioms:

1. *Linearity*: $\varphi(\alpha v + \beta w) = \alpha \varphi(v) + \beta \varphi(w)$ for all $v, w \in \Gamma$ and $\alpha, \beta \in \mathbb{R}$.
2. *Null-player property*: if $i \in \Omega$ is a "null-player" in a game v, i.e. $v(R \cup \{i\}) = v(R)$ for each $R \subseteq \Omega$, then $\varphi_i(v) = 0$.
3. *Efficiency*: $\sum_i \varphi_i(v) = v(\Omega)$ for all games v.
4. *Symmetry* (sometimes called *anonymity*): $\varphi(\pi \cdot v) = \pi \cdot \varphi(v)$ for every permutation π of Ω, where the game $\pi \cdot v$ is defined by $(\pi \cdot v)(R) := v(\pi^{-1}(R))$ for any $R \subseteq \Omega$.

The value defined by these axioms is called *Shapley value*. Axioms 1-4 are independent. Gilles [11] and Schmeidler [5] give examples of values satisfying any 3 of them and not the 4th.

Any game value satisfying axioms 1, 2 and 3 is called a *quasi-value*. In the original economic interpretation, the fourth axiom (Symmetry) is an expression of equity of all the participating players. It can be formulated in a more elegant way by the commutativity of the following diagram.

$$
\begin{array}{ccc}
\Gamma & \xrightarrow{\ \varphi\ } & \Gamma_1 \\
\downarrow{\scriptstyle \pi} & & \downarrow{\scriptstyle \pi} \\
\Gamma & \xrightarrow{\ \varphi\ } & \Gamma_1
\end{array}
\tag{1}
$$

Axiom 4 requires that it commutes for each permutation of players π.

The following definition introduces the main object of our study.

Definition 2. *Let G by a group of permutations of Ω. A G-symmetric quasi-value is a game value that satisfies axioms 1, 2, 3 and such that $\varphi(\pi \cdot v) = \pi \cdot \varphi(v)$ for every permutation $\pi \in G$. In other words, diagram (1) commutes for all $\pi \in G$.*

Throughout this work, we will need the following basis of the space of cooperative games, introduced in Shapley's original paper [22].

Definition 3. *The* unanimity basis *is the basis $\{u_R\}_{\emptyset \neq R \subseteq \Omega}$ of the vector space of all cooperative games over the set Ω, defined by $u_R(S) = 1$ if $R \subseteq S$ and 0 otherwise.*

2.2   Group Theory

We say that a group G *acts* on the set X, if G is a subgroup of the group S_X of permutations of X. Any set $G \cdot x$ is called an *orbit*, or a G-orbit of x. The set of all G-orbits is denoted by X/G. The action of G on X is *transitive*, if for each $x, y \in X$, there exists a $g \in G$ such that $g \cdot x = y$. The *stabilizer* of a subset $A \subseteq X$ is the subgroup G_A of all elements $g \in G$ such that $g \cdot A \subseteq A$. For a subgroup H of G, $g \cdot H$ denotes a *left* and $H \cdot g$ a *right coset* of H and any group $H' = g^{-1}Hg$ is *conjugate* to H.

We introduce here a definition that will help us to describe a property of permutation groups we will need later.

Definition 4. *Let G be a group acting on a set X. We say that the action is a* supertransitive action, *if the stabilizer G_A of any subset $A \subseteq X$ acts transitively on A. A permutation group $G \subseteq S_n$ is* supertransitive, *if the stabilizer G_A acts transitively on each $A \subseteq \{1, \ldots, n\}$.*

For any n, S_{n-1} may be embedded into S_n as a set of permutations preserving one element. However, for $n = 6$, there exists an embedding of S_5 into S_6 different from the standard one. This embedding $S_5 \hookrightarrow S_6$ may be realized as the action of the projective linear group $PGL(2, 5)$ on the projective line over $\mathbb{Z}_5$. The reader may find the details in the literature [7, p. 60-61], [4]. We will call this embedding an *exotic embedding*. It is well known that such a nonstandard embedding is only one up to conjugation by an element of S_6. In this paper, we only need the property that the image of the exotic embedding is a supertransitive subgroup of S_6. This is proved in the appendix.

3   Dimension of G-Symmetric Quasi-Values

If a quasi-value is symmetric with respect to a set of permutations, it is also symmetric with respect to any permutation they generate in S_Ω, hence the set of all symmetries of a quasi-value is always a group. For a finite set Ω and a group $G \subseteq S_\Omega$ of permutations, we denote by $\mathcal{A}_G$ the set of all G-symmetric quasi-values.

We will represent $\mathcal{A}_G$ as a space of matrices. Each game value φ can be represented as a map from Γ to $\mathbb{R}^\Omega$ by the natural identification $\Gamma_1 \simeq \mathbb{R}^\Omega$. Choosing the unanimity basis on Γ (Def. 3) and the canonical basis $(e_i)_{i \in \Omega}$ on $\mathbb{R}^\Omega$, we may represent linear game values as matrices of the size $|\Omega| \times (2^{|\Omega|} - 1)$. The null player property applied to the unanimity basis implies $\varphi(u_R)(\{i\}) = 0$ for each $i \notin R$, because such player i doesn't contribute to any coalition in the game u_R. As a consequence, a matrix A with elements $(a_{iR})_{i \in \Omega, \emptyset \neq R \subseteq \Omega}$ corresponds to a linear game value satisfying the null-player-property iff $a_{iR} = 0$ for all pairs (i, R) such that $i \notin R$. Further, the game value satisfies the efficiency axiom iff for any nonempty $R \subseteq \Omega$, $\varphi(u_R)(\Omega) = 1$, which translates to a constraint on matrix coefficients $\sum_{i \in R} a_{iR} = 1$ for each $\emptyset \neq R \subseteq \Omega$. The G-symmetry of a game value requires $\varphi(g \cdot v) = g \cdot (\varphi(v))$ for any game v and permutation $g \in G$, the action of G on Γ defined by $(g \cdot v)(R) = v(g^{-1}R)$. An element u_R from the unanimity basis satisfies $(g \cdot u_R)(S) = u_R(g^{-1}(S)) = u_{gR}(S)$, so the unanimity basis is invariant with respect to the group action and $g \cdot u_R = u_{gR}$. The symmetry axiom is equivalent to

$$((g \cdot \varphi)(u_R))(\{i\}) = (\varphi(u_{gR}))(\{i\}),$$

for all $i \in \Omega$ and $\emptyset \neq R \subseteq \Omega$. The left-hand side is equal to $\varphi(u_R)(\{g^{-1}i\})$. So, in the matrix representation of φ, the symmetry axiom translates to the condition $a_{(g^{-1}i)\,R} = a_{i\,(gR)}$, or simply $a_{iR} = a_{(gi)\,(gR)}$ for all $i \in \Omega$, $\emptyset \neq R \subseteq \Omega$ and $g \in G$.

Summarizing this, we have the following.

Lemma 1. *Choosing the unanimity basis of Γ and the canonical basis of $\mathbb{R}^{\Omega} \simeq \Gamma_1$, $\mathcal{A}_G$ may be identified with a set of matrices $A = (a_{iR})$ with elements satisfying the following equations:*

- $a_{iR} = 0$ *if* $i \notin R$,
- *The sum of elements in each column is 1,*
- *Matrix elements a_{iR} are constant on the orbits of the G-action $g \cdot (i, R) = (gi, gR)$.*

All these conditions are linear equations for matrix elements a_{iR} and they are all satisfied by the Shapley value. So, $\mathcal{A}_G$ is a nonempty affine space.

Theorem 1. *Let $X = \{(i, R); i \in R \subseteq \Omega\}$, $\chi = \{R; \emptyset \neq R \subseteq \Omega\}$ and let $G \subseteq S_G$ be a group of permutations acting on sets X and χ, extending naturally its action on Ω. Then the dimension of $\mathcal{A}_G$ is $|X/G| - |\chi/G|$. Explicitly it can also be expressed as*

$$\dim \mathcal{A}_G = (\frac{dZ_G}{dx_1} - Z_G)|_{(2,2...2)} + 1 \tag{2}$$

where Z_G is the cycle index of the group G

$$Z_G(x_1...x_n) = \frac{1}{|G|} \sum_{\pi \in G} x_1^{j_1(\pi)} \cdots x_n^{j_n(\pi)}, \tag{3}$$

$j_k(\pi)$ *denotes the number of cycles of length k in the permutation π [8, p. 85].*

Proof. We will identify elements of $\mathcal{A}_G$ with matrices as described in Lemma 1. Let $p : X \to \chi$ be the map $(i, R) \to R$. For any $x = (i, R) \in X$ and $g \in G$, $p(gx) = g(p(x))$. For $\emptyset \neq R \subseteq \Omega$, the stabilizer G_R acts on R and R splits into k_R orbits $\{R_1, \ldots, R_{k_R}\}$ with respect to this action. If $R' = gR$, then the stabilizer of R' is gG_Rg^{-1} and g maps each G_R-orbit $R_i \subseteq R$ bijectively onto a $G_{R'}$-orbit $R'_i \subseteq R'$. So, $k_R = k_{R'}$ and $|R_i| = |R'_i|$ for $i = 1, \ldots, k_R$. For $m \in \chi/G$, we define $k_m := k_R$ for any $R \in m$ and $l_{mi} = |R_i|$ for $i = 1, \ldots, k_m$. These numbers are independent on the choice of R.

We will say that $m \in \chi/G$ *contains* an orbit $Gx \in X/G$, if $p(x) \in m$. Each $m \in \chi/G$ contains k_m orbits $\{o_1, \ldots, o_{k_m}\} \subseteq X/G$ and we may choose real numbers c_{mi} such that $\sum_{i=1}^{k_m} c_{mi}l_{mi} = 1$ with $k_m - 1$ degrees of freedom. Choosing such numbers c_{mi} for all $m \in \chi/G$ gives

$$\sum_{m \in \chi/G} (k_m - 1) = \sum_{m \in M} k_m - |\chi/G| = |X/G| - |\chi/G|$$

degrees of freedom. Any such choice of c_{mi} defines a matrix of game value

$$a_{iR} = \begin{cases} c_{mi} \text{ if } i \in R_i \subseteq R \in m \\ 0 \text{ if } i \notin R \end{cases}$$

These are exactly matrices A constant on the orbits of X satisfying $\sum_i a_{iR} = 1$ for all R and $a_{iR} = 0$ for all $i \notin R$. The number of degrees of freedom for the choice of c_{mi} is equal to the dimension of $\mathcal{A}_G$. This proves the first part.

Burnside lemma [21, p. 58] enables to express the number of orbits of a group action in an explicit way. If a finite group H acts on a finite set Y, then

$$|Y/H| = \frac{1}{|H|} \sum_{h \in H} |\{y \in Y \mid h(y) = y\}|. \qquad (4)$$

A permutation $\pi \in G$ fixes those sets $R \subseteq \Omega$ that don't split any cycle of π. There exists $2^{\#\,cycles(\pi)}$ such sets, $2^{\#\,cycles(\pi)} - 1$ of them nonempty. So,

$$|\chi/G| = \left(\frac{1}{|G|} \sum_{\pi \in G} 2^{\#\,cycles(\pi)} \right) - 1.$$

Elements of X fixed by π are pairs (i, R) such that $i \in R$, $\pi(i) = i$ and $\pi(R) = R$. There exists $\#\,fixedpoints(\pi)*2^{\#\,cycles(\pi)-1}$ such pairs. We derived the following equation:

$$\dim \mathcal{A}_G = \frac{1}{|G|} \left(\sum_{\pi \in G} (\#\text{fixedpoints}(\pi) * 2^{\#cycles(\pi)-1}) - \sum_{\pi \in G} 2^{\#cycles(\pi)} \right) + 1.$$

The statement of the theorem follows from this by a direct computation. $\square$

The cycle index Z_G is known in a more explicit form than (3) for many subgroups of S_n and it has also been generalized and computed for finite classical groups [9].

Further, we will show for which groups G the dimension of $\mathcal{A}_G$ is zero, i.e. for which G the only G-symmetric quasi-value is the Shapley value. In Section 2.2, we defined a group $G \subseteq S_\Omega$ to be supertransitive, if the stabilizer G_R acts transitively on R for each subset $R \subseteq \Omega$. In other words, if for each R and each $i, j \in R$, there exists a $g \in G$ such that $g(R) = R$ and $g \cdot i = j$. We will show that this condition is equivalent to the existence of a unique G-symmetric quasi-value.

Theorem 2. *Let Ω be finite and $G \leq S_\Omega$. There exists a unique G-symmetric quasi-value if and only if G acts supertransitively on Ω. Equivalently, this is if and only one of the following conditions is satisfied:*

- *$G = S_\Omega$, the full symmetric group*
- *$|\Omega| > 3$ and $G = A_\Omega$, the alternating group*
- *$|\Omega| = 6$ and G is the image of an exotic embedding $S_5 \hookrightarrow S_6$ (see Section 2.2).*

Proof. We will work with the matrix representation of $\mathcal{A}_G$, described in Lemma 1. Let (a_{iR}) be a matrix representing a value in $\mathcal{A}_G$.

If the action of G on Ω is supertransitive, then for each $\emptyset \neq R \subseteq \Omega$, all elements $\{(i, R); i \in R\}$ lie on the same G-orbit, so all the corresponding matrix elements a_{iR} are equal. The null-player property implies that $a_{iR} = 0$ for $i \notin R$

and together with the efficiency condition we obtain that for each $i \in R$, $a_{iR} = 1/|R|$. This implies uniqueness.

If the action of G on Ω is not supertransitive, then there exists a nonempty subset $\tilde{R} \subseteq \Omega$ such that the stabilizer $G_{\tilde{R}}$ has not a transitive action on $\tilde{R}$. So, $\tilde{R}$ contains at least two $G_{\tilde{R}}$-orbits. We may define the matrix a_{iR} as follows. In the matrix column corresponding to $\tilde{R}$ we choose $a_{i\tilde{R}} = 0$ if $i \notin \tilde{R}$ and the other elements $a_{j\tilde{R}}$ arbitrary, constant on $G_{\tilde{R}}$-orbits and such that $\sum_j a_{j\tilde{R}} = 1$. For all R' on the G-orbit of R, we define the coefficients $a_{iR'}$ in a unique way so that they are constant on the G-orbits and the remaining matrix elements may be equal to elements of the original Shapley matrix. In this way, we may construct an infinite number of different G-symmetric quasi-values which proves that $\dim \mathcal{A}_G \geq 1$.

For the classification part, it remains to prove that the groups listed in the theorem are exactly the groups acting supertransitively on $\{1, \ldots, n\}$. The proof of this is technical and we postpone it to the Appendix (Chapter 5). $\square$

4 Consequences

4.1 Examples

First we give some examples of groups and G-symmetric quasi-values. In all these examples, we assume that the player set Ω consists of n players.

Example 1. Let $G_1 = \{\text{id}\}$ be the trivial group. In this case, any quasi-value is G_1-symmetric. Consider a selector $\gamma : 2^\Omega \to \Omega$ with $\gamma(R) \in R$ for all $\emptyset \neq R \subseteq \Omega$. Now we define the value φ as

$$\varphi_i(v) = \sum_{i=\gamma(R)} \Delta_v(R) \tag{5}$$

where $\Delta_v(R) \in \mathbb{R}$ is a *Harsanyi dividend* of the coalition $R \subseteq \Omega$ defined by $\Delta_v(R) = \sum_{T \subseteq R} (-1)^{|R|-|T|} v(T)$. It was shown in [6] that such values satisfy the axioms for quasi-values. [1] The cycle index of the trivial group is $Z(x_1) = x_1^n$ and substituting into (2) yields $\dim \mathcal{A}_{G_1} = n2^{n-1} - 2^n + 1$. However, the number of selectors $\gamma : 2^\Omega \to \Omega$ is much larger, so many of the quasi-values defined by (5) are affine dependent.[2]

Example 2. ("Caste system") The set Ω is split into k nonempty disjoint subsets ("castes") $\Omega_1, \ldots, \Omega_k$ and G_2 is chosen so that it guarantees equity within each Ω_i. Formally, $G_2 = \{\pi \in S_\Omega \mid \forall i \ \pi(\Omega_i) = \Omega_i\}$.

Some examples of G_2-symmetric quasivalues have been described in the literature. The *Owen value*, defined in [20], can be obtained as the expected value

---

[1] In the matrix representation, such values correspond to matrices $a_{i\,R} = \delta_{i\gamma(R)}$.
[2] For $n \geq 4$, $\dim \mathcal{A}_{G_1}$ is strictly smaller than $n! - 1$ which implies that the set of marginal operators (defined in Section 4.2) is also affine dependent.

of *marginal operators* (see Section 4.2), if we first randomly choose an order of the castes and then the order of the players within each caste. Another related concept is the *weighted Shapley value*, studied by Kalai and Samet in [14]. Here an order of the castes is given and within each caste, the profit is diveded among players proportional to their *weights*. In the case of equal weights of all players, the weighted Shapley value is symmetric with respect to all G_2-permutations.

The cycle index is $Z_{G_2} = \prod_{r=1}^{k} Z_{S_{\Omega_r}}$. We know from the proof of Theorem 2 that $|\chi/G| = \frac{1}{|G|}\sum_g 2^{\#cycles(g)}$ for each set χ with a G-action. In particular, for $G = S_n$, $|2^\Omega/G| = n+1$, because S_n-orbits of 2^Ω are $O_s = \{R \subseteq \Omega \,|\, |R| = s\}$ for $s = 0, 1, \dots, n$. This enables as to calculate

$$Z_{S_n}|_{(2,\dots,2)} = \frac{1}{n!} \sum_{\pi \in S_n} 2^{j_1(\pi)+\dots+j_n(\pi)} = \frac{1}{n!} \sum_{\pi \in S_n} 2^{\#cycles(\pi)} = |2^\Omega/S_n| = n+1.$$

If $G = S_n$, then the Shapley value is the only game value, so it follows from Theorem 2 that $(\frac{dZ_{S_n}}{dx_1} - Z_{S_n})|_{(2,\dots,2)} + 1 = 0$ and $\frac{dZ_{S_n}}{dx_1}|_{(2,\dots,2)} = n$. So, for $G_2 = \prod_{r=1}^{k} S_{\Omega_r}$

$$\frac{dZ_{G_2}}{dx_1}|_{(2,2\dots2)} = \Big(\sum_{r=1}^{k} \frac{dZ_{S_{\Omega_r}}}{dx_1} \prod_{s\neq r} Z_{S_{\Omega_s}}\Big)|_{(2,2\dots2)} = \sum_{r=1}^{k} |\Omega_r| \prod_{s\neq r} (1+|\Omega_s|)$$

and

$$\dim \mathcal{A}_{G_2} = \Big(\sum_{r=1}^{k} \frac{|\Omega_r|}{1+|\Omega_r|} - 1\Big) \prod_{r=1}^{k} (1+|\Omega_r|) + 1.$$

For the case of two castes $k = 2$ this simplifies to $|\Omega_1| \times |\Omega_2|$.

Example 3. (Cyclic group) This example illustrates that transitive group action does not imply a unique G-symmetric quasi-value. If G_3 is the cyclic group $C_n \subseteq S_n$, the cycle index is $Z_{C_n} = \frac{1}{n}\sum_{f|n} \phi(f)x_f^{n/f}$, where $\phi(f)$ is the Euler totient function $\phi(f) = p_1^{k_1-1}(p_1 - 1)\dots p_r^{k_r-1}(p_r - 1)$, where $f = p_1^{k_1}\dots p_r^{k_r}$ is the prime number decomposition.[8, p. 86]. Substituting into the formula in Theorem 2 gives

$$\dim \mathcal{A}_{G_3} = 2^{n-1} - \frac{1}{n}\sum_{f|n} \phi(f)2^{n/f} + 1.$$

In the case of $n = 3$, the dimension turns out to be $2^2 - \frac{1}{3}(2^3 + 2 \times 2) + 1 = 1$, so there exists a one-dimensional space of quasi-values symmetric with respect to cyclic permutations of players.

4.2 Shapley-Value as an Expected Value of Non-uniformly Distributed Marginal Vectors

Suppose that $\Omega = \{1, 2, \dots, n\}$, i.e. an order is given on the set of player. For a game $v \in \Gamma$ and a permutation $\pi \in S_n$, we may define a quasi-value m_π by $(m_\pi)(v)_{\pi(1)} = v(\pi(1))$ and

$$(m_\pi(v))_{\pi(i)} = v(\{\pi(1), \pi(2), \ldots, \pi(i)\}) - v(\{\pi(1), \pi(2), \ldots, \pi(i-1)\})$$

for $i = 2, \ldots, n$. We call m_π the *marginal operator* and $m_\pi(v)$ the *marginal vector* [2, p. 19]. It corresponds to a situation where the players arrive in the order $\pi(1), \pi(2), \ldots, \pi(n)$ and each player is assigned the value of his or her contribution to the coalition of all players that have arrived before. The evaluation of m_π on a game u_R from the unanimity basis is $m_\pi(u_R)(\{\pi(i)\}) = u_R(\pi(1), \ldots, \pi(i)) - u_R(\pi(1), \ldots, \pi(i-1))$ which is equal to 1 if and only if $\pi(i) \in R$ and $\pi(j) \notin R$ for $j > i$ and 0 otherwise. After the identification 1, we can represent m_π is as a matrix

$$(m_\pi)_{iR} = \begin{cases} 1 \text{ iff } i \in R \text{ and } \pi^{-1}(i) = \max \pi^{-1}(R) \\ 0 \text{ otherwise.} \end{cases}$$

A theorem of Weber [23] shows that if π is a random permutation taken from a uniform distribution on S_n then for any game v, the expected value of a marginal operator m_π is the Shapley value. This can be generalized to the following statement.

Proposition 1. *Let G be a subgroup of S_n and A^π be a probability distributioin on S_n constant on the right cosets $\{G \cdot \pi\}_\pi$, i.e. $A^\pi = A^{g\pi}$ for all $g \in G$ and $\pi \in S_n$. Then $\sum A^\pi m_\pi$ is a G-symmetric quasi-value.*

Proof. We will show that the identity holds if evaluated on games from the unanimity basis of Γ. For the game u_R (Definition 3), we start with the following equation:

$$(g \cdot m_\pi)(u_R) = m_{g\pi}(u_{gR}). \tag{6}$$

To prove this, we evaluate both sides on $\{i\}$ and rewrite the left-hand side to the equivalent equation

$$(m_\pi(u_R))(\{g^{-1}(i)\}) = (m_{g\pi}(u_{gR}))(\{i\}).$$

Both sides are equal to 1 if and only if $\pi^{-1}(g^{-1}(i)) = \max \pi^{-1}(R)$ and 0 otherwise, which proves (6) for all $R \subseteq \Omega$, $i \in \Omega$ and $g \in G$. The G-symmetry of $\sum_{\pi \in S_n} A^\pi m_\pi$ follows from

$$(g \cdot \sum_{\pi \in S_n} A^\pi m_\pi)(u_R) = \sum_{\pi \in S_n} A^\pi(g \cdot m_\pi)(u_R) = \sum_{\pi \in S_n} A^\pi m_{g\pi}(u_{gR}) =$$

$$= \sum_{\pi \in S_n} A^{g\pi} m_{g\pi}(u_{gR}) = \sum_{g\pi = \pi' \in S_n} A^{\pi'} m_{\pi'}(g \cdot u_R) = ((\sum_{\pi' \in S_n} A^{\pi'} m_{\pi'}) \cdot g)(u_R)$$

where we used (6) in the second and $A^\pi = A^{g\pi}$ in the third equality. $\square$

An immediate consequence of the classification Theorem 2 is that for $|\Omega| > 3$ any quasi-value symmetric with respect to the alternating group A_n is already the Shapley value. It follows from the last proposition that $\sum_\pi A^\pi m_\pi$ is the

Shapley value not only for $A^\pi = \frac{1}{n!}$ but also for $A^\pi = \frac{s}{n!}$ for π even and $A^\pi = \frac{2-s}{n!}$ for π odd, $s \in [0,2]$. In fact, there are many other possibilities how to express the Shapley value as a convex combination of marginal operators. The space of all quasi-values on Ω is $(n2^{n-1} - 2^n + 1)$-dimensional and the set of all probability distributions on S_n is a $(n! - 1)$-dimensional convex region in $\mathbb{R}^{n!}$, so there are at least $n! - n2^n + 2^{n-1} - 2$ degrees of freedom for the choice of a distribution A^π such that $\sum_\pi A^\pi m_\pi = $ Shapley.

Exponentially many (with respect to n) of these probability distributions A^π can be constructed as follows. Choose $\Omega_0 \subseteq \Omega$, $|\Omega_0| > 3$ and define S_0 to be a group of all permutations π acting identically on $\Omega \setminus \Omega_0$. Choose $\alpha \in (0,2)$ and define a probability distribution on S_n by

$$
A^\pi(\Omega_0) = \begin{cases} \frac{1}{n!} & \text{if } \pi \notin S_0 \\ \frac{\alpha}{n!} & \text{if } \pi \in S_0 \text{ and } \pi \text{ is even} \\ \frac{2-\alpha}{n!} & \text{if } \pi \in S_0 \text{ and } \pi \text{ is odd} \end{cases}
$$

One can verify that the corresponding expected value of marginal operators m_π is the Shapley value. For a set $\{\Omega_1, \Omega_2, \ldots, \Omega_k\}$ s.t. $\Omega_i \nsubseteq \Omega_j$ for all i and j, the vectors $(A^\pi(\Omega_i) - \frac{1}{n!})_i \in \mathbb{R}^{n!}$ are linearly independent and the distributions $(A^\pi(\Omega_i))_i$ are affine independent.

5 Appendix

Here we finish the proof of Theorem 2 by the classification of supertransitive groups. Our proof is based on a classification of set-transitive permutation groups given by Beamont and Peterson in 1955 [1]. Another proof of the supertransitive groups classification was given by Michal Jordan on mathoverflow [13].

Theorem 3. *G is a supertransitive subgroup of S_n if and only if one of the following conditions holds:*

- *G is the full symmetric group S_n for some n,*
- *G is the alternating group A_n for $n > 3$,*
- *G is conjugate to the image of an exotic embedding of S_5 to S_6.*

Proof. Let $G \subseteq S_n$ be a group of permutations acting supertransitively on $\{1, \ldots, n\}$. This means that the stabilizer of each $A \subseteq \{1, \ldots, n\}$ acts transitively on A. Let $B \subseteq \{1, \ldots, n\}$ and $i, j \notin B$. Then G acts transitively on $B \cup \{i, j\}$ and there exists a permutation $\pi \in G$ taking $B \cup \{i\}$ to $B \cup \{j\}$ such that $\pi(i) = j$. This implies that for each A and B s.t. $|A| = |B| > 1$, there exists a permutation $\pi \in G$ s.t. $\pi(A) = B$. If $|A| = |B| = 1$, the same is true because supertransitivity implies transitivity. We have shown that if the action of G is supertransitive, it is also set-transitive.

If G has a supertransitive action on $\{1, \ldots, n\}$, then its order has to be divisible by each $k \le n$, because each k-element set A is isomorphic to G/G_A, hence $|G| = |A| \times |G_A|$. So, G has to be divisible by the least common multiple of $\{1, \ldots, n\}$.

Beamont and Petrson classified all set-transitive permutation groups in [1]. It follows that such subgroups of S_n are exactly the full symmetric group S_n for any n, the alternating group A_n for $n > 2$ and 5 exceptions. The first and second exceptions are subgroups of S_5 of order 10, resp. 20. These groups cannot have a supertransitive action on $\{1, \ldots, 5\}$, because the lowest common multiple of $\{1, \ldots, 5\}$ is 60. Two other exceptions in Beamont's classification are subgroups of S_9 of orders 504 and 1512. These numbers are not divisible by the lowest common multiple of $\{1, \ldots, 9\}$ so we can exclude them as well. The last exception is a subgroup of S_6 of order 120. This groups is equivalent to the exotic embedding of S_5 to S_6 and we will show that it acts supertransitively on S_6.

In [12], the authors realize this group action on $\{1, \ldots, 6\}$ as the conjugate action of S_5 on its six Sylow 5-subgroups. Using this realisation, we may show that this action is supertransitive by direct calculation. Let as denote the Sylow 5-subgroups by $I = \langle (12345) \rangle$, $II = \langle (12354) \rangle$, $III = \langle (12435) \rangle$, $IV = \langle (12453) \rangle$, $V = \langle (12534) \rangle$ and $VI = \langle (12543) \rangle$. An elementary calculation shows that the image of a transposition in S_5 is the product of three disjoint transpositions in S_6, e.g. $(1, 2) \in S_5 \mapsto (I, VI)(II, IV)(III, V)$ in the above realisation. Together with the set-transitivity of this S_5-action, this implies 2-supertransitivity. The image of a 3-cycle in S_5 is a product of two disjoint 3-cycles in S_6, which implies 3-supertransitivity. Similarly, the image of a 4-, resp. 5-cycle in S_5 is a 4-, resp. 5-cycle in S_6, which implies 4- and 5-supertransitivity.

It remains to prove that A_n is supertransitive if and only if $n > 3$. First note that $A_2 = \{id\}$, reps. $A_3 = \langle (123) \rangle$ are not supertransitive, because no element of these groups takes 1 to 2 and preserves $\{1, 2\}$. Let $n > 3$ and $A \subseteq \{1, \ldots, n\}$ be a k-set. If $k < n - 1$, then any permutation of A can be extended to an even permutation of $\{1, \ldots, n\}$. If $k = n - 1 > 2$, then for each $i, j \in A$, there exists an even permutation of A taking i to j. This can be extended to an even permutation of $\{1, \ldots, n\}$, acting identically on the complement of A. $\qquad \square$

Acknowledgements. We would like to thank to Michal Jordan for his mathematical remarks and discussion on mathoverflow. This work was supported by MŠMT project number OC10048 and by the institutional research plan AV0Z100300504 and by the Excelence project P402/12/G097 DYME Dynamic Models in Economics of GAČR.

References

1. Beaumont, R., Peterson, R.: Set-transitive permutation groups. Canadian Journal of Mathematics 7(1), 35–42 (1955)
2. Brânzei, R., Dimitrov, D., Tijs, S.: Models in Cooperative Game Theory: Crisp, Fuzzy, and Multi-Choice Games. Lecture Notes in Economics and Mathematical Systems. Springer (2005)
3. Brink, R.: An axiomatization of the shapley value using a fairness property. International Journal of Game Theory 30, 309–319 (2002)
4. Carnahan, S.: Small finite sets (2007)

5. David, S.: The nucleolus of a characteristic function game. Siam Journal on Applied Mathematics 17(6), 1163–1166 (1969)
6. Derks, J., Haller, H., Peters, H.: The selectope for cooperative games. Open access publications from maastricht university, Maastricht University (2000)
7. Dixon, J., Mortimer, B.: Permutation groups. Springer (1996)
8. Flajolet, P., Sedgewick, R.: Analytic combinatorics. Cambridge University Press (2009)
9. Fulman, J.: Cycle indices for the finite classical groups (1997)
10. Gilboa, I., Monderer, D.: Quasi-value on subspaces. International Journal of Game Theory 19(4), 353–363 (1991)
11. Gilles, R.: The Cooperative Game Theory of Networks and Hierarchies. Theory and decision library: Game theory, mathematical programming, and operations research. Springer (2010)
12. Janusz, G., Rotman, J.: Outer automorphisms of S_6. The American Mathematical Monthly 89(6), 407–410 (1982)
13. Jordan, M.: Super-transitive group action (mathoverflow contribution), http://mathoverflow.net/questions/71917
14. Kalai, E., Samet, D.: On Weighted Shapley Values. International Journal of Game Theory 16(3), 205–222 (1987)
15. Kar, A.: Axiomatization of the shapley value on minimum cost spanning tree games. Games and Economic Behavior 38(2), 265–277 (2002)
16. Monderer, D., Ruckle, W.H.: On the Symmetry Axiom for Values of Nonatomic Games. Int. Journal of Math. and Math. Sci. 13(1), 165–170 (1990)
17. Neumann, J., Morgenstern, O., Rubinstein, A., Kuhn, H.: Theory of Games and Economic Behavior. Princeton Classic Editions. Princeton University Press (2007)
18. Neyman, A.: Uniqueness of the shapley value. Games and Economic Behavior 1(1), 116–118 (1989)
19. Neyman, A.: Values of Games with Infinitely Many Players. Handbook of Game Theory with Economic Applications, vol. 3, ch. 56, pp. 2121–2167. Elsevier (2002)
20. Owen, G.: Values of Games with A Priori Unions. In: Henn, R., Moeschlin, O. (eds.) Mathematical Economics and Game Theory. Lecture Notes in Economics and Mathematical Systems, vol. 141, pp. 76–88. Springer, Heidelberg (1977)
21. Rotman, J.: An introduction to the theory of groups. Springer (1995)
22. Shapley, L.S.: A value for n-person games. Annals of Mathematics Studies 2(28), 307–317 (1953)
23. Weber, R.: Probabilistic Values of Games. In: Roth, A. (ed.) The Shapley Value: Essays in Honor of Lloyd S. Shapley, pp. 101–120. Cambridge Univ. Press (1988)
24. Young, H.P.: Monotonic solutions of cooperative games. International Journal of Game Theory 14, 65–72 (1985)

Dividing Connected Chores Fairly

Sandy Heydrich[1] and Rob van Stee[2]

[1] Universität des Saarlandes, Saarbrücken, Germany
s9saheyd@stud.uni-saarland.de
[2] Max Planck Institute for Informatics, Saarbrücken, Germany
vanstee@mpi-inf.mpg.de

Abstract. In this paper we consider the fair division of chores (tasks that need to be performed by agents, with negative utility for them), and study the loss in social welfare due to fairness. Previous work has been done on this so-called price of fairness, concerning fair division of cakes and chores with non-connected pieces and of cakes with connected pieces. We provide tight or nearly tight bounds on the price of fairness in situations where each player has to receive one connected piece of the chores. We also give the first proof of the existence of equitable divisions for chores with connected pieces.

1 Introduction

Motivated by the fact that social interaction often requires dividing goods, researchers in economics, law and computer science dealt with fair division since the 1940's, and already the ancient Greeks knew the problem. In fair division, one tries to divide some desirable or undesirable good between a number of people that all have individual preferences and dislikes, while satisfying some fairness condition. We will only focus on the case where the goods are divisible, i.e. can be cut in arbitrary pieces; dividing indivisible goods is a much harder problem. The typical analogy for fair division when considering desirable goods is *cake cutting* [1], meaning that we want to divide a cake that has various sections with different toppings, whereas in the *chore division problem* [2] one tries to minimize the discontent of the players when dividing work. Many algorithms found for cake cutting also apply to the division of chores, but interestingly, as we will see in this work their theoretical properties differ in several cases.

Of course one has to decide how to define *fairness*, and the three criteria *proportionality, envy-freeness* and *equitability* considered in many earlier papers (e.g. [3],[4],[5]) will also be considered here. Informal definitions for these are given in the next paragraph. Apart from achieving a division which is fair, another goal is optimizing the *social welfare*, and the natural question arises what the trade-off between those two goals is. Caragiannis et al. [5] and Aumann and Dombb [4] examined this trade-off for the division of cakes and chores; Caragiannis et al. found bounds for this trade-off, called the *price of fairness*, for both cakes and chores, but without any restriction on the number of pieces each player receives. This may lead to the undesirable situation that players receive a huge number

B. Vöcking (Ed.): SAGT 2013, LNCS 8146, pp. 171–182, 2013.

of small pieces, e.g. a bunch of crumbs in the cake analogy. Therefore, Aumann and Dombb [4] examined the price of fairness for connected pieces, requiring that every player receives exactly one connected part of the cake; however, they did not consider division of chores. To close the gap, in this paper we give bounds on the price of fairness with connected pieces in division of chores. An analogy for this could be that a group of gardeners needs to maintain a garden and each of them wants to be responsible for one connected area.

Model. In our model, the chores are represented by the real interval $[0,1]$ and we consider n players. Each player has a *disutility function* over this interval that gives his discontent for a particular piece. These functions are required to be non-atomic measures, i.e. they are non-negative and additive and if an interval is valued strictly positive, it must have a subinterval that has a strictly less but still strictly positive value. Furthermore the functions are normalized, so the disutility for the whole chores is 1. The disutility of a player in a division is then the disutility of this player for the piece he receives. The *utilitarian welfare* for a division is defined as the sum over the disutilities of all players, while *egalitarian welfare* is the greatest disutility among all players (i.e. the disutility of the worst-off player). A division is called optimal if it minimizes the welfare. We call a division *proportional* if every player thinks that he receives his fair share, we call it *envy-free* if no player thinks that another player receives less than him, and *equitable* if the disutilities of all players are equal.

To quantify the loss in welfare due to fairness we use the notion of *price of fairness*. We define the price of fairness as the ratio between the welfare of the best fair division and the welfare of the optimal division.

1.1 Related Work

Modern mathematicians started working on the topic in the 1940's with Banach, Steinhaus and Knaster giving the "Last Diminisher" mechanism for proportional divisions [1]. In the following years, research mainly focused on finding algorithms for achieving fair divisions ([6],[7],[8],[9]), also trying to bound the number of cuts required. Furthermore, Dubins and Spanier as well as Stromquist gave existence theorems for certain fair divisions [8],[9].

On the problem of fair division of chores however, much less work has been done. The problem was first mentioned by Gardner in [2], and Oskui [3, p. 73] gave the first three person envy-free chore division algorithm. Peterson and Su gave envy-free protocols for four and later for n players [10],[11]. The existence proof for proportional cake divisions of Steinhaus [1] can also be applied to chores. Su [12] has proven that envy-free divisions of chores with connected pieces also always exist. For the existence of equitable divisions with connected pieces, as far as we know no proof was given so far.

The problem of the efficiency of fair divisions was first addressed by Caragiannis et al. [5]. Their work considered the price of fairness for utilitarian welfare and the three fairness notions proportionality, envy-freeness and equitability, and examined bounds for these for divisible and indivisible cakes as well as chores. Aumann and Dombb [4] gave bounds for the price of fairness for utilitarian and egalitarian welfare, restricted to the case that only connected pieces are allowed to be given to the players so they do not end up with arbitrarily many pieces, but they only considered cake cutting. In this work we give bounds for the remaining case of chore division with connected pieces.

Following the work of Caragiannis et al. and Aumann and Dombb, Cohler et al. [13] provided a polynomial time approximation scheme for computing envy-free cake divisions that are optimal w.r.t. utilitarian welfare. Based on this work, Bei et al. [14] give a PTAS for computing optimal proportional cake divisions with connected pieces.

Brams et al. [15] connected the topic of efficient fair divisions with the sphere of Pareto-optimal divisions, i.e. divisions in which it is not possible to give one player a strictly higher utility while giving no player a lower utility. They examined whether we can always find fair divisions maximizing the (utilitarian) social welfare that are also Pareto-optimal and showed that for a special class of evaluation functions, the optimal (w.r.t. utilitarian welfare) equitable division has never a higher social welfare than the optimal envy-free division.

1.2   Overview of Results

We examine the price of fairness for utilitarian and egalitarian welfare and the three fairness notions *proportionality, envy-freeness* and *equitability* as a function of the number of players n. We give tight bounds for all cases except for the utilitarian price of proportionality, where there is still a small gap between the lower and the upper bound. All results are summarized and compared to the results by Caragiannis et al. [5] and Aumann and Dombb [4] in Table 1.

For utilitarian welfare, we show that the price of proportionality is linear (between $n/2$ and n) in the number of players for $n > 2$. This matches the $\Theta(n)$ bounds for chore division with non-connected pieces by Caragiannis et al. [5]. For egalitarian welfare we show that there is no trade-off between proportionality and egalitarian welfare, which is the same result as shown by Aumann and Dombb [4] for cake cutting.

When considering the price of envy-freeness and more than two players, we show how to construct instances that have an arbitrarily high price of fairness. We hence see that for this fairness notion there is an inherent difference between cakes and chores (Aumann and Dombb [4] as well as Caragiannis et al. [5] found bounds for the price of envy-freeness for cake cutting).

Our proof is the first for the existence of equitable divisions of chores with connected pieces. We prove that the egalitarian price of equitability is 1 by constructing an equitable division starting with an egalitarian-optimal one. We also give a tight bound of n for the utilitarian price of equitability. For this, Aumann and Dombb [4] gave an upper bound of n and a lower bound of $n-1+\frac{1}{n}$, and for non-connected chores, Caragiannis et al. [5] gave a tight bound of n.

Essentially the same results, but with a tight bound of n for the utilitarian price of proportionality, were achieved independently by Hoffmann et al. [16].

Table 1. Results of this work, compared to [5] and [4]. Some results only hold for $n \geq 3$. See the text for the case $n = 2$.

Chores:	connected (this work)		egalitarian	non-connected ([5])	
	utilitarian			utilitarian	
	lower	upper		lower	upper
Proportionality	$n/2$	n	1	$\frac{(n+1)^2}{4n}$	n
Envy-Freeness	∞			$\frac{(n+1)^2}{4n}$	∞
Equitability	n		1	n	
Cakes:	connected ([4])		egalitarian	non-connected ([5])	
	utilitarian			utilitarian	
	lower	upper		lower	upper
Proportionality	$\frac{\sqrt{n}}{2}$	$\frac{\sqrt{n}}{2} + 1 - o(1)$	1	$\Omega(\sqrt{n})$	$O(\sqrt{n})$
Envy-Freeness	$\frac{\sqrt{n}}{2}$	$\frac{\sqrt{n}}{2} + 1 - o(1)$	$n/2$	$\Omega(\sqrt{n})$	$n - 1/2$
Equitability	$n - 1 + 1/n$	n	1	$\frac{(n+1)^2}{4n}$	n

2   Definitions

In this section we formally define the chores division problem itself, the notions of fairness and social welfare used in this work and finally the *price of fairness*, the measure for the trade-off between fairness and social welfare.

The chores are represented by the real interval $[0, 1]$ and our players are denoted by $1, \ldots, n$. Each player i has a certain valuation function $v_i(\cdot)$, that maps any possible subset of the chores to a real valuation between 0 and 1. This valuation function needs to be a non-atomic measure (i.e. non-negative, zero for the empty interval, additive and each interval with positive value must have a sub-interval with strictly less but strictly positive value) with $v_i(0, 1) = 1$.

Definition 1. *A division x of the chores is a vector $x = (x_1, \ldots, x_{n-1}, \pi) \in [0, 1]^{n-1} \times S_n$. The point x_i denotes the position of the i-th cut, we define $x_0 := 0, x_n := 1$, and the cuts are sorted: $x_0 \leq x_1 \leq \ldots \leq x_{n-1} \leq x_n$. π is a permutation that denotes the assignment of the pieces to the players: Player i receives the interval $(x_{\pi(i)-1}, x_{\pi(i)})$. By X we denote the set of all possible divisions.*

The unhappiness of the players with a certain division is given by the notion of *disutility*.

Definition 2. *The disutility of a division x for a player i is $d_i(x) = v_i(x_{\pi(i)-1}, x_{\pi(i)})$.*

In this work, the following three different notions of fairness are considered.

Definition 3. *A* *division* x *is* **proportional** *if* $d_i(x) \leq \frac{1}{n}$ *for every player* i.
A *division* x *is* **envy-free** *if* $v_i(x_{\pi(i)-1}, x_{\pi(i)}) \leq v_i(x_{\pi(j)-1}, x_{\pi(j)})$ *for every pair of players* i, j.
A *division* x *is* **equitable** *if* $d_i(x) = d_j(x)$ *for every pair of players* i, j.

Intuitively, a division is proportional if all players get a portion they consider their fair share of the chores (or less). A division is envy-free if no player envies any other player, in the sense that he dislikes the other player's piece less than his own piece. Note that every envy-free division is proportional. Finally, a division is equitable if the disutilities of all players are equal (by their own valuations).

The social welfare of a division can be defined in two ways: In utilitarian welfare, the total disutility of all players is considered, whereas egalitarian welfare refers to the disutility of the worst-off player.

Definition 4. *A* *division* x *has* *utilitarian social welfare* $u(x) = \sum_{i=1,\ldots,n} d_i(x)$
and egalitarian social welfare $eg(x) = \max_{i=1,\ldots,n} d_i(x)$.

To quantify the amount of social welfare one has to sacrifice to achieve fairness, we define the *price of fairness*:

Definition 5. *The price of fairness (price of proportionality, respectively envy-freeness, equitability) is the minimal welfare achievable in fair (proportional, respectively envy-free, equitable) divisions divided by the minimal welfare achievable in arbitrary divisions.*

For example the price of envy-freeness with egalitarian welfare is $\frac{\min_{x \in X_{EF}} eg(x)}{\min_{x \in X} eg(x)}$, where X_{EF} denotes the set of all connected envy-free divisions.

3   The Price of Proportionality

We start with bounds for the price of proportionality. For utilitarian welfare, the results do not differ much from the results for non-connected chores by Caragiannis et al. [5], although the lower bound is slightly better. Concerning egalitarian welfare, we can use a proof analogous to the proof by Aumann and Dombb for the price of proportionality with connected cakes [4].

3.1   Utilitarian Welfare

Theorem 1. *The utilitarian price of proportionality for the division of chores with connected pieces is lower-bounded by* $\frac{n}{2}$ *for* $n > 2$.

We give no full proof here; it can be found in the full version. The idea is to construct an instance where one player, who dislikes the chores uniformly, receives a piece slightly greater than $\frac{1}{n}$ in the optimal division, and where it is very costly to give some part of this piece to any other player. Intuitively one could say that in this scenario one player "sacrifices" himself to do more work than his fair share (in terms of proportionality), as he himself does not dislike this work as much as the other players.

For two players, we only state the result; the proof is given in the full version.

Theorem 2. *The utilitarian price of proportionality for $n = 2$ players is lower-bounded by 2.*

For an upper bound on the utilitarian price of proportionality, we refer the reader to the proof by Caragiannis et al. [5], as it also applies to connected chores.

Theorem 3. *The utilitarian price of proportionality for the division of chores with connected pieces is upper-bounded by n.*

In the case with two players, envy-free and proportional divisions coincide, hence these results immediately imply the following:

Corollary 1. *The utilitarian price of envy-freeness for $n = 2$ players is 2.*

3.2   Egalitarian Welfare

For the egalitarian price of proportionality, we can again apply the result of Aumann and Dombb [4].

Theorem 4. *Every egalitarian-optimal division of chores with connected pieces is proportional, and therefore the egalitarian price of proportionality in this case is 1. For $n = 2$ players, this again also holds for the price of envy-freeness.*

4   The Price of Equitability

The price of equitability is a more interesting case than proportionality, as so far no proof was given for the existence of equitable divisions of chores with connected pieces. In Theorem 5, we show that we can transform any egalitarian-optimal division into an equitable one with the same welfare, and with this, we give this existence proof and prove that no trade-off between equitability and egalitarian welfare exists. The construction relies on the fact that optimality with respect to egalitarian welfare and the non-atomicity of the evaluation measures imply that we can make pieces that are adjacent to a piece with maximal dislike (among all pieces) also maximal.

Afterwards we give proofs for a tight bound of n for utilitarian welfare. This matches the bound for non-connected chores given by Caragiannis et al. [5].

4.1   Egalitarian Welfare

Theorem 5. *For every instance of the chores division problem, there exists an equitable division with connected pieces. Furthermore, the egalitarian price of equitability for the division of chores with connected pieces is 1.*

Proof. We need some more terminology for this proof:

- The *value* of a piece is the dislike that is assigned to this piece by the player who receives it.
- Let $m = \min_{x \in X} eg(x)$ be the optimal egalitarian welfare. Pieces that have a value of m are called *maximal pieces*.
- A *block of maximal pieces* is sequence of one or more adjacent maximal pieces $p_1, \ldots, p_k$ where the left neighbor of p_1 and the right neighbor of p_k are non-maximal (i.e. a maximal sequence of maximal pieces). Those non-maximal neighbors are called *neighbors of this block*.

Consider an egalitarian-optimal division x that has the minimal number of maximal pieces among all egalitarian-optimal divisions. We want to make all pieces in x maximal by moving cuts, and for this we need a lemma, which is proven in the full version of the paper:

Lemma 1. *Consider a block of k maximal pieces. Then we can either make the right neighbor p' of this block maximal as well or we can make all pieces in the block as well as p' non-maximal by only moving the cuts inside the block and the cut between the block and p' to the left.*

The lemma can also be shown for the left neighbor symmetrically.

Now consider our optimal division x and look at its left-most block of maximal pieces. Note that this block must exists, as at least one maximal piece must exist (by definition of maximal piece and optimality of x). By the lemma we can make its right neighbor maximal, as otherwise we could make the block non-maximal, contradicting the assumption that x has the minimal number of maximal pieces. By applying the lemma again and again, we can make all pieces to the right of this (steadily growing) left-most block maximal. Note that every time, if we find that we can make the entire block non-maximal, we find a division with less maximal pieces than x. If we reach a piece that is already maximal during this process, we just add it to the block without moving cuts. We can then apply the lemma symmetrically for the pieces to the left of this block (which is now the only block of maximal pieces in x) and make all of them maximal too. Finally we have a division where all pieces are maximal. □

4.2 Utilitarian Welfare

While achieving equitability does not influence the egalitarian optimality, it has an impact on the utilitarian welfare, as shown in the next three theorems. The idea of the lower bound proof is to make sure that one indifferent player has to receive at least a piece of a certain value in both fair and unfair divisions, which leads to a price of fairness of n, as in equitable divisions all players have to receive this certain disutility, while the indifferent player is the only one to receive any disliked piece in the utilitarian-optimal division.

Theorem 6. *The utilitarian price of equitability is lower-bounded by n.*

Proof. We construct an instance of the chores division problem that has a utilitarian price of equitability of at least n as follows:

Let $\epsilon > 0$ be arbitrarily small. We create $(n-1)^2$ so-called "disliked pieces" $p_1, \ldots, p_{(n-1)^2}$, where p_i is located at $(\frac{i}{(n-1)^2+1} - \epsilon, \frac{i}{(n-1)^2+1} + \epsilon)$.

We divide those pieces into $(n-1)$ blocks of $n-1$ pieces each, and each block contains one piece for every player $\{1, \ldots, n-1\}$. The first piece of the first block is associated with player 1, the second with player 2 and so on, until the last piece of the first block is associated with player $n-1$. The pieces of the second block are then associated with players $2, 3, \ldots, n-1, 1$ (in this order), and so on. Generally, the pieces of the i-th block are associated with players $i, i+1, \ldots, n-1, 1, \ldots, i-1$. Each player dislikes each piece associated with him as $\frac{1}{n-1}$ and the rest of the chores as 0, which sums up to a total valuation of 1 for the entire chores for each of the first $n-1$ players. Finally player n dislikes the entire chores uniformly. This construction is shown in figure 1.

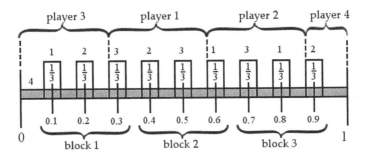

Fig. 1. Example construction for $n = 4$ players. The numbers above the columns denote the player this piece is associated with. Above the interval the optimal division is shown.

First, we want to upper bound the welfare in an optimal division of this instance (note that this is enough to give a lower bound on the price of fairness) and for this, take a look at the following division: we give the piece $(0, \frac{n-1}{(n-1)^2+1} - \epsilon)$ to player $n-1$, the piece $(\frac{i \cdot (n-1)}{(n-1)^2+1} - \epsilon, \frac{(i+1) \cdot (n-1)}{(n-1)^2+1} - \epsilon)$ to player i for $i = 1, \ldots, n-2$ and finally the piece $(\frac{(n-1)^2}{(n-1)^2+1} - \epsilon, 1)$ to player n.

We observe the following: For $i = 1, \ldots, n-2$ the i-th disliked piece of player i is at $(\frac{(i-1) \cdot (n-1)+1}{(n-1)^2+1} - \epsilon, \frac{(i-1) \cdot (n-1)+1}{(n-1)^2+1} + \epsilon)$ and the $i+1$-st disliked piece of player i is at $(\frac{(i+1) \cdot (n-1)}{(n-1)^2+1} - \epsilon, \frac{(i+1) \cdot (n-1)}{(n-1)^2+1} + \epsilon)$. This follows by construction, as these pieces are the first piece of the i-th block, or the last piece of the $i+1$-st block, respectively.

We conclude that the piece that is assigned to player $i = 1, \ldots, n-2$ as above is between the i-th and $i+1$-st piece of this player. The piece player $n-1$ receives is before his first disliked piece. Hence, the disutility of the players $1, \ldots, n-1$ are all 0, as they all do not receive any of their disliked pieces. Player n's disutility is exactly the physical size of the piece he receives, so the utilitarian welfare in this division is $u(x) = d_n(x) = 1 - (\frac{(n-1)^2}{(n-1)^2+1} - \epsilon) = \frac{1}{(n-1)^2+1} + \epsilon$. This division for the example with $n = 4$ players can be seen in figure 1.

Now, we claim that the disutility of player n in any equitable division of the chores is at least $\frac{1}{(n-1)^2+1} - 2\epsilon$. From this it follows that the utilitarian welfare in equitable divisions is at least n times as high, as all players must have the same disutility. Thus, the price of equitability is at least $\frac{\frac{n}{(n-1)^2+1} - 2n\epsilon}{\frac{1}{(n-1)^2+1} + \epsilon}$. The bound follows as ϵ approaches 0.

It remains to show that indeed player n has to receive a disutility of at least $\frac{1}{(n-1)^2+1} - 2\epsilon$ in every equitable division. The idea behind this claim is that no player $1, \ldots, n-1$ can receive a whole block of pieces without getting some positive disutility (as in every block there is one disliked piece of every player), and hence player n must get at least the slot between either two disliked pieces or the first (last) disliked piece and the left (right) end of the chores. The details of the proof can be found in the full version of the paper.   □

Proving a matching upper bound is simple when re-using Theorem 5, as we can use it to show that switching from an arbitrary to an equitable division does not increase the egalitarian welfare and hence the trivial bound $eg(x) \leq u(x) \leq n \cdot eg(x)$ can be applied. See the full version for the detailed proof.

Theorem 7. *The utilitarian price of equitability is upper-bounded by n.*

5   The Price of Envy-Freeness

Finally, we take a look at the price of envy-freeness. For this fairness notion, we get the most interesting deviation from former results on connected cakes and non-connected chores, as we can prove unboundedness of the price of fairness here (for more than two players). In contrast to the previous theorems, the arbitrary high price of fairness now does not result from giving an indifferent player more than his fair share in the optimal division, but from the fact that in the optimal division for the concrete instance given below, it is optimal to give the indifferent player *no* piece of the chores. But a situation where one player does not receive any piece in the optimal division has a negative effect on the price of envy-freeness, as every other player receiving a positive disutility will envy this player. By choosing the preferences in a certain way, we can make the price of envy-freeness arbitrarily high.

Theorem 8. *The price of envy-freeness for the division of chores with connected pieces is unbounded for both utilitarian and egalitarian welfare for $n > 2$ players.*

Proof. We use the same construction we used in the proof of Theorem 6 for the utilitarian price of equitability with other valuation functions. Again, let $0 < \epsilon < \frac{1}{(n-1)^2}$ and consider $(n-1)^2$ disliked pieces arranged in $(n-1)$ blocks as before, where piece p_i is located at $(\frac{i}{(n-1)^2+1} - \epsilon, \frac{i}{(n-1)^2+1} + \epsilon)$ for $i = 1, \ldots, (n-1)^2$.

Call the first piece of each block "type A" piece, the other pieces "type B" pieces. The pieces are associated with players as before.

Each player dislikes the only type A piece that is associated with him as $1 - (n-2)\epsilon$, the $(n-2)$ type B pieces associated with him as ϵ and the rest of the chores as 0. Player n assigns a dislike of $\frac{1}{(n-1)^2}$ to every disliked piece of either type and 0 to the rest. An example with 4 players can be seen in figure 2.

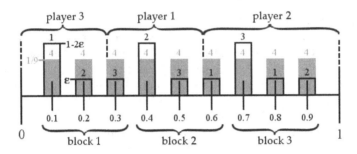

Fig. 2. Example construction for $n = 4$. The numbers above the columns denote the player this piece is associated with. The gray pieces denote the valuation of player 4. Above the interval the optimal division is shown.

An upper bound for the optimal division is constructed as follows: Similarly to the division given in the proof for Theorem 6, we give the piece $(0, \frac{n-1}{(n-1)^2+1} - \epsilon)$ to player $n-1$, the piece $(\frac{i \cdot (n-1)}{(n-1)^2+1} - \epsilon, \frac{(i+1) \cdot (n-1)}{(n-1)^2+1} - \epsilon)$ to player i for $i = 1, \ldots, n-3$, but we give the whole remaining piece $(\frac{(n-2)(n-1)}{(n-1)^2+1} - \epsilon, 1)$ to player $n-2$. Player n does not receive any piece of the chores (see also figure 2).

We observe the same facts as before:

- For $i = 1, \ldots, n-2$ the i-th disliked piece of player i is at $(\frac{(i-1) \cdot (n-1)+1}{(n-1)^2+1} - \epsilon, \frac{(i-1) \cdot (n-1)+1}{(n-1)^2+1} + \epsilon)$.
- For $i = 1, \ldots, n-2$ the $i+1$-st disliked piece of player i is at $(\frac{(i+1) \cdot (n-1)}{(n-1)^2+1} - \epsilon, \frac{(i+1) \cdot (n-1)}{(n-1)^2+1} + \epsilon)$.

Therefore, the piece player $i = 1, \ldots, n-3$ receives is between the i-th and $i+1$-st piece of this player. The piece player $n-1$ receives is before his first disliked piece (which starts at $\frac{n-1}{(n-1)^2+1} - \epsilon$). Hence, all players except player $n-2$ do not receive any of their disliked pieces and therefore have disutility 0. Only player $n-2$ receives one type B piece, therefore the maximal disutility, the egalitarian welfare and the utilitarian welfare of the optimal division is ϵ. Note that the division from Theorem 6 yields in this case a higher disutility, namely $\frac{1}{(n-1)^2}$ (which is strictly greater than ϵ according to our assumptions).

This division however is not envy-free, as player $n-2$ envies player n, for the empty piece is preferred by every player. Analogous to the argumentation in the last step of Theorem 6, we can argue that it is impossible to divide the chores between the first $n-1$ players entirely without giving one player a piece he dislikes. We are always left with at least one disliked piece p that has to be assigned to one player among players $1, \ldots, n-1$ who dislikes it > 0. Furthermore, this means we also have to give some piece to player n, as otherwise all players receiving a disutility > 0 envy n, and assigning n less than one half of p makes the player who receives the rest of p envy n. Thus, we can show that we cannot do better than giving n one half of p and another player i who also dislikes p the other half. Therefore, to achieve an optimal envy-free division we give players $1, \ldots, n-3, n-1$ the same pieces as in the non-envy-free division, the piece $\left(\frac{(n-2)(n-1)}{(n-1)^2+1} - \epsilon, \frac{(n-1)^2}{(n-1)^2+1} \right)$ to player $n-2$ and the piece $\left(\frac{(n-1)^2}{(n-1)^2+1}, 1 \right)$ to player n, i.e. we split the type B piece that player $n-2$ received in the optimal division between players $n-2$ and n. In this division x, $d_{n-2}(x) = \frac{\epsilon}{2}$ and $d_n(x) = \frac{1}{2(n-1)^2}$. As both get the same amount of player $n-2$'s last disliked piece, $n-2$ does not envy n any more. Furthermore, player n receives half of a disliked piece, whereas every other player receives more than one such piece, hence n does also not envy any other player. Players $1, \ldots, n-3, n-1$ have 0 disutility and therefore also envy no other player. Finally, $d_n(x) > d_{n-2}(x)$ for $\epsilon < \frac{1}{(n-1)^2}$, hence $eg(x) = d_n(x) = \frac{1}{2(n-1)^2}$, and $u(x) = d_{n-2}(x) + d_n(x) = \frac{\epsilon}{2} + \frac{1}{2(n-1)^2}$.

So finally we have a utilitarian price of envy-freeness of $\frac{1}{2} + \frac{1}{2(n-1)^2\epsilon}$ and an egalitarian price of envy-freeness of $\frac{1}{2(n-1)^2\epsilon}$, which both becomes arbitrarily large when ϵ approaches 0. $\square$

6 Conclusion

In this work we examined the decrease of social welfare due to fairness when dividing chores so that every player receives exactly one connected piece of the chores. We considered three important fairness criteria and two different social welfare functions and found tight bounds for almost all cases. For utilitarian welfare and proportionality or equitability the bounds are in $\Theta(n)$, for egalitarian welfare there is no trade-off for these two fairness criteria. For envy-freeness however, no bound exists for both welfare functions except for 2 players.

Upon finding that the price of envy-freeness for the division of chores is the only case that is unbounded, one could ask the question why there is such a fundamental difference between envy-freeness and the other two fairness notions, and why this difference does not appear when considering cakes (Aumann and Dombb [4] and Caragiannis et al. [5] gave bounds for this case). The answer for the first question lies in the inherent difference between envy-freeness and the other two fairness notions, namely that only the first relies on the valuation of a player for pieces other than his own. The difference between chores and cakes seems to arise from the different nature of the two problems: Infinite envy always results from one player receiving no piece of the cake/chores, which is - at least in some instances - desirable in chores division (see proof of Theorem 8), but undesirable when dividing cakes. This difference can also be seen in the results for indivisible cakes and chores by Caragiannis et al. [5], where the price of envy-freeness is bounded for cakes and unbounded for chores.

Some questions still remain open. For the setting where non-connected pieces are allowed, Caragiannis et al. [5] only considered utilitarian welfare. Bounding the egalitarian price of fairness for non-connected divisions of cakes and chores could be examined. Furthermore, Caragiannis et al. [5] provided an analysis of the price of fairness for indivisible cakes and chores, but again they only considered utilitarian welfare. Further research could investigate the impact of fairness on egalitarian welfare for this setting.

References

1. Steinhaus, H.: The problem of fair division. Econometrica 16, 101–104 (1948)
2. Gardner, M.: Aha! Insight. W. F. Freeman and Co. (1978)
3. Robertson, J.M., Webb, W.A.: Cake-cutting algorithms - be fair if you can. A K Peters (1998)
4. Aumann, Y., Dombb, Y.: The efficiency of fair division with connected pieces. In: Saberi, A. (ed.) WINE 2010. LNCS, vol. 6484, pp. 26–37. Springer, Heidelberg (2010)
5. Caragiannis, I., Kaklamanis, C., Kanellopoulos, P., Kyropoulou, M.: The efficiency of fair division. In: Leonardi, S. (ed.) WINE 2009. LNCS, vol. 5929, pp. 475–482. Springer, Heidelberg (2009)
6. Brams, S.J., Taylor, A.D.: An envy-free cake division protocol. American Mathematical Monthly 102(1), 9–18 (1995)
7. Brams, S.J., Taylor, A.D., Zwicker, W.: A moving-knife solution to the four-person envy-free cake division. Proceedings of the American Mathematical Society 125, 547–554 (1997)
8. Dubins, L.E., Spanier, E.H.: How to cut a cake fairly. American Mathematical Monthly 68(1), 1–17 (1961)
9. Stromquist, W.: How to cut a cake fairly. American Mathematical Monthly 87(8), 640–644 (1980)
10. Peterson, E., Su, F.E.: Four-person envy-free chore division. Mathematics Magazine 75(2), 117–122 (2002)
11. Peterson, E., Su, F.E.: N-person envy-free chore division. arXiv.org:0909.0303 (2009)
12. Su, F.E.: Rental harmony: Sperner's lemma in fair division. American Mathematical Monthly 106(10), 930–942 (1999)
13. Cohler, Y.J., Lai, J.K., Parkes, D.C., Procaccia, A.D.: Optimal envy-free cake cutting. In: Proceedings of the Twenty-Fifth AAAI Conference on Artificial Intelligence (2011)
14. Bei, X., Chen, N., Hua, X., Tao, B., Yang, E.: Optimal proportional cake cutting with connected pieces. In: [17]
15. Brams, S.J., Feldman, M., Lai, J.K., Morgenstern, J., Procaccia, A.D.: On maxsum fair cake divisions. In: [17]
16. Hoffman, Y., Aumann, Y., Dombb, Y.: Private communication
17. Hoffmann, J., Selman, B. (eds.): Proceedings of the Twenty-Sixth AAAI Conference on Artificial Intelligence, Toronto, Ontario, Canada, July 22-26. AAAI Press (2012)

On Popular Random Assignments

Haris Aziz[1], Felix Brandt[2], and Paul Stursberg[3]

[1] NICTA and University of New South Wales, Sydney, NSW 2033, Australia
haris.aziz@nicta.com.au
[2] Institut für Informatik, Technische Universität München, 85748 München, Germany
brandtf@in.tum.de
[3] Zentrum Mathematik, Technische Universität München, 85747 München, Germany
stursb@ma.tum.de

Abstract. One of the most fundamental and ubiquitous problems in microeconomics and operations research is how to assign objects to agents based on their individual preferences. An assignment is called popular if there is no other assignment that is preferred by a majority of the agents. Popular assignments need not exist, but the minimax theorem implies the existence of a popular *random* assignment. In this paper, we study the compatibility of popularity with other properties that have been considered in the literature on random assignments, namely efficiency, equal treatment of equals, envy-freeness, and strategyproofness.

1 Introduction

One of the most fundamental and ubiquitous problems in microeconomics and operations research is how to assign objects to agents based on their individual preferences (see, e.g., [21, 4, 5]). In its simplest form, the problem is known as the *assignment problem*, the *house allocation problem*, or *two-sided matching with one-sided preferences*. Formally, the *assignment problem* concerns a set of agents $A = \{a_1, \ldots, a_n\}$ and a set of houses $H = \{h_1, \ldots h_n\}$. Each agent has preferences over the elements of H and the goal is to assign or allocate exactly one house to each agent in an efficient and fair manner. An important assumption in this setting is that monetary transfers between the agents are not permitted.[1] The assignment problem has numerous applications in a variety of settings such the assignment of dormitories to students, jobs to applicants, rooms to housemates, processor time slots to jobs, parking spaces to employees, offices to workers, kidneys to patients, school seats to student applicants, etc. Clearly, deterministic assignments may fail to satisfy even extremely mild fairness criteria such as equal treatment of equals. It is therefore an established practice to restore (*ex ante*) fairness by introducing randomization. Random assignments

---

[1] Monetary transfers may be impossible or highly undesirable, as is the case if houses are public facilities provided to low-income people. There are a number of settings such as voting, kidney-exchange, or school choice in which money cannot be used as compensation due to practical, ethical, or legal constraints (see, e.g., [20]).

B. Vöcking (Ed.): SAGT 2013, LNCS 8146, pp. 183–194, 2013.

are strongly related to fractional assignments and random assignment rules can also be used to fractionally allocate resources to agents.

A deterministic assignment (or matching) is deemed *popular* if there exists no other assignment that a majority of agents prefers to the former (see, e.g., [1, 3, 17, 12]). Popular assignments were first considered by Gärdenfors [10]. While popular assignments can be computed in polynomial time [1], they unfortunately may not exist. Taking cue from this observation, McCutchen [17] proposed two quantities—the *unpopularity margin* and the *unpopularity factor*—to measure the unpopularity of an assignment and defined the notion of a *least unpopular* assignment, which is guaranteed to exist.[2] However, computing least unpopular assignments turned out to be NP-hard. Alternatively, Kavitha et al. [13] suggested the notion of popular *random* assignments. A random assignment p is popular if there is no other assignment q such that the expected number of agents who prefer the outcome of q to that of p is greater than $n/2$. Kavitha et al. [13] showed that popular random assignments not only exist due to the minimax theorem but can also be computed in polynomial time via linear programming. To the best of our knowledge, axiomatic properties of popular random assignments have not been studied so far. In this paper, we aim at improving our understanding of popular random assignments by investigating which common axiomatic properties are compatible with popularity.

Contributions. We first point out that popular random assignments can be viewed as a special case of *maximal lotteries*, which were proposed in the context of social choice by Fishburn [8].[3] Assignment can be seen as a restricted domain of social choice in which each alternative corresponds to an assignment. Preferences over houses can be easily extended to preferences over assignments by assuming that each agent only cares about the house assigned to himself and is indifferent between all assignments in which he is assigned the same house. We prove the following statements.

- Every popular assignment is efficient.
- There always exists a popular assignment that satisfies equal treatment of equals. Such an assignment can furthermore be computed in polynomial time.
- Popularity and envy-freeness are incompatible if $n \geq 3$. If a popular and envy-free assignment exists, it can be computed in polynomial time.
- There are no strategyproof popular random assignment rules if $n \geq 3$.

---

[2] The unpopularity margin of a matching is the maximum majority difference by which it is dominated by any other matching. The unpopularity factor of a matching is the maximum factor by which it is dominated by any other matching.

[3] Maximal lotteries were first considered by Kreweras [14] and independently rediscovered and studied in detail by Fishburn [8]. Interestingly, maximal lotteries or variants thereof have been rediscovered again by economists, mathematicians, political scientists, and computer scientists [15, 7, 9, 19]. Strategyproofness and efficiency of maximal lotteries were recently analyzed by Aziz et al. [2].

Related work. Random assignment rules have received enormous attention in recent years. Most notable among these rules are *random serial dictatorship (RSD)* (see e.g., [4, 16]) and the *probabilistic serial rule (PS)* [4]. Each of these rules has its own merits. However, it can be easily shown that the assignment returned by any of these rules may not be popular.

Perhaps closest to our work are the papers by Kavitha et al. [13], who introduced popular random assignments, and Bogomolnaia and Moulin [4], who outlined a systematic way of studying the properties of random assignments and random assignment rules. In particular, Bogomolnaia and Moulin [4] popularized the use of first-order stochastic dominance to formalize various notions of envy-freeness, efficiency, and strategyproofness that we also consider in this paper.

2   Preliminaries

An assignment problem is a triple $(A, H, \succsim)$ such that $A = \{a_1, \ldots, a_n\}$ is a set of agents, $H = \{h_1, \ldots, h_n\}$ is a set of houses, and $\succsim = (\succsim_1, \ldots, \succsim_n)$ is a preference profile in which $\succsim_i$ denotes an antisymmetric, complete, and transitive relation on H representing the preferences of agent i over the houses in H.[4]

A *deterministic assignment* (or *pure matching*) $M \subset A \times H = \mathcal{M}$ is a subset of non-adjacent arcs in the bipartite graph $G = (A \,\dot\cup\, H, A \times H)$. If $(i, h) \in M$, we write $M(i) = h$. A matrix $p = (p_{ih})_{(i,h) \in A \times H}$ with $p_{ih} \geq 0$, $\sum_{i \in A} p_{ih} = 1$ for all $h \in H$ and $\sum_{h \in H} p_{ih} = 1$ for all $a_i \in A$, $h \in H$ is called a *random assignment* (or *mixed matching*). Note that the entries $p_i = (p_{i1}, \ldots, p_{in})$ corresponding to arcs incident with some agent i constitute a random allocation for this agent. Further note that every random assignment may be represented by a (not necessarily unique) lottery over deterministic assignments and that in turn, every lottery over deterministic assignments induces a unique random assignment. This is known as the Birkhoff-Von Neumann theorem (see, e.g., [13]).

A natural way to compare random assignments is by means of *stochastic dominance (SD)*. Given two random assignments p and q, $p_i \succsim_i^{SD} q_i$ i.e., agent i *SD-prefers* p_i to q_i iff

$$\sum_{\substack{h \in H \\ h \succsim_i h^*}} p_{ih} \geq \sum_{\substack{h \in H \\ h \succsim_i h^*}} q_{ih} \text{ for all } h^* \in H.$$

This preference extension is of particular importance because one random assignment is SD-preferred to another iff, for any utility representation consistent with the ordinal preferences, the former yields at least as much *expected utility* as the latter (see, e.g., [11, 6]). Since for all $i \in A$, agent i compares assignment p with assignment q only with respect to his allocations p_i and q_i, we will sometimes abuse the notation by writing $p \succsim_i^{SD} q$ instead of $p_i \succsim_i^{SD} q_i$.

---

[4] Although we assume strict preferences for the ease of exposition, all our positive results hold for arbitrary preferences and our negative results even hold for strict preferences.

Finally, a *random assignment rule* f is a function which for each input $(A, H, \succsim)$ returns a random assignment p. When A and H are clear from the context, we simply write $f(\succsim)$ for $f(A, H, \succsim)$.

3 Desirable Properties of Random Assignment Rules

In this section, we define a number of desirable properties for random assignments and random assignment rules. Properties of assignments naturally translate to properties of assignment rules: We say that a random assignment rule f satisfies property P if every assignment p returned by f satisfies P.

Popularity. In order to define popularity, we first associate a function ϕ_i with each preference relation $\succsim_i$ on H by letting $\phi_i : H \times H \to \{-1, 0, 1\}$ such that for all $h, h' \in H$,

$$\phi_i(h, h') = \begin{cases} +1 & \text{if } h \succ_i h', \\ -1 & \text{if } h' \succ_i h, \text{ and} \\ 0 & \text{otherwise.} \end{cases}$$

Now consider the natural extension of ϕ_i to random assignments and take the sum over all agents. To this end, we define

$$\phi(p, q) := \sum_{a_i \in A} \sum_{h, h' \in H} p_{ih} q_{ih'} \, \phi_i(h, h')$$

and say that p is *more popular* than q if $\phi(p, q) > 0$. A random assignment p is *popular* if there is no assignment q more popular than p. It can be easily shown that both PS and RSD fail to satisfy popularity.

Efficiency. A deterministic assignment M is *Pareto efficient* if there exists no other deterministic assignment M' such that $M'(a_i) \succsim_i M(a_i)$ for all $a_i \in A$, and there exists an agent $a_i \in A$ such that $M'(a_i) \succ_i M(a_i)$. A random assignment is *ex post efficient* if it can be represented as a probability distribution over Pareto efficient deterministic assignments. Finally, a random assignment p is *SD-efficient* if there exists no assignment q such that q stochastically dominates p, i.e. $q_i \succsim_i^{SD} p_i$ for all $a_i \in A$ and $q_i \succ_i^{SD} p_i$ for some $a_i \in A$. It can be shown that SD-efficiency implies *ex post* efficiency. Furthermore, while PS satisfies SD-efficiency, RSD is only *ex post* efficient [4].

Fairness. A random assignment p satisfies *equal treatment of equals* if agents with identical preferences receive identical random allocations, i.e., $\succsim_i = \succsim_j$ implies that $p_i = p_j$ for any pair of agents i and j. Equal treatment of equals is considered as one of the most fundamental requirements in resource allocation and a "minimal test for fairness" [18]. A random assignment satisfies *SD-envy-freeness* if each agent (weakly) SD-prefers his allocation to that of any other agent. A random assignment satisfies *weak SD-envy-freeness* if no agent strictly SD-prefers someone elses allocation to his. SD-envy-freeness implies equal treatment of equals while weak SD-envy-freeness does not. PS is known to satisfy SD-envy-freeness whereas RSD only satisfies weak SD-envy-freeness [4].

Strategyproofness. In contrast to the previous conditions, strategyproofness can only meaningfully be defined as the property of an assignment *rule* rather than that of an assignment. A random assignment rule f is *SD-strategyproof* if for every preference profile $\succsim$, and for all $a_i \in A$ and $\succsim_i'$, $f(\succsim_i, \succsim_{-i}) \succsim_i^{SD} f(\succsim_i', \succsim_{-i})$. A random assignment rule f is *weakly SD-strategyproof* if for every preference profile $\succsim$, there exists no $\succsim_i'$ for some agent $a_i \in A$ such that $f(\succsim_i', \succsim_{-i}) \succ_i^{SD} f(\succsim_i, \succsim_{-i})$. RSD is SD-strategyproof whereas PS is only weakly SD-strategyproof. (When also allowing ties in the preferences, RSD remains SD-strategyproof whereas PS fails to be even weakly SD-strategyproof.)

In the remainder of this paper, we investigate whether and to which extent popularity is compatible with efficiency, fairness, and strategyproofness.

4 Efficiency

It is easy to see that popular assignments are *ex post* efficient. For the sake of contradiction let us assume that there is a deterministic assignment which is in the support of a lottery representation of some popular random assignment but which is not Pareto optimal. This implies that the deterministic assignment is Pareto dominated by another deterministic assignment and hence cannot be in the support of the popular random assignment (as replacing it by the assignment that dominates it would yield a more popular assignment).

We address SD-efficiency by first observing that popular random assignments are a special case of maximal lotteries in general social choice [8]. A lottery p is a *maximal lottery* if there exists no other lottery q for which the expected number of agents who prefer q over p is more than the expected number of agents who prefer p over q.

An assignment problem $(A, H, \succsim)$ may also be seen as a social choice problem where A is the set of agents and the alternatives to choose from are all the different (deterministic) assignments between agents in A and houses in H. The preferences of the agents over these alternatives can naturally be defined according to their preferences over the houses allocated to them (which means that agents will be indifferent between assignments that assign the same house to them). As Kavitha et al. [13] note, popularity of a random assignment p may also be defined in terms of its representation as a lottery over deterministic assignments. Furthermore, for every possible such representation the "unpopularity margin" is *equal to* that of the original assignment p. This means that every maximal lottery induces a popular random assignment, and every lottery that represents a popular assignment is maximal.

We now show that popular assignments are not only *ex post* efficient but even SD-efficient.

Lemma 1. *Let $L_p = [p^1 : M_1, \ldots, p^k : M_k]$ and $L_q = [q^1 : N_1, \ldots, q^k : N_l]$ be lotteries over deterministic assignments that induce the fractional assignments p and q. Then, $p \succsim_i^{SD} q$ iff $L_p \succsim_i^{SD} L_q$.*

Proof. For reasons of notational convenience, we write

$$p(M) = \begin{cases} p^j & M = M_j \in \operatorname{supp}(L_p) \\ 0 & M \notin \operatorname{supp}(L_p). \end{cases}$$

For every agent i and $h^* \in H$ we can pick some assignment N with $N(i) = h$ to obtain

$$\sum_{\substack{h \in H \\ h \succsim_i h^*}} p_{ih} = \sum_{\substack{M \in \mathcal{M} \\ M(i) \succsim_i h^*}} p(M) = \sum_{\substack{M \in \mathcal{M} \\ M \succsim_i N}} p(M).$$

Analogously, for every agent i and assignment N, we have

$$\sum_{\substack{M \in \mathcal{M} \\ M \succsim_i N}} p(M) = \sum_{\substack{M \in \mathcal{M} \\ M(i) \succsim_i N(i)}} p(M) = \sum_{\substack{h \in H \\ h \succsim_i N(i)}} p_{ih}.$$

This means that

$$\forall h \in H : \sum_{\substack{h \in H \\ h \succsim_i h^*}} p_{ih} \geq \sum_{\substack{h \in H \\ h \succsim_i h^*}} q_{ih} \quad \text{iff} \quad \forall N \in \mathcal{M} : \sum_{\substack{M \in \mathcal{M} \\ M \succsim_i N}} p(M) \geq \sum_{\substack{M \in \mathcal{M} \\ M \succsim_i N}} q(M),$$

i.e., $p \succsim_i^{SD} q$ iff $L_p \succsim_i^{SD} L_q$. $\square$

Theorem 1. *Every popular assignment is SD-efficient.*

Proof. Let p be a popular assignment. Suppose that p is SD-dominated by some assignment q. Let L_p be a lottery representation of p and L_q a lottery representation of q. Then Lemma 1 implies that L_q SD-dominates L_p. But, as argued above, L_p is a maximal lottery which is a contradiction to the fact that maximal lotteries satisfy SD-efficiency (see [2]).

$\square$

5   Equal Treatment of Equals

Even though popular assignments satisfy fairness in the sense of respecting majorities of agents, they can be highly unfair on the individual level. In fact, popular assignments may not even satisfy equal treatment of equals. This can be seen by considering the extremely simple case of two agents with identical preferences in which *every* random assignment is popular.

We will now show that a popular assignment that satisfies equal treatment of equals always exists and that it can be computed in polynomial time. To this end, we introduce the notion of an S-leveling:

Definition 1. *Let x be a random assignment for (A, H) and $S \subset A$. The S-leveling of x is the random assignment y given by*

$$y_{ah} = \begin{cases} x_{ah} & a \notin S \\ \frac{1}{|S|} \sum_{a \in S} x_{ah} & a \in S. \end{cases}$$

It is easy to see that the S-leveling of a random assignment is again a random assignment, as the sum over all edges incident to any house or agent remains unchanged.

Lemma 2. *Let x and z be random assignments for (A, H) and $S \subset A$ such that all $a \in S$ have identical preferences. Let furthermore y be the S-leveling of x. Now, the S-leveling z' of z satisfies*

$$\phi(x, z') = \phi(y, z).$$

Proof. We begin by showing that $\sum_{a \in S} (x_{ah} z'_{ah'} - y_{ah} z_{ah'}) = 0$ for all $h, h' \in H$. Let $h, h' \in H$. Then

$$\sum_{a \in S} (x_{ah} z'_{ah'} - y_{ah} z_{ah'}) = \sum_{a \in S} x_{ah} z'_{ah'} - \sum_{a \in S} y_{ah} z_{ah'} \tag{1}$$

$$= \sum_{a \in S} x_{ah} \left(\frac{1}{|S|} \sum_{a \in S} z_{ah'} \right) - \sum_{a \in S} \left(\frac{1}{|S|} \sum_{a \in S} x_{ah} \right) z_{ah'} \tag{2}$$

$$= \frac{1}{|S|} \sum_{a \in S} z_{ah'} \sum_{a \in S} x_{ah} - \frac{1}{|S|} \sum_{a \in S} x_{ah} \sum_{a \in S} z_{ah'} \tag{3}$$

$$= 0, \tag{4}$$

where we use the definition of S-leveling in (2) and the fact that one of the factors in each sum does not depend on a in (3). We use the definition of function ϕ as defined in Section 3.

Now, we define $\phi^* := \phi_a$ for an arbitrary agent $a \in S$ and have $\phi^* = \phi_a$ for all agents in S due to their identical preferences. Using this notation, we show, that $\phi(x, z') - \phi(y, z) = 0$:

$$\phi(x, z') - \phi(y, z) = \sum_{a \in A} \sum_{h, h' \in H} x_{ah} z'_{ah'} \phi_a(h, h') - \sum_{a \in A} \sum_{h, h' \in H} y_{ah} z_{ah'} \phi_a(h, h')$$
$$\tag{5}$$

$$= \sum_{a \in A} \sum_{h, h' \in H} \phi_a(h, h') (x_{ah} z'_{ah'} - y_{ah} z_{ah'}) \tag{6}$$

$$= \sum_{a \in S} \sum_{h, h' \in H} \phi_a(h, h') (x_{ah} z'_{ah'} - y_{ah} z_{ah'}) \tag{7}$$

$$= \sum_{h, h' \in H} \phi^*(h, h') \sum_{a \in S} (x_{ah} z'_{ah'} - y_{ah} z_{ah'}) \tag{8}$$

$$= 0 \tag{9}$$

using the fact that x and y as well as z and z' coincide on $A \setminus S$ in equation (7), the identical preferences of agents in S in (8) and finally our first claim in (9). □

Theorem 2. *There always exists a popular random assignment that satisfies equal treatment of equals. Such an assignment can furthermore be computed in polynomial time.*

Proof. Let x be a popular random assignment (the existence of which is guaranteed due to the minimax theorem) that does not satisfy equal treatment of equals for a subset A' of A and $S \subset A'$ a set of agents with identical preferences. Denote by y the S-leveling of x, which obviously has the property of treating these agents with identical preferences equally.

Suppose for contradiction that there is a random assignment z more popular than y, that is $\phi(z, y) > 0$. Using Lemma 2, we obtain a random assignment z' with $\phi(z', x) > 0$. Hence, z' is more popular than x which yields a contradiction to our assumption that x was popular.

We thus obtain a random assignment (y) that does not satisfy equal treatment of equals for a strictly smaller subset $A' \setminus S$ of A. Applying this argument iteratively, we finally obtain a random assignment that satisfies equal treatment of equals. □

To efficiently compute a popular assignment that satisfies equal treatment of equals, consider LP3 by Kavitha et al. [13] which computes a popular random assignment. With at most $O(n^2)$ extra constraints, it can be ensured that agents with same preferences get the same allocations. For each a_i, a_j such that $\succsim_i = \succsim_j$, we can impose the condition that $x(a_i, h_k) = x(a_j, h_k)$ for all $h_k \in H$. This ensures the equal treatment to equals condition.

6   Envy-Freeness

In this section, we investigate to which extent popularity is compatible with envy-freeness. There are popular assignments that fail to satisfy even weak SD-envy-freeness (again, consider the case with two agents who have identical preferences). The question that we are interested in is whether, for every preference profile, there exists at least one popular assignment that satisfies SD-envy-freeness or weak SD-envy-freeness.

Theorem 3. *There exists an instance of a random assignment problem with $n = 3$ for which no popular assignment satisfies SD-envy-freeness.*

Proof. Consider the following assignment problem with three agents and three houses.

$$a_1 : h_1, h_2, h_3$$
$$a_2 : h_1, h_2, h_3$$
$$b \ \ : h_2, h_1, h_3$$

As noted in Section 3, any assignment that satisfies SD-envy-freeness must also satisfy equal treatment of equals. We now show that the unique popular assignment that satisfies equal treatment of equals is as follows:

$$p_{a_1 h_1} = 1/2, \qquad p_{a_1 h_2} = 0, \qquad p_{a_1 h_3} = 1/2,$$
$$p_{a_2 h_1} = 1/2, \qquad p_{a_2 h_2} = 0, \qquad p_{a_2 h_3} = 1/2,$$
$$p_{b h_1} = 0, \qquad p_{b h_2} = 1, \qquad p_{b h_3} = 0.$$

Consider an assignment p which satisfies equal treatment of equals. Denote by $p_1 := p_{a_1h_1} = p_{a_2h_1}$ and $p_2 := p_{a_1h_2} = p_{a_2h_2}$. Note that, in particular, as p is popular it has to be at least as popular as the pure assignment $M_1 = \{(b, h_2), (a_1, h_1), (a_2, h_3)\}$. Hence, p must fulfil $\phi(M_1, p) = 1 + p_2 - 2p_1 \leq 0$ which means that $p_1 \geq 1/2 + p_2/2$. Secondly, $1 \geq p_{a_1h_1} + p_{a_2h_1} = 2p_1$ which means that $p_1 \leq 1/2$.

The only assignment that satisfies the constraints $p_1 \geq 1/2 + p_2/2$, $p_1 \leq 1/2$, $p_1 \geq 0$, and $p_2 \geq 0$ is the one for which $p_1 = 1/2$ and $p_2 = 0$. In this assignment p, the allocations of a_1, a_2 do not SD-dominate the allocation of b according to the preference of a_1 and a_2. Therefore the only popular assignment satisfying equal treatment of equals does not satisfy SD-envy-freeness.   □

Despite this negative result, an SD-envy-free popular random assignment can be computed in polynomial time whenever it exists. For each pair of agents a, b, we need the constraint that $p_a \succsim_a^{SD} p_b$. This can be encoded easily by considering at most as many partial sums as the number of houses n.

$$\sum_{\substack{h \in H \\ h \succsim_a h^*}} p_{ah} \geq \sum_{\substack{h \in H \\ h \succsim_a h^*}} p_{bh} \text{ for all } h^* \in H.$$

There are $O(n^2)$ such constraints.

Regarding *weak* SD-envy-freeness, the alternative characterization of the SD relation in terms of utility functions mentioned in Section 2 might help. This characterization allows us to ensure weak SD-envy-freeness by adding constraints to the linear program used to compute popular assignments as follows: An assignment p is *not* strictly preferred to an assignment q by agent i, if there exists some utility function u for which the expected utility of q is greater than that of p. This can be expressed by adding variables to represent the utility function u (for each agent). However, we have shown that the resulting feasible region is non-convex, which implies that this representation hardly leads to an efficient algorithm to compute such an assignment. This assessment does of course not preclude the *existence* of such an assignment.

7   Strategyproofness

Finally, we examine how popular assignment rules fare in terms of strategyproofness. It turns out that popularity is incompatible with SD-strategyproofness.

Theorem 4. *For $n \geq 3$, there are no SD-strategyproof popular randomized assignment rules.*

Proof. Consider an assignment problem with three agents and three houses and the following preferences.

$$a_1 : h_1, h_3, h_2$$
$$a_2 : h_1, h_2, h_3$$
$$a_3 : h_1, h_2, h_3$$

We show that there exists some utility function for agent a_1, compatible with his preferences, which allows him to obtain a higher expected utility if he misreports his preferences. In light of the equivalence mentioned in Section 2, this means that agent a_1 does not SD-prefer his original outcome to that which she may achieve by misreporting.

The set of all deterministic assignments is as follows:

$$M_{123} = \{\{a_1, h_1\}, \{a_2, h_2\}, \{a_3, h_3\}\}, \quad M_{312} = \{\{a_1, h_3\}, \{a_2, h_1\}, \{a_3, h_2\}\},$$
$$M_{231} = \{\{a_1, h_2\}, \{a_2, h_3\}, \{a_3, h_1\}\}, \quad M_{132} = \{\{a_1, h_1\}, \{a_2, h_3\}, \{a_3, h_2\}\},$$
$$M_{321} = \{\{a_1, h_3\}, \{a_2, h_2\}, \{a_3, h_1\}\}, \quad M_{213} = \{\{a_1, h_2\}, \{a_2, h_1\}, \{a_3, h_3\}\}.$$

Then consider the matrix corresponding to the pairwise weighted majority comparisons. An entry in the matrix denotes the number of agents who prefer the row assignment to the column assignment minus number of agents who prefer the column assignment to the row assignment. An assignment is popular if and only if it is a maximin strategy of the symmetric two-player zero-sum game represented by the matrix. It can be checked using an LP solver that each maximin strategy only randomizes over M_{312} and M_{321}.

Since a_1 gets h_3 in both M_{312} and M_{321}, a_1 gets h_3 with probability one in every popular assignment.

	M_{123}	M_{312}	M_{231}	M_{132}	M_{321}	M_{213}
M_{123}	0	−1	+1	0	0	0
M_{312}	+1	0	+1	0	0	+2
M_{231}	−1	−1	0	+2	−2	0
M_{132}	0	0	−2	0	−1	+1
M_{321}	0	0	+2	+1	0	+1
M_{213}	0	−2	0	−1	−1	0

Now if a_1 misreports his preferences as h_1, h_2, h_3, the new preference profile is as follows.

$$a_1 : h_1, h_2, h_3$$
$$a_2 : h_1, h_2, h_3$$
$$a_3 : h_1, h_2, h_3$$

Then, the pairwise majority margins are shown in the matrix below.

	M_{123}	M_{312}	M_{231}	M_{132}	M_{321}	M_{213}
M_{123}	0	−1	+1	0	0	0
M_{312}	+1	0	−1	0	0	0
M_{231}	−1	+1	0	0	0	0
M_{132}	0	0	0	0	−1	+1
M_{321}	0	0	0	+1	0	−1
M_{213}	0	0	0	−1	+1	0

It can be shown that a maximin strategy is a probability distribution over the following two strategies $[M_{123} : 1/3; M_{312} : 1/3; M_{231} : 1/3]$ and $[M_{132} : 1/3; M_{321} : 1/3; M_{213} : 1/3]$. Hence the induced popular assignment for any lottery corresponding to a maximin strategy is one which specifies a probability of $1/3$ of each agent getting each object. Thus a_1 gets h_1, h_2, and h_3 each with probability $1/3$. Now, let us assume that $u_{a_1}(h_1) + u_{a_1}(h_2) > 2u_{a_1}(h_3)$. Then a_1 gets utility $(u_{a_1}(h_1) + u_{a_1}(h_2) + u_{a_1}(h_3))/3 > 3u_{a_1}(h_3)/3 = u_{a_1}(h_3)$. $\quad\square$

An important open question is whether there are *weakly* SD-strategyproof popular random assignment rules. Related questions have recently also been analyzed in the more general context of social choice where it was shown that popularity is incompatible with weak SD-strategyproofness, but compatible with a significantly weaker version of weak SD-strategyproofness called *weak ST-strategyproofness* [2].

8 Conclusion

Kavitha et al. [13] have recently shown that every assignment problem admits a popular random assignment which can furthermore be computed in polynomial time using linear programming. In this paper, we investigated which common axiomatic properties are compatible with popularity. Results were mixed. It turned out that a particularly desirable aspect of popularity is that many conditions can be formalized as linear constraints that can be simply plugged into the linear program for computing popular random assignments. Furthermore, all properties considered in this paper (including popularity) do *not* require the asymmetry or transitivity of the agents' preferences. By contrast, two of the most studied random assignment rules, PS and RSD, are only defined for transitive preferences and many axiomatic results concerning these rules even require linear preferences.

A number of interesting questions arise from our study. Two of the most important ones are whether there always exists a weakly SD-envy-free popular random assignment and whether there exists a popular random assignment rule that satisfies weak SD-strategyproofness.

Acknowledgments. This material is based upon work supported by the Australian Government's Department of Broadband, Communications and the Digital Economy, the Australian Research Council, the Asian Office of Aerospace Research and Development through grant AOARD-124056, and the Deutsche Forschungsgemeinschaft under grants BR 2312/7-1 and BR 2312/7-2.

References

1. Abraham, D.K., Irving, R.W., Kavitha, T., Mehlhorn, K.: Popular matchings. SIAM Journal on Computing 37(4), 1030–1034 (2007)

2. Aziz, H., Brandt, F., Brill, M.: On the tradeoff between economic efficiency and strategyproofness in randomized social choice. In: Proc. of 12th AAMAS Conference, pp. 455–462. IFAAMAS (2013)
3. Biró, P., Irving, R.W., Manlove, D.F.: Popular matchings in the marriage and roommates problems. In: Calamoneri, T., Diaz, J. (eds.) CIAC 2010. LNCS, vol. 6078, pp. 97–108. Springer, Heidelberg (2010)
4. Bogomolnaia, A., Moulin, H.: A new solution to the random assignment problem. Journal of Economic Theory 100(2), 295–328 (2001)
5. Budish, E., Che, Y.-K., Kojima, F., Milgrom, P.: Designing random allocation mechanisms: Theory and applications. American Economic Review (forthcoming, 2013)
6. Cho, W.J.: Probabilistic assignment: A two-fold axiomatic approach (unpublished manuscript, 2012)
7. Felsenthal, D.S., Machover, M.: After two centuries should Condorcet's voting procedure be implemented? Behavioral Science 37(4), 250–274 (1992)
8. Fishburn, P.C.: Probabilistic social choice based on simple voting comparisons. Review of Economic Studies 51(167), 683–692 (1984)
9. Fisher, D.C., Ryan, J.: Tournament games and positive tournaments. Journal of Graph Theory 19(2), 217–236 (1995)
10. Gärdenfors, P.: Match making: Assignments based on bilateral preferences. Behavioral Science 20(3), 166–173 (1975)
11. Katta, A.-K., Sethuraman, J.: A solution to the random assignment problem on the full preference domain. Journal of Economic Theory 131(1), 231–250 (2006)
12. Kavitha, T.: Popularity vs maximum cardinality in the stable marriage setting. In: Proc. of 23rd SODA, pp. 123–134. ACM Press (2012)
13. Kavitha, T., Mestre, J., Nasre, M.: Popular mixed matchings. Theoretical Computer Science 412(24), 2679–2690 (2011)
14. Kreweras, G.: Aggregation of preference orderings. In: Sternberg, S., Capecchi, V., Kloek, T., Leenders, C. (eds.) Mathematics and Social Sciences I: Proceedings of the Seminars of Menthon-Saint-Bernard, France (July 1-27, 1960), and of Gösing, Austria (July 3-27, 1962), pp. 73–79 (1965)
15. Laffond, G., Laslier, J.-F., Le Breton, M.: The bipartisan set of a tournament game. Games and Economic Behavior 5, 182–201 (1993)
16. Manea, M.: Serial dictatorship and Pareto optimality. Games and Economic Behavior 61, 316–330 (2007)
17. McCutchen, R.M.: The least-unpopularity-factor and least-unpopularity-margin criteria for matching problems with one-sided preferences. In: Laber, E.S., Bornstein, C., Nogueira, L.T., Faria, L. (eds.) LATIN 2008. LNCS, vol. 4957, pp. 593–604. Springer, Heidelberg (2008)
18. Moulin, H.: Fair Division and Collective Welfare. The MIT Press (2003)
19. Rivest, R.L., Shen, E.: An optimal single-winner preferential voting system based on game theory. In: Proc. of 3rd International Workshop on Computational Social Choice, pp. 399–410 (2010)
20. Roth, A.E.: Repugnance as a constraint on markets. Journal of Economic Perspectives 21(3), 37–58 (2007)
21. Young, H.P.: Dividing the indivisible. American Behavioral Scientist 38, 904–920 (1995)

Scheduling a Cascade with Opposing Influences*

MohammadTaghi Hajiaghayi, Hamid Mahini, and Anshul Sawant

University of Maryland at College Park
{hajiagha,hmahini,asawant}@cs.umd.edu

Abstract. Adoption or rejection of ideas, products, and technologies in a society is often governed by simultaneous propagation of positive and negative influences. Consider a planner trying to introduce an idea in different parts of a society at different times. How should the planner design a schedule considering this fact that positive reaction to the idea in early areas has a positive impact on probability of success in later areas, whereas a flopped reaction has exactly the opposite impact? We generalize a well-known economic model which has been recently used by Chierichetti, Kleinberg, and Panconesi (ACM EC'12). In this model the reaction of each area is determined by its initial preference and the reaction of early areas. We model the society by a graph where each node represents a group of people with the same preferences. We consider a full propagation setting where news and influences propagate between every two areas. We generalize previous works by studying the problem when people in different areas have various behaviors.

We first prove, independent of the planner's schedule, influences help (resp., hurt) the planner to propagate her idea if it is an appealing (resp., unappealing) idea. We also study the problem of designing the optimal non-adaptive spreading strategy. In the non-adaptive spreading strategy, the schedule is fixed at the beginning and is never changed. Whereas, in adaptive spreading strategy the planner decides about the next move based on the current state of the cascade. We demonstrate that it is hard to propose a non-adaptive spreading strategy in general. Nevertheless, we propose an algorithm to find the best non-adaptive spreading strategy when probabilities of different behaviors of people in various areas drawn i.i.d from an unknown distribution. Then, we consider the influence propagation phenomenon when the underlying influence network can be any arbitrary graph. We show it is $\#P$-complete to compute the expected number of adopters for a given spreading strategy. However, we design a polynomial-time algorithm for the problem of computing the expected number of adopters for a given schedule in the full propagation setting. Last but not least, we give a polynomial-time algorithm for designing an optimal adaptive spreading strategy in the full propagation setting.

Keywords: Influence maximization, Scheduling, Spreading strategy, Algorithm.

1 Introduction

People's opinions are usually formed by their friends' opinions. Whenever a new concept is introduced into a society, the high correlation between people's reactions

* Supported in part by NSF CAREER award 1053605, NSF grant CCF-1161626, ONR YIP award N000141110662, and a DARPA/AFOSR grant FA9550-12-1-0423.

B. Vöcking (Ed.): SAGT 2013, LNCS 8146, pp. 195–206, 2013.

initiates an influence propagation. Under this propagation, the problem of promoting a product or an opinion depends on the problem of directing the flow of influences. As a result, a planner can develop a new idea by controlling the flow of influences in a desired way. Although there have been many attempts to understand the behavior of influence propagation in a social network, the topic is still controversial due to lack of reliable information and complex behavior of this phenomenon. For example, one compelling approach is "seeding" which was introduced by the seminal work of Kempe, Kleinberg, and Trados [1] and is well-studied in the literature [1, 2, 3]. The idea is to influence a group of people in the initial investment period and spread the desired opinion in the ultimate exploitation phase. Another approach is to use time-varying and customer-specific prices to propagate the product (see e.g., [4, 5, 6]). All of these papers investigate the influence propagation problem when only positive influences spread into the network. However, in many real world applications people are affected by both positive and negative influences, e.g., when both consenting and dissenting opinions broadcast simultaneously.

We generalize a well-known economic model introduced by Arthur [7]. This model has been recently used by Chierichetti, Kleinberg, and Panconesi [8]. Assume an organization is going to develop a new idea in a society where the people in the society are grouped into n different areas. Each area consists of people living near each other with almost the same preferences. The planner schedules to introduce a new idea in different areas at different times. Each area may accept or reject the original idea. Since areas are varied and effects of early decisions boost during the diffusion, a schedule-based strategy affects the spread of influences. This framework closely matches to various applications from economics to social science to public health where the original idea could be a new product, a new technology, or a new belief.

Consider the spread of two opposing influences simultaneously. Both positive and adverse reactions to a single idea originate different flows of influences simultaneously. In this model, each area has an *initial preference* of $\mathcal{Y}$ or $\mathcal{N}$. The initial preference of $\mathcal{Y}$ ($\mathcal{N}$) means the area will accept (decline) the original idea when there are no network externalities. Let c_i be a non-negative number indicating how reaction of people in area i depends on the others'. We call c_i the *threshold* of area i. Assume the planner introduced the idea in area i at time s. Let $m_{\mathcal{Y}}$ and $m_{\mathcal{N}}$ be the number of areas which accept or reject the idea before time s. If $|m_{\mathcal{Y}} - m_{\mathcal{N}}| \geq c_i$ the people in area i decide based on the majority of previous adopters. It means they adopt the idea if $m_{\mathcal{Y}} - m_{\mathcal{N}} \geq c_i$ and drop it if $m_{\mathcal{N}} - m_{\mathcal{Y}} \geq c_i$. Otherwise, if $|m_{\mathcal{Y}} - m_{\mathcal{N}}| < c_i$ the people in area i accept or reject the idea if the initial preference of area i is $\mathcal{Y}$ or $\mathcal{N}$ respectively. The planner does not know exact initial preferences and has only prior knowledge about them. Formally speaking, for area i the planner knows the initial preference of area i will be $\mathcal{Y}$ with probability p_i and will be $\mathcal{N}$ with probability $1 - p_i$. We call p_i the *initial acceptance probability* of area i.

We consider the problem when the planner classifies different areas into various types. The classification is based on the planner's knowledge about the reaction of people living in each area. Hence, the classification is based on different features, e.g., preferences, beliefs, education, and age such that people in areas with the same type react almost the same to the new idea. It means all areas of the same type have the same threshold c_i and the same initial acceptance probability p_i. It is worth mentioning previous works only consider the problem when all areas have the same type, i.e., all p_i's and c_i's are the same [7, 8]. The planner wants to manage the flow of influences, and her *spreading strategy* is a permutation π over different areas. Her goal is to find a spreading strategy π which maximizes the expected number of adopters. We consider both *adaptive* and *non-adaptive* spreading strategies in this paper. In the adaptive spreading strategy, the planner can see results of earlier areas for further decisions. On the other hand, in the non-adaptive spreading strategy the planner decides about the permutation in advance. We show the effect of a spreading strategy on the number of adopters with an example in the full version of the paper.

1.1 Related Work

We are motivated by a series of well-known studies in economics and politics literature in order to model people's behavior [7, 9, 10, 11]. Arthur first proposed a framework to analyze people's behavior in a scenario with two competing products [7]. In this model people are going to decide about one of two competing products alternatively. He studied the problem when people are affected by all previous customers, and the planner has the same prior knowledge about people's behavior, i.e., people have the same types. He demonstrated that a cascade of influences is formed when products have positive network externalities, and early decisions determine the ultimate outcome of the market. It has been showed the same cascade arises when people look at earlier decisions, not because of network externalities, but because they have limited information themselves or even have bounded rationality to process all available data [9, 10].

Chierichetti, Kleinberg, and Panconesi argued when relations between people form an arbitrary network, the outcome of an influence propagation highly depends on the order in which people make their decisions [8]. In this setting, a potential spreading strategy is an ordering of decision makers. They studied the problem of finding a spreading strategy which maximizes the expected number of adopters when people have the same type, i.e., people have the same threshold c and the same initial acceptance probability p. They proved for any n-node graph there is an adaptive spreading strategy with at least $O(np^c)$ adopters. They also showed for any n-node graph all non-adaptive spreading strategies result in at least (resp. at most) $\frac{n}{2}$ if initial acceptance probability is less (resp. greater) than $\frac{1}{2}$. They considered the problem on an arbitrary graph when nodes have the same type. While we mainly study the problem on a complete graph when nodes have various types, we improve their result in our setting and show the expected number of adopters for all adaptive spreading strategies is at least (resp. at most) np if initial acceptance probability is $p \geq \frac{1}{2}$ (resp. $p \leq \frac{1}{2}$). We also show the problem of designing the best spreading strategy is hard on an arbitrary graph with several types of customers. We prove it is $\#P$-complete to compute the expected number of adopters for a given spreading strategy.

The problem of designing an appropriate marketing strategy based on network externalities has been studied extensively in the computer science literature. For example, Kempe, Kleinberg, and Tardos [1] studied the following question in their seminal work: How can we influence a group of people in an investment phase in order to propagate an idea in the exploitation phase? This question was introduced by Domingos and Richardson [12]. The answer to this question leads to a marketing strategy based on seeding. There are several papers that study the same problem from an algorithmic point of view, e.g., [2, 3, 13]. Hartline, Mirrokni, and Sundararajan [6] also proposed another marketing strategy based on scheduling for selling a product. Their marketing strategy is a permutation π over customers and price p_i for customer i. The seller offers the product with price p_i to customer i at time t where $t = \pi^{-1}(i)$. The goal is to find a marketing strategy which maximizes the profit of the seller. This approach is followed by several works, e.g., [4, 5, 14]. These papers study the behavior of an influence propagation when there is only one flow on influences in the network. In this paper, we study the problem of designing a spreading strategy when both negative and positive influences propagate simultaneously.

The propagation of competitive influences has been studied in the literature (See [15] and its references). These works studied the influence propagation problem in the presence of competing influences, i.e., when two or more competing firms try to propagate their products at the same time. However we study the problem of influence propagation when there exist both positive and negative reactions to the same idea. There are also studies which consider the influence propagation problem in the presence of positive and negative influences [16, 17]. Che et al. [16] use a variant of the independent cascade model introduced in [1]. They model negative influences by allowing each person to flips her idea with a given probability q. Li et al. [17] model the negative influences by negative edges in the graph. Although they study the same problem, we use different models to capture behavior of people.

1.2   Our Results

We analyze an influence propagation phenomenon where two opposing flows of influences propagate through a social network. As a result, a mistake in the selection of early areas may result in propagation of negative influences. Therefore a good understanding of influence propagation dynamics seems necessary to analyze the properties of a spreading strategy. Besides the previous papers which have studied the problem with just one type [7, 8], we consider the scheduling problem with various types. Also, we mainly study the problem in a *full propagation* setting as it matches well to our motivations. In the full propagation setting news and influences propagate between every two areas. One can imagine how internet, media, and electronic devices broadcast news and influences from everywhere to everywhere. In the *partial propagation* setting news and influences do not necessarily propagate between every two areas. In the partial propagation setting the society can be modeled with a graph, where there is an edge from area i to area j if and only if influences propagate from area i to area j.

Our main focus is to analyze the problem when the planner chooses a non-adaptive spreading strategy. Consider an arbitrary non-adaptive spreading strategy when initial preferences of all areas are p. The expected number of adopters is exactly np if all areas decide independently. We demonstrate that in the presence of network influences, the expected number of adopters is greater/less than np if initial acceptance probability p is greater/less than $\frac{1}{2}$. These results have a bold message: **The influence propagation is an amplifier for an appealing idea and an attenuator for an unappealing idea.**

Chierichetti, Kleinberg, and Panconesi [8] studied the problem on an arbitrary graph with only one type. They proved the number of adopters is greater/less than $\frac{n}{2}$ if initial acceptance probability p is greater/less than $\frac{1}{2}$. Theorem 1 improves their result from $\frac{n}{2}$ to np in our setting. All missing proofs are in the full version of the paper.

Theorem 1. *Consider an arbitrary non-adaptive spreading strategy π in the full propagation setting. Assume all initial acceptance probabilities are equal to p. If $p \geq \frac{1}{2}$, then the expected number of adopters is at least np. Furthermore, If $p \leq \frac{1}{2}$, then the expected number of adopters is at most np.*

Chierichetti, Kleinberg, and Panconesi [8] studied the problem of designing an optimum spreading strategy in the partial propagation setting. They design an approximation algorithm for the problem when the planner has the same prior knowledge about all areas, i.e., all areas have the same type. We study the same problem with more than one type. We first consider the problem in the full propagation setting. One approach is to consider a non-adaptive spreading strategy with a constant number of switches between different types. The planner has the same prior knowledge about areas with the same type. It means areas with the same type are identical for the planner. Thus any spreading strategy can be specified by types of areas rather than areas themselves. Let $\tau(i)$ be the type of area i and $\tau(\pi)$ be the sequence of types for spreading strategy π. For a given spreading strategy π a *switch* is a position k in the sequence such that $\tau(\pi(k)) \neq \tau(\pi(k+1))$. As an example consider a society with 4 areas. Areas 1 and 2 are of type 1. Areas 3 and 4 are of type 2. Then spreading strategy $\pi_1 = (1, 2, 3, 4)$ with $\tau(\pi_1) = (1, 1, 2, 2)$ has a switch at position 2 and spreading strategy $\pi_2 = (1, 3, 2, 4)$ with $\tau(\pi_2) = (1, 2, 1, 2)$ has switches at positions 1, 2, and 3.

Theorem 2. *A σ-switch spreading strategy is a spreading strategy with at most σ switches. For any constant σ, there exists a society with areas of two types such that no σ-switch spreading strategy is optimal.*

We construct a society with n areas with $\frac{n}{2}$ areas of type 1 and $\frac{n}{2}$ areas of type 2. We demonstrate an optimal non-adaptive spreading strategy should switch at least $\Omega(n)$ times. It means no switch-based non-adaptive spreading strategy can be optimal.

On the positive side, we analyze the problem when thresholds are drawn independently from an unknown distribution and initial acceptance probabilities are arbitrary numbers. We characterize the optimal non-adaptive spreading strategy in this case.

Theorem 3. *Assume that the planner's prior knowledge about all values of c_i's is the same, i.e., all c_i's are drawn independently from the same but unknown distribution. Let initial acceptance probabilities be arbitrary numbers. Then, the best non-adaptive spreading strategy is to order all areas in non-increasing order of their initial acceptance probabilities.*

We also study the problem of designing the optimum spreading strategy in the partial propagation setting with more than one types. We show it is hard to determine the expected number of adopters for a given spreading strategy. Formally speaking, we show it is $\#P$-complete to compute the expected number of adopters for a given spreading strategy π in the partial propagation setting with more than one type. This is another evidence to show the influence propagation is more complicated with more than one type.

We prove Theorem 4 based on a reduction from a variation of the *network reliability* problem.

Theorem 4. *In the partial propagation setting, it is #P-complete to compute the expected number of adopters for a given non-adaptive spreading strategy π.*

We also present a polynomial-time algorithm to compute the expected number of adopters for a given non-adaptive spreading strategy in a full propagation setting. We design an algorithm in order to simulate the amount of propagation for a given spreading strategy.

Theorem 5. *Consider a full propagation setting. The expected number of adopter can be computed in polynomial time for a given non-adaptive spreading strategy π.*

At last we study the problem of designing the best adaptive spreading strategy. We overcome the hardness of the problem and design a polynomial-time algorithm to find the best adaptive marketing strategy in the following theorem.

Theorem 6. *A polynomial-time algorithm finds the best adaptive spreading strategy for a society with a constant number of types.*

2   Notation and Preliminaries

In this section we define basic concepts and notation used throughout this paper. We first formally define the spread of influence through a network as a stochastic process and then give the intuition behind the formal notation. We are given a graph $G = (V, E)$ with thresholds, $c_v \in \mathbb{Z}_{>0}, \forall v \in V$ and initial acceptance probabilities $p_v \in [0, 1], \forall v \in V$. Let $|V| = n$. Let d_v be the degree of vertex v. Let $N(v)$ be the set of neighboring vertices of v. Let $\boldsymbol{c}$ be the vector $(c_1, \ldots, c_n)$ and $\boldsymbol{p}$ be the vector $(p_1, \ldots, p_n)$. Given a graph $G = (V, E)$ and a permutation $\pi : V \mapsto V$, we define a discrete stochastic process, *IS* (Influence Spread) as an ordered set of random variables $(X^1, X^2, \ldots, X^n)$, where $X^t \in \Omega = \{-1, 0, 1\}^n, \forall t \in \{1, \ldots, n\}$. The random variable X_v^t denotes decision of area v at time t. If it has not yet been scheduled, $X_v^t = 0$. If it accepts the idea then $X_v^t = 1$, and if it rejects the idea then $X_v^t = -1$. Note that $X_v^t = 0$ iff $t < \pi^{-1}(v)$. Let $D(v) = \sum_{u \in N(v)} X_u^{\pi^{-1}(v)}$ be the sum of decision's of v's neighbors. For simplicity in notation, we denote X_v^n by X_v.

We now briefly explain the intuition behind the notation. The input graph models the influence network of areas on which we want to schedule a cascade, with each vertex representing an area. There is an edge between two vertices if two corresponding areas influence each others decision. The influence spread process models the spread of idea acceptance and rejection for a given spreading strategy. The permutation π maps a position in spreading strategy to an area in V. For example, $\pi(1) = v$ implies that v is the first area to be scheduled. Once the area v is given a chance to accept or reject the idea at time $\pi^{-1}(v)$, $X_v^{\pi^{-1}(v)}$ is assigned a value based on v's decision and at all times t after $\pi^{-1}(v)$, $X_v^t = X_v^{\pi^{-1}(v)}$. The random variable X_v denotes whether an area v accepted or rejected the idea. We note that $X_v^t = X_v, \forall t \geq \pi^{-1}(v)$. The random variable X^t is complete snapshot of the cascade process at time t. The variable $D(v)$ is the decision variable for v. It denotes the sum of decisions of v's neighbors at the time v is scheduled in the cascade and it determines whether v decides to follow the majority decision or whether v decides based on its initial acceptance probability. The random variable I_t is the sum of decisions of all areas at time t. Thus, I_n is the variable we are interested in as it denotes the difference between number of people who accept the idea and people who reject the idea. Let $v = \pi(t)$. Given X^{t-1}, X^t is defined as follows:

- Every area decides to accept or reject the idea exactly once when it is scheduled and its decision remains the same at all later times. Therefore $\forall i \neq \pi(t)$:
  - $X_i^t = X_i^{t-1}$
- Decision of area v is based on decision of previous areas if its threshold is reached.
  - $X_v^t = 1$ if $D(v) \geq c_v$
  - $X_v^t = -1$ if $D(v) \leq -c_v$
- If threshold of area v is not reached, then it decides to accept the idea with probability p_v, its initial acceptance probability, and decides to reject it with probability $1 - p_v$.

In partial propagation setting, we represent such a stochastic process by tuple $IS = (G, c, p, \pi)$. For full propagation setting, the underlying graph is a complete graph and hence we can denote the process by (c, p, π). When c and p are clear from context, we denote the process simply by spreading strategy, π. We define random variable $I_t = \sum_{v \in V} X_v^t$. We denote by $q_v = 1 - p_v$ the probability that v rejects the idea based on initial preference. We denote by $Pr(A; IS)$, the probability of event A occurring under stochastic process IS. Similarly, we denote by $E(z; IS)$, the expected value of random variable z under the stochastic process IS.

3   A Bound on Spread of Appealing and Unappealing Ideas

Lets call an idea unappealing if its initial acceptance probability for all areas is p for some $p \leq \frac{1}{2}$. We prove in this section, that for such ideas, no strategy can boost the acceptance probability for any area above p. We note that exactly the opposite argument can be made when $p \geq \frac{1}{2}$ is the initial acceptance probability of all areas, i.e., any spreading strategy guarantees that every area accepts the idea with probability of at least p.

Theorem 1. *Consider an arbitrary non-adaptive spreading strategy π in the full propagation setting. Assume all initial acceptance probabilities are equal to p. If $p \geq \frac{1}{2}$, then the expected number of adopters is at least np. Furthermore, If $p \leq \frac{1}{2}$, then the expected number of adopters is at most np.*

Proof. We prove this result for the case when $p \leq \frac{1}{2}$. The other case ($p \leq \frac{1}{2}$) follows from symmetry. To avoid confusion, we let $p_0 = p$ and use p_0 instead of the real number p throughout this proof. If we prove that any given area accepts the idea with probability of at most p_0, then from linearity of expectation, we are done. Consider an area v scheduled at time $t + 1$. The probability that the area accepts or rejects the idea is given by

$$Pr(X_v = 1) = p_0(1 - Pr(I_t \geq c_v) - Pr(I_t \leq -c_v)) + Pr(I_t \geq c_v),$$
$$Pr(X_v = -1) = (1 - p_0)(1 - Pr(I_t \geq c_v) - Pr(I_t \leq -c_v)) + Pr(I_t \leq -c_v).$$

Since $Pr(X_v = 1) + Pr(X_v = -1) = 1$, if we prove that $\frac{Pr(X_v = 1)}{Pr(X_v = -1)} \leq \frac{p_0}{1-p_0}$, then we have $Pr(X_v = 1) \leq p_0$. We have

$$\frac{Pr(X_v = 1)}{Pr(X_v = -1)} = \frac{p_0(1 - Pr(I_t \geq c_v) - Pr(I_t \leq -c_v)) + Pr(I_t \geq c_v)}{(1 - p_0)(1 - Pr(I_t \geq c_v) - Pr(I_t \leq -c_v)) + Pr(I_t \leq -c_v)}.$$

We have:

$$\frac{p_0(1 - Pr(I_t \geq c_v) - Pr(I_t \leq -c_v))}{(1 - p_0)(1 - Pr(I_t \geq c_v) - Pr(I_t \leq -c_v))} = \frac{p_0}{1 - p_0}.$$

We know that for any $a, b, c, d, e \in \mathbb{R}_{>0}$, if $\frac{a}{b} \leq e$ and $\frac{c}{d} \leq e$ then

$$\frac{a + c}{b + d} \leq e. \tag{1}$$

Therefore, if we prove that $\frac{Pr(I_t \geq c_v)}{Pr(I_t \leq -c_v)} \leq \frac{p_0}{1-p_0}$, we are done. Thus, we can prove this theorem by proving that $\frac{Pr(I_k \geq x)}{Pr(I_k \leq -x)} \leq \frac{p_0}{1-p_0}$ for all $x \in \{1 \ldots k\}, k \in \{1 \ldots n\}$. We prove this by induction on number of areas. If there is just one area, then that area decides to accept with probability p_0 (as all initial acceptance probabilities are equal to p_0). Assume if the number of areas is less than or equal to n, then $\frac{Pr(I_k \geq x)}{Pr(I_k \leq -x)} \leq \frac{p_0}{1-p_0}$ for all $x \in \{1 \ldots k\}, k \in \{1 \ldots n\}$. We prove the statement when there are $n + 1$ areas. Let $par(n, x) : \mathbb{N} \times \mathbb{N} \mapsto \{0, 1\}$ be a function which is 0 if n and x have the same parity, 1 otherwise. Let v be the area scheduled at time $n + 1$. Let $\nu = par(n, x)$. We now consider the following three cases.

Case 1: $1 \leq x \leq n - 2$. The event $I_{n+1} \geq x + 1$ is the union of the following two disjoint events:

1. $I_n \geq x + 2$, and whatever the nth area decides, I_{n+1} is at least $x + 1$.
2. $I_n = x + \nu$ and $n + 1^{th}$ area decides to accept.

Similarly, the event $I_{n+1} \leq -x - 1$ is the union of the event $I_n \leq -x - 2$ and the event — $I_n = -x - \nu$ and the $n + 1^{th}$ area rejects the idea. We note that we require the par function because only one of the events $I_n = x$ and $I_n = x + 1$ can occur w.p.p. depending on parities of n and x. Thus

$$Pr(I_{n+1} \geq x + 1) = Pr(I_n \geq x + 2) + Pr(X_v = 1 | I_n = x + \nu)Pr(I_n = x + \nu),$$
$$Pr(I_{n+1} \leq -x - 1) = Pr(I_n \leq -x - 2)$$
$$+ Pr(X_v = -1 | I_n = -x - \nu)Pr(I_n = -x - \nu).$$

Now, if $x + \nu \geq c_v$, then $Pr(X_v = 1 | I_n = x + \nu) = Pr(X_v = -1 | I_n = -x - \nu) = 1$, otherwise $Pr(X_v = 1 | I_n = x + \nu) = p_0 < 1 - p_0 = Pr(X_v = -1 | I_n = -x - \nu)$. Therefore, $Pr(X_v = 1 | I_n = x + \nu) \leq Pr(X_v = -1 | I_n = -x - \nu)$. Let $\beta = Pr(X_v = -1 | I_n = -x - \nu)$. Using the above, we have

$$Pr(I_{n+1} \geq x+1) \leq Pr(I_n \geq x+2) + \beta Pr(I_n = x+\nu),$$
$$Pr(I_{n+1} \leq -x-1) = Pr(I_n \leq -x-2) + \beta Pr(I_n = -x-\nu).$$

From above, we have

$$f(\beta) = \frac{Pr(I_n \geq x+2) + \beta Pr(I_n = x+\nu)}{Pr(I_n \leq -x-2) + \beta Pr(I_n = -x-\nu)} \geq \frac{Pr(I_{n+1} \geq x+1)}{Pr(I_{n+1} \geq -x-1)}. \quad (2)$$

The function $f(\beta)$ is either increasing or decreasing and hence has extrema at end points of its range. The maxima is $\leq \max\{\frac{Pr(I_n \geq x+2)}{Pr(I_n \leq -x-2)}, \frac{Pr(I_n \geq x+2) + Pr(I_n = x+\nu)}{Pr(I_n \leq -x-2) + Pr(I_n = -x-\nu)}\}$ because $\beta \in [0,1]$. Now $Pr(I_n \geq x+2) + Pr(I_n = x+1) + Pr(I_n = x) = Pr(I_n \geq x)$ and $Pr(I_n \leq -x-2) + Pr(I_n = -x-\nu) = Pr(I_n \leq -x)$. Thus $f \leq \max\{\frac{Pr(I_n \geq x+2)}{Pr(I_n \leq -x-2)}, \frac{Pr(I_n \geq x)}{Pr(I_n \leq -x)}\} \leq \frac{p_0}{1-p_0}$ (from induction hypothesis). From above and (2), $\frac{Pr(I_{n+1} \geq x+1)}{Pr(I_{n+1} \leq -x-1)} \leq \frac{p_0}{1-p_0}$.

Case 2: $x = 0$. If n is odd then $Pr(I_{n+1} \geq 1) = Pr(I_{n+1} \geq 2)$ and $Pr(I_{n+1} \leq -1) = Pr(I_{n+1} \leq -2)$ and this case is the same as $x = 1$ and hence considered above. Thus, assume that n is even. Thus

$$Pr(I_{n+1} \geq 1) = Pr(I_n \geq 2) + Pr(X_v = 1|I_n = 0)Pr(I_n = 0), \quad (3)$$
$$Pr(I_{n+1} \leq -1) = Pr(I_n \leq -2) + Pr(X_v = -1|I_n = 0)Pr(I_n = 0). \quad (4)$$

Since, if $I_n = 0$, then areas decide based on the initial acceptance probability. We have $Pr(X_v = 1|I_n = 0) = p_0$ and $Pr(X_v = -1|I_n = 0) = 1 - p_0$. Using this fact, by dividing (3) and (4), we have

$$\frac{Pr(I_{n+1} \geq 1)}{Pr(I_{n+1} \leq -1)} \leq \frac{Pr(I_n \geq 2) + p_0 Pr(I_n = 0)}{Pr(I_n \leq -2) + (1-p_0)Pr(I_n = 0)}.$$

From induction hypothesis, $\frac{Pr(I_n \geq 2)}{Pr(I_n \leq -2)} \leq \frac{p_0}{1-p_0}$. Thus, we conclude $\frac{Pr(I_{n+1} \geq 1)}{Pr(I_{n+1} \leq -1)} \leq \frac{p_0}{1-p_0}$ based on (1).

Case 3: $x \in \{n-1, n\}$. In this case $Pr(I_n \geq x+2) = 0$, since the number of adopters can never be more than the number of total areas. Also, I_{n+1} cannot be equal to n because n and $n+1$ don't have the same parity. Therefore, $Pr(I_{n+1} \geq n) = Pr(I_{n+1} \geq n+1)$ and $Pr(I_{n+1} \leq -n) = Pr(I_{n+1} \leq -n-1)$. Thus, it is enough to analyze the case $x = n$. We have

$$Pr(I_{n+1} \geq n+1) = Pr(X_v = 1|I_n = n)Pr(I_n = n),$$
$$Pr(I_{n+1} \leq n+1) = Pr(X_v = -1|I_n = -n)Pr(I_n = -n).$$

Since either both decisions are made based on thresholds with probability 1 or both are made based on initial probabilities and initial acceptance probability is less than the initial rejection probability, We know that $Pr(X_v = 1|I_n = n) \leq Pr(X_v = -1|I_n = -n)$. Therefore $\frac{Pr(I_{n+1} \geq n+1)}{Pr(I_{n+1} \leq n+1)} \leq \frac{Pr(I_n = n)}{Pr(I_n = -n)}$. Now, since $Pr(I_n = n) = Pr(I_n \geq n)$ and $Pr(I_n = -n) = Pr(I_n \leq -n)$, from induction hypothesis, we have $\frac{Pr(I_{n+1} \geq n+1)}{Pr(I_{n+1} \leq n+1)} \leq \frac{p_0}{1-p_0}$ and we are done.

4 Non-adaptive Marketing Strategy with Random Thresholds

We consider the problem of designing a non-adaptive spreading strategy when the thresholds are drawn independently from the same but unknown distribution. We show the best spreading strategy is to schedule areas in a non-increasing order of initial acceptance probabilities. We prove the optimality of the algorithm using a coupling argument. First we state the following lemma which will be useful in proving Theorem 3.

Lemma 1. *Let π and π' be two spreading strategies. If $\exists k$ such that $\pi(i) = \pi'(i)$, $\forall i \geq k$ and $Pr(I_k \geq x; \pi) \geq Pr(I_k \geq x; \pi')$, $\forall x \in \mathbb{Z}$, then $E(I_n; \pi) \geq E(I_n; \pi')$.*

Theorem 3. *Assume that the planner's prior knowledge about all values of c_i's is the same, i.e., all c_i's are drawn independently from the same but unknown distribution. Let initial acceptance probabilities be arbitrary numbers. Then, the best non-adaptive spreading strategy is to order all areas in non-increasing order of their initial acceptance probabilities.*

Proof. Let π' be a spreading strategy where areas are scheduled in an order that is not non-increasing. Thus, there exists k such that $p_{\pi'(k)} < p_{\pi'(k+1)}$. We prove that if a new spreading strategy π is created by exchanging position of areas $\pi'(k)$ and $\pi'(k + 1)$, then the expected number of people who accept the idea cannot decrease. It means the best spreading strategy is non-increasing in the initial acceptance probabilites.

To prove the theorem, we will prove that $Pr(I_{k+1} \geq x; \pi) \geq Pr(I_{k+1} \geq x; \pi')$ and the result then follows from Lemma 1. Since, the two spreading strategies are identical till time $k - 1$ and therefore the random variable I_{k-1} has identical distribution under both the strategies, we can prove the above by proving that $Pr(I_{k+1} \geq I_{k-1} + y | I_{k-1}; \pi) \geq Pr(I_{k+1} \geq I_{k-1} + y | I_{k-1}; \pi')$ for all $y \in \mathbb{Z}$. We note that the only feasible values for y are in $\{-2, 0, 2\}$. Hence, if $y > 2$ then both sides of the above inequality are equal to 1 and the inequality holds. Similarly, if $y <= -2$ both sides of the inequality are equal to 1 and the inequality holds. Thus, we only need to analyze the values $y = 0$ and $y = 2$.

Now we define some notation to help with rest of the proof. Let $u = \pi'(k + 1)$, $v = \pi'(k)$, and $q_i = 1 - p_i$. It means $p_v < p_u$. Let $\chi(i, j)$ be the event where i and j are indicators of decision of areas scheduled at time k and $k + 1$ respectively, e.g., $\chi(1, 1)$ means that areas scheduled at time k and $k + 1$ accepted the idea, whereas $\chi(1, -1)$ implies that area scheduled at time k accepted the idea, while the area scheduled at time $k + 1$ rejected the idea. Let $B(y)$ be the event $I_{k+1} \geq I_{k-1} + y | I_{k-1} = z$ for some arbitrary $z \in \mathbb{Z}$. We consider the cases $I_{k-1} > 0$, $I_{k-1} < 0$ and $I_{k-1} = 0$ separately.

Case 1: $I_{k-1} = z, z > 0$. We have, $B(0) = \chi(1, 1) \cup \chi(1, -1) \cup \chi(-1, 1)$ which is equal to the complement of $\chi(-1, -1)$. Since we assume $z > 0$, the thresholds $-c_u$ and $-c_v$ cannot be hit. Thus, $\chi(-1, -1)$ occurs only when both areas decide to reject the idea based on their respective initial acceptance probabilities. Thus, from chain rule of probability, it is the product of following four terms:

1. $Pr(z < c_u)$, i.e, the threshold rule does not apply and u decides based on initial acceptance probabilities.
2. u rejects the idea based on initial probability of rejection, q_u.

3. $Pr(z - 1 < c_v)$. Given u rejected the idea, $D(v)$, the decision variable for v becomes $z - 1$ and the threshold rule does not apply and v decides based on initial acceptance probabilities.
4. v rejects the idea based on initial probability of rejection, q_v.

Therefore, $Pr(\chi(-1, -1)) = Pr(z < c_u)q_u Pr(z - 1 < c_v)q_v$. Thus, $Pr(B(0); \pi) = 1 - Pr(z < c_u)q_u Pr(z - 1 < c_v)q_v$. Since, c_u and c_v are i.i.d random variables, we can write any probability of form $Pr(z \gtrless c_u)$ or $Pr(z \gtrless c_v)$ as $Pr(z \gtrless x)$, where x is an independent random variable with the same distribution as c_u and c_v. Thus

$$Pr(B(0); \pi) = 1 - Pr(z < x)q_u Pr(z - 1 < x)q_v. \tag{5}$$

Now, $Pr(\chi(1, 1)) = Pr(X_u = 1 | I_{k-1} = z)Pr(X_v = 1 | I_k = z + 1)$. Event $X_u = 1$ is the union of following two non-overlapping events:

1. $z \geq c_u$; u accepts the idea because of the threshold rule.
2. $z < c_u$ and u accepts the idea based on initial acceptance probability, p_u.

Thus, $Pr(X_u = 1 | I_{k-1} = z) = Pr(z \geq c_u) + Pr(z < c_u)p_u$. Similarly, $Pr(X_v = 1 | I_k = z + 1) = Pr(z + 1 \geq c_v) + Pr(z + 1 < c_v)p_v$. Therefore

$$Pr(B(2); \pi) = (Pr(z \geq x) + Pr(z < x)p_u) \\ \times (Pr(z + 1 \geq x) + Pr(z + 1 < x)p_v). \tag{6}$$

where we have replaced c_u and c_v by x because they are i.i.d. random variables. We can obtain corresponding probabilities for process π' by exchanging p_u and p_v. Thus, $Pr(B(0); \pi) = Pr(B(0); \pi') = 1 - Pr(z < x)q_u Pr(z - 1 < x)q_v$. We can write $Pr(B(2); \pi')$ as follows.

$$Pr(B(2); \pi') = (Pr(z \geq x) + Pr(z < x)p_v) \\ \times (Pr(z + 1 \geq x) + Pr(z + 1 < x)p_u). \tag{7}$$

On the other hand $Pr(z < x) \geq Pr(z + 1 < x)$ and $Pr(z + 1 \geq x) \geq Pr(z \geq x)$. Comparing (6) and (7) along with these facts that $p_v < p_u$ and $Pr(z < x)Pr(z + 1 \geq x) \geq Pr(z \geq x)Pr(z + 1 < x)$, we get $Pr(B(2); \pi) \geq Pr(B(2); \pi')$.

Case 2: $I_{k-1} = -z, z > 0$. A similar analysis can be applied to get the result.

Case 3: $I_{k-1} = 0$. We have

$$Pr(B(2); \pi) = p_u(Pr(x > 1)p_v + Pr(x = 1)), \tag{8}$$
$$Pr(B(0); \pi) = p_u + q_u Pr(x > 1)p_v, \tag{9}$$
$$Pr(B(2); \pi') = p_v(Pr(x > 1)p_u + Pr(x = 1)), \tag{10}$$
$$Pr(B(0); \pi') = p_v + q_v Pr(x > 1)p_u. \tag{11}$$

By comparing (8) with (10) and (9) with (11), we see that $Pr(B(2); \pi) \geq Pr(B(2); \pi')$ and $Pr(B(0); \pi) \geq Pr(B(0); \pi')$ respectively. Thus, $Pr(I_{k+1} \geq I_{k-1} + x | I_{k-1}; \pi) \geq Pr(I_{k+1} \geq I_{k-1} + x | I_{k-1}; \pi'), \forall x \in \mathbb{Z}$.

Acknowledgments. Authors would like to thank Jon Kleinberg for his useful comments about the motivation of our problem.

References

[1] Kempe, D., Kleinberg, J., Tardos, É: Maximizing the spread of influence through a social network. In: KDD, pp. 137–146 (2003)

[2] Kempe, D., Kleinberg, J., Tardos, É.: Influential nodes in a diffusion model for social networks. In: Caires, L., Italiano, G.F., Monteiro, L., Palamidessi, C., Yung, M. (eds.) ICALP 2005. LNCS, vol. 3580, pp. 1127–1138. Springer, Heidelberg (2005)

[3] Mossel, E., Roch, S.: On the submodularity of influence in social networks. In: STOC, pp. 128–134 (2007)

[4] Anari, N., Ehsani, S., Ghodsi, M., Haghpanah, N., Immorlica, N., Mahini, H., Mirrokni, V.S.: Equilibrium pricing with positive externalities (Extended abstract). In: Saberi, A. (ed.) WINE 2010. LNCS, vol. 6484, pp. 424–431. Springer, Heidelberg (2010)

[5] Akhlaghpour, H., Ghodsi, M., Haghpanah, N., Mirrokni, V.S., Mahini, H., Nikzad, A.: Optimal iterative pricing over social networks (Extended abstract). In: Saberi, A. (ed.) WINE 2010. LNCS, vol. 6484, pp. 415–423. Springer, Heidelberg (2010)

[6] Hartline, J., Mirrokni, V.S., Sundararajan, M.: Optimal marketing strategies over social networks. In: WWW, pp. 189–198 (2008)

[7] Arthur, W.B.: Competing technologies, increasing returns, and lock-in by historical events. The Economic Journal 99(394), 116–131 (1989)

[8] Chierichetti, F., Kleinberg, J., Panconesi, A.: How to schedule a cascade in an arbitrary graph. In: EC, pp. 355–368 (2012)

[9] Banerjee, A.V.: A simple model of herd behavior. The Quarterly Journal of Economics 107(3), 797–817 (1992)

[10] Bikhchandani, S., Hirshleifer, D., Welch, I.: A theory of fads, fashion, custom, and cultural change in informational cascades. Journal of Political Economy 100(5), 992–1026 (1992)

[11] Granovetter, M.: Threshold models of collective behavior. American Journal of Sociology 83(6), 1420–1443 (1978)

[12] Domingos, P., Richardson, M.: Mining the network value of customers. In: KDD, pp. 57–66 (2001)

[13] Chen, W., Wang, Y., Yang, S.: Efficient influence maximization in social networks. In: KDD, pp. 199–208 (2009)

[14] Arthur, D., Motwani, R., Sharma, A., Xu, Y.: Pricing strategies for viral marketing on social networks. In: Leonardi, S. (ed.) WINE 2009. LNCS, vol. 5929, pp. 101–112. Springer, Heidelberg (2009)

[15] Goyal, S., Kearns, M.: Competitive contagion in networks. In: STOC, pp. 759–774 (2012)

[16] Chen, W., Collins, A., Cummings, R., Ke, T., Liu, Z., Rincon, D., Sun, X., Wang, Y., Wei, W., Yuan, Y.: Influence maximization in social networks when negative opinions may emerge and propagate. In: ICDM, pp. 379–390 (2011)

[17] Li, Y., Chen, W., Wang, Y., Zhang, Z.L.: Influence diffusion dynamics and influence maximization in social networks with friend and foe relationships. In: WSDM, pp. 657–666 (2013)

Designing Budget-Balanced Best-Response Mechanisms for Network Coordination Games

Bruno Escoffier[1], Diodato Ferraioli[1], Laurent Gourvès[2], and Stefano Moretti[2]

[1] PSL*, Université Paris Dauphine, Lamsade and CNRS UMR 7243
{bruno.escoffier,diodato.ferraioli}@dauphine.fr
[2] CNRS UMR 7243 and PSL*, Université Paris Dauphine, Lamsade
{laurent.gourves,stefano.moretti}@dauphine.fr

Abstract. Network coordination games (NCGs) have recently received a lot of attention since they model several kinds of interaction problems in social networks. However, the performance of these games at equilibrium may be very bad. This motivates the adoption of mechanisms for inducing a socially optimal state. Many settings are naturally dynamical and thus we believe it is worth to consider the design of incentive compatible best-response mechanisms (Nisan, Schapira, Valiant, Zohar, 2011) for NCGs. Specifically, we would like to assign to players special fees in order to induce the optimum profile of an NCG. Moreover, we would like the mechanism to be budget-balanced, i.e., implementable with no cost.

We show that a budget-balanced and incentive compatible best-response mechanism for inducing the optimal profile of a two-strategy NCG always exists. Moreover, for such a mechanism, we investigate other properties inspired by envy-freeness, collusion-resistance and fairness.

1 Introduction

A Motivating Example: French Academics Pools. Let us introduce the following example, drawn from a (simplified version of a) real case occurring in France. The Paris academic system is constituted by a myriad of institutions, including 15 universities and dozens of engineering and business schools. For several (strategic, scientific, maybe political) reasons, former administration proposed to group some of these institutions into large and geographically coherent pools. Rather quickly, several proposals came out, and let us focus on two important projects of pools in Paris region: PSL (Paris-Sciences-Lettres), located inside Paris, and UPSa (University Paris-Saclay), located in the suburbs.

This setting can be easily modeled as follows. There is a set $N = \{1, 2, \ldots, n\}$ of institutions and each institution may decide to join PSL or UPSa. For an institute i, there is a personal cost to join one particular pool that does not depend on the institutions already in the pool (in particular, they arise from the necessity of changing location, since the pools shall be geographically coherent). Also, for some pairs of institutions i and j, there is an interest of being in the same pool (for instance, more scientific cooperation or common teaching programs).

B. Vöcking (Ed.): SAGT 2013, LNCS 8146, pp. 207–218, 2013.
© Springer-Verlag Berlin Heidelberg 2013

The introduction of these pools is motivated by the necessity to improve the welfare of the Paris university system through an improved classification in international ranking systems and a more accurate distribution of public funds. Nevertheless, the composition of the two pools was not fixed, neither the universities have been forced to join in a specific pool (for example, by setting a penalty whenever an university join in the undesired pool). As a matter of fact some institutes decided over time to join in one of the pools. Often, this decision is made after a long bargaining with the administration and between institutions.

Then, it is natural to ask if and how the administration can implement the desired outcome in such an inherently sequential environment.

Network Coordination Games. In order to address the above question, we consider the class of *network games*. These games are used to represent systems consisting of a network of interconnected components. These components, or *agents*, interact only with their neighbors and the relationship between them can be modeled as simple two-player games. Examples of network games can be found in Economics [1], in Biology [2], in Physics [3], and in Computer Science [4].

Among network games, particular interest has been given to *Network Coordination Games (NCGs)* (see Section 3 for a formal definition). In these games, the agents prefer to coordinate with their neighbors rather than conflicting with them. NCGs have been adopted, for example, in the study of the ferromagnetic Ising model [3], for modeling the diffusion of innovations [5] or the formation of opinions [6]. The academic pools' example also can be modeled by an NCG.

A specific feature of many network games is given by the twofold nature of cost functions: in addition to costs arising from the relationship with neighbors on the network, adopting a strategy may also have a cost that does not depend on what other players are doing. For example, in the Ising model this cost is given by external magnetic fields; in the formation of academic pools, by the change of location; in opinion games, it instead embeds the personal belief of players. Thus, these costs model the personal "feeling" of agents or their interaction with the environment. They are then fixed costs attached to specific strategies.

Unluckily, the analysis of NCGs shows that the self-interested behavior of agents may worsen the performance of the system. Indeed, the *Price of Anarchy* (PoA) [7] for NCGs can be unbounded [6]. Moreover, the optimum is not necessarily a Nash equilibrium as shown in the following example. Consider two players, both with strategy set $\{0,1\}$, involved in the (network) coordination game specified by the following cost matrix:

	0	1
0	$0, \varepsilon$	$1 - \varepsilon, 1 - \varepsilon$
1	$1 - \varepsilon, 1 - \varepsilon$	$\varepsilon, 0$

Assume moreover that the row player incurs a preference cost of $b \in \{0,1\}$ for playing strategy b while the column player's preference cost for playing strategy b is $1 - b$. Thus, for $0 < \varepsilon < \frac{1}{3}$, the two configurations of minimal total cost are $(0,0)$ and $(1,1)$ but the unique Nash equilibrium of the game is $(0,1)$.

However, we will show (see Theorem 2) that if each player has exactly two strategies, then there is a polynomial centralized algorithm to find the profile that optimizes the performance of these systems. Nevertheless, it is often impossible to force agents to play according to such optimal solution because the optimum may not be an equilibrium. And even if the optimum is an equilibrium, the intrinsic dynamics of these settings (see, for example, the case of Paris academic pools) can push the system into a sub-optimal outcome. The occurrence of these events, large PoA, efficient centralized optimization and inability to force central solutions, suggests to design mechanisms able to influence the players' behavior towards the desired direction. Several and different mechanisms of this kind have been proposed, including *taxes* [8,9], *Stackelberg strategies* [10], *mechanisms via creditability* [11] and *coordination mechanisms* [12].

Best-Response Mechanisms. In this work we focus on *incentive compatible best-response mechanisms*. This is a class of indirect mechanisms introduced in [13] in which agents repeatedly play a base game and at each time step they are pre-scribed to choose the best-response to the strategies currently selected by other agents. This class of mechanisms takes advantage of the dynamical nature of many systems (as, for example, the intrinsic dynamics underlying the formation of Paris academic pools) to induce the desired outcome.

In [13] it is shown that for a specific class of games, namely *NBR-solvable games with clear outcomes* (roughly speaking, these are games for which a Nash equilibrium can be computed by iterated elimination of "useless" strategies; see below for a detailed definition), players have no incentive to deviate from this prescribed behavior and the mechanism converges to an equilibrium. Thus, for inducing the optimum of an NCG, it is sufficient to modify the players' cost functions so that the NCG becomes an NBR-solvable game with clear outcomes.

However, several constraints should be satisfied. First, the new cost functions should not worsen the performance of the desired profile. Indeed, it does not make sense to induce the profile that minimizes the social cost and then ask to players more than this cost. Second, any cost function should include the players' preference costs. Indeed, as suggested above, these often are fixed costs depending only on personal or environmental features and they cannot be influenced by neither other players nor any external authority (for example, in the case of academic pools, there is no way of avoiding the change of location). Finally we would like to model an authority that does not assign penalties or taxes for undesired strategies, but influences only the costs faced by the agents when they adopt the desired strategy. This is motivated by the example of Paris academic pools, where the "penalty" approach has been largely avoided.

Thus, we assume that the mechanism may assign to players playing the desired strategy special *fees* (possibly negative) in place of the costs arising from their relationships. Note we will assume these fees depend only on the strategy of the agent with which we are bargaining, and not on the entire strategy profile. For example, if in the setting of the example above, we would like to induce the optimal profile $(0,0)$, then the mechanism may offer to the column player a fee $-\delta$, with $\delta > \varepsilon$, whenever she plays strategy 0. That is, the players face this new game (the payoffs below do not include preference costs):

	0	1
0	$0, -\delta$	$1 - \varepsilon, 1 - \varepsilon$
1	$1 - \varepsilon, -\delta$	$\varepsilon, 0$

It is easy to see that, by considering the preference costs described in the example above, $(0,0)$ is the unique Nash equilibrium, its cost is lower than in the original game and, moreover, the new game is solvable by iterated elimination of dominated strategies (a subclass of NBR-solvable games): playing 0 is a dominant strategy for the row player, and then (given that the row player plays 0) playing 0 is a dominant strategy for the column player.

Implementing this mechanism has a cost $\delta + \varepsilon$. Indeed, it should be necessary not only to pay δ to the column player, but also to pay the communication costs of ε in her place. In this work we focus on *budget-balanced* mechanisms, i.e. on mechanisms that can be implemented with no cost. This means that whenever inducing a player to play the target strategy has a cost, it should be possible to recover this cost from other players (see below for a detailed definition). For example, in the above setting, the mechanism may offer to the row player a fee of $\delta + \varepsilon$, with $\varepsilon < \delta < 1$, for playing 0; then, after she payed this fee, it may offer to the column player a fee of $-\delta$ for playing 0. The resulting game is

	0	1
0	$\delta + \varepsilon, -\delta$	$\delta + \varepsilon, 1 - \varepsilon$
1	$1 - \varepsilon, -\delta$	$\varepsilon, 0$

As above the payoffs in the matrix do not include preference costs, and by considering the ones described in the example above, $(0,0)$ is the unique Nash equilibrium and the game is solvable by iterated elimination of dominated strategies. However, now the mechanism can be implemented with no cost.[1]

Our Contribution. The focus of this work is on designing budget-balanced incentive compatible best-response mechanisms for NCGs through the assignment of special fees to players in case they play the desired strategy. We consider NCGs in which each player has two strategies. As stated above, the optimal profile can be efficiently computed for these games. Then, we show that it is always possible to design a budget-balanced best-response mechanism for inducing this optimal profile (Theorem 3). Thus, an authority can always find policies that allow to exploit the dynamical nature of a system to induce the desired outcome.

Given this positive result, we investigate other desired properties that the mechanism may satisfy. The first property we consider, named *order-freeness*, deals with the possibility that several budget-balanced best-response mechanisms can be adopted for inducing the optimal profile. If these mechanisms treat one player in different ways, then this player will care about which mechanism is actually implemented. A best-response mechanism is order-free if no player prefers that another mechanism is adopted. We will show that an order-free

---

[1] The definition of budget-balanced mechanisms reminds the 0-implementation definition in [11]. However, they are different: indeed, we can enforce payments from agents, but our new cost functions should satisfy several other requirements.

best-response mechanism always exists (Theorem 4). However, we also show that verifying if a mechanism satisfies this property is hard (Theorem 5).

The second property on which we focus is the *collusion-resistance*. We would like that no coalition has any incentive to leave the induced profile even if side-payments are allowed. We show that a budget-balanced best-response mechanism that is collusion-resistant always exists (Theorem 6). We also characterize collusion-resistance in terms of solutions of a suitable cooperative game.

Finally, we look for *fair* mechanisms. We give different definitions of fairness, that focus on different aspects of the problem. The first definition is based on the cooperative game characterization previously discussed. In this area, the Shapley value is uniformly recognized as a standard for fairness and, for this reason, we look at the extent to which we can adopt this concept in our setting. It turns out that the Shapley value corresponds to a best-response mechanism only for special NCGs (Theorem 7). For the other fairness definitions, we specify an ideal fair mechanism which we would like to be as close to as possible. Unfortunately, it will turn out that for each of the ideal mechanisms we considered, it is hard to compute the closest budget-balanced best-response mechanism (Theorem 8).

Notation. Throughout this paper we use bold symbols for vectors, i.e. $\mathbf{x} = (x_1, x_2, \ldots, x_n)$. Given a vector $\mathbf{x}$ of size n and a set $A \subseteq [n]$, we will write $\mathbf{x}_A$ for the vector $(x_i)_{i \in A}$ and $\mathbf{x}_{-A}$ for the vector $(x_i)_{i \notin A}$. We also use $\mathbf{x}_{-i}$ as a shorthand for $\mathbf{x}_{-\{i\}}$. Finally, for any $b \in \{0, 1\}$ we write $\bar{b}$ for $1 - b$.

Due to space limits, proofs are omitted (see the full version of the paper [14]).

2   The Model

Network Games. In an n-player *network game* $\mathcal{G}$, each player $i \in [n]$ has strategy set S_i and is represented by a vertex of a *graph* $G = (V, E)$. For each player i and each strategy $s_i \in S_i$, we denote by $p_i(s_i) \geq 0$ the *preference cost* for i for the strategy s_i. To each edge $e = (i, j) \in E$ is linked a two player game $\mathcal{G}_e$ in which the set of strategies of the endpoints are exactly S_i and S_j. We denote by $c_i^e(s_i, s_j) \geq 0$ the *communication cost* for player i in the game $\mathcal{G}_e$, with $e = (i, j) \in E$, when i chooses strategy $s_i \in S_i$ and j selects strategy $s_j \in S_j$. Given a strategy profile $\mathbf{x} \in S_1 \times \cdots \times S_n$, the total cost of player i in the network game $\mathcal{G}$ is given by $c_i(\mathbf{x}) = p_i(x_i) + \sum_{e=(i,j)} c_i^e(x_i, x_j)$.

Best-Response Mechanisms. Nisan et al. [13] studied a class of indirect mechanisms, termed *repeated-response mechanisms*: starting from a given profile, at each time step t, some player i is selected and she announces a strategy $s_i^t \in S_i$. A *best-response mechanism* is a repeated-response mechanism in which the pre-scribed behavior for each player is to always choose a best-response to the strategies currently played by other players. A repeated-response mechanism *converges* to the target profile $\mathbf{x}$ if the players eventually play according to this strategy profile, i.e. there is $t^\star > 0$ such that $\mathbf{x}^t = \mathbf{x}$ for any $t \geq t^\star$, where $\mathbf{x}^t$ is the strategy profile after that the players selected at time step t announced their strategies. For a player i enrolled in a repeated-response mechanism, let us denote by $z_i(\mathbf{x}^t)$ the cost of i in the profile $\mathbf{x}^t$. If the mechanism converges to $\mathbf{x}$, we say that the *total cost* of i is $Z_i = z_i(\mathbf{x})$, otherwise we say the total cost of i is $Z_i = \limsup_{t \to \infty} z_i(\mathbf{x}^t)$. A repeated-response mechanism is *incentive compatible* if any player behaving as prescribed by the mechanism achieves a total cost that is at most as high as the total cost achieved by deviating from the prescribed behavior, given that the other players play as prescribed. Specifically, a best-response mechanism is incentive compatible if always choosing the best-response is a pure Nash equilibrium of the n-player game whose player's strategies are all possible behaviors and player's costs are their total costs. It is useful to remark that in [13] (and in this work), players are only interested in minimizing the cost in the profile at which the mechanism converges and they do not care about the cost in the intermediate profiles generated by the dynamics.

NBR-solvable Games. A strategy $s_i \in S_i$ is a *never-best-response (NBR) strategy* if for every $\mathbf{x}_{-i}$ there exists $s_i' \neq s_i$ such that $z_i(s_i, \mathbf{x}_{-i}) > z_i(s_i', \mathbf{x}_{-i})$. A game is *NBR-solvable* if there exists a sequence of eliminations of NBR strategies that results in a single strategy profile[2]. Let us denote by $\mathbf{y}^\star$ the unique profile to which the sequence of eliminations of an NBR-solvable game converges. For an NBR-solvable game $\mathcal{G}$ with a sequence of eliminations of length k, we denote by $\mathcal{G}_j$, with $j = 0, 1, \ldots, k$, the sub-game resulting from the first j eliminations in the sequence (observe that $\mathcal{G}_0 = \mathcal{G}$ and $\mathcal{G}_k$ allows only the profile $\mathbf{y}^\star$). Then an NBR-solvable game is said to have *clear outcomes* if for each player i, there exists a (player-specific) sequence of elimination of NBR strategies such that at the first step τ_i in which a strategy of S_i is eliminated, player i "likes" $\mathbf{y}^\star$ at least as much as any other profile in $\mathcal{G}_{\tau_i}$, i.e. $z_i(\mathbf{y}^\star) \leq z_i(\mathbf{y})$ for every $\mathbf{y} \in \mathcal{G}_{\tau_i}$. Next theorem shows how this class of games relates to best-response mechanisms.

Theorem 1 ([13]). *A best-response mechanism for an NBR-solvable game with clear outcomes is incentive compatible and converges to $\mathbf{y}^\star$, regardless of the starting state and the order in which players are selected.*

We highlight that the above theorem holds for any possible schedule of players such that each player appear in the schedule infinitely many times.

Our Mechanism. We aim to design an incentive compatible best-response mechanism for inducing a profile $\mathbf{y}$. It is then sufficient to compute a new cost function c_i' for any player i so that the new game $\mathcal{G}'$ is NBR-solvable with clear outcomes and $\mathbf{y}$ is the Nash equilibrium.

In our setting we force the new cost functions to satisfy several constraints. First, the social cost of the induced profile should not increase, i.e. $\sum_i c_i'(\mathbf{y}) \leq \sum_i c_i(\mathbf{y})$, where $\mathbf{y}$ is the target profile. Moreover, we defined preference costs as being fixed, independent of the network and attached to strategies. That is, whenever a player adopts a strategy, he must pay the corresponding preference cost. Hence, the functions c_i' must include these costs. Finally, we would like that the new cost function affects only profiles in which players play the desired strategy, i.e. $c_i'(\mathbf{x}) = c_i(\mathbf{x})$ if $x_i \neq y_i$ and $\mathbf{y}$ is the target profile. This constraint is motivated by the impossibility in many setting (for practical or ethical reasons)

---

[2] In case of ties, we assume that an opportune tie-breaking rule is adopted.

to force the actions of autonomous agents by introducing large penalties or taxes on the undesired strategies. Thus, we have the following mechanism:

Definition 1 (The mechanism). *Let j_i be the i-th player appearing in the elimination sequence defining $\mathcal{G}'$ and let i_j be the first time the player j appears in this elimination sequence. Consider an arbitrary schedule of players. For $i = 1, \ldots, \ell$, where ℓ is the length of the elimination sequence of $\mathcal{G}'$, let t_i be such that j_i is scheduled at t_i, and before t_i, players $j_1, \ldots, j_{i-1}$ are scheduled at least once in this order. Then, at each time step t the mechanism assigns to player j selected at time t a cost $c_j(\mathbf{x}^{t-1})$ if $t < t_{i_j}$ and a cost $c_j'(\mathbf{x}^{t-1})$ otherwise.*

Then, since $\mathcal{G}'$ is NBR-solvable with clear outcomes, by Theorem 1 at each time step players prefer to play according to the best-response and this dynamics converges to the target profile $\mathbf{y}$. Note that the time necessary for convergence depends on how players are scheduled for announcing their strategies. However, if we only consider schedules in which no player is "adversarially" delayed for arbitrarily long time, then the mechanism converges to the target profile quickly.

In this work we focus on a special way of building the new cost functions. Formally, we consider a *vector of fees* $\boldsymbol{\gamma} = (\gamma(i))_{i \in [n]}$ and say that $\boldsymbol{\gamma}$, a network game $\mathcal{G}$ and a strategy profile $\mathbf{y}$ *define* a game $\mathcal{G}_{\mathbf{y}, \boldsymbol{\gamma}}$ in which the cost function c_i' of player i is as follows: if $x_i = y_i$, then $c_i'(\mathbf{x}) = c_i'(y_i) = \gamma(i) + p_i(y_i)$, otherwise $c_i'(\mathbf{x}) = c_i(\mathbf{x})$. Then, our aim becomes to compute a vector of fees $\boldsymbol{\gamma}$ such that the game $\mathcal{G}_{\mathbf{y}, \boldsymbol{\gamma}}$ defined by $\mathcal{G}, \mathbf{y}$ and $\boldsymbol{\gamma}$ is NBR-solvable with clear outcomes.

Budget-Balancedness. Fix a network game $\mathcal{G}$, a profile $\mathbf{y}$, and a vector of fees $\boldsymbol{\gamma}$. The *cost* of $\boldsymbol{\gamma}$ is defined as $\sum_i (c_i(\mathbf{y}) - c_i'(\mathbf{y})) = C(\mathbf{y}) - \sum_i \gamma(i)$, where $C(\mathbf{y}) = \sum_i \sum_{e=(i,j)} c_i^e(y_i, y_j)$. Note that the cost of $\boldsymbol{\gamma}$ is always non-negative. We are interested in designing incentive compatible best-response mechanisms that are *budget-balanced*, i.e. mechanisms for which $\boldsymbol{\gamma}$ has *null cost*.

However, a mechanism can run out of money by paying a fee to some players before that an equivalent amount of money has been collected from other players. We would like to avoid the occurrence even of this "temporary deficit". That is, we would like that the mechanism can schedule the players so that it is able to pay a player i with money collected from players scheduled before i. Specifically, let an *order of players* be a permutation π on the set of players. Then, we say that a vector of fees $\boldsymbol{\gamma}$ is *deficit-free according to* π if $\gamma(i) + \sum_{j \in N_\pi(i)} \gamma(j) \geq 0$ for each player i, where $N_\pi(i)$ is the set of players that are scheduled before i in π, i.e. $N_\pi(i) = \{j : \pi(j) < \pi(i)\}$. A vector of fees $\boldsymbol{\gamma}$ is *deficit-free* if there exists at least one order π of players such that $\boldsymbol{\gamma}$ is deficit-free according to π.

Deficit-freeness implies that the mechanism has a non-negative cost even in case of non-convergence. That is, the cost of the fees will always be non-negative no matter what the players will play and not only in the target profile.

3 Two-Strategy Network Coordination Games

Here, we consider the following subclass of network games, named *two-strategy NCGs*, where each player has only two strategies, 0 and 1, and for every edge $e = (i, j)$ the game $\mathcal{G}_e$ is given by the following cost matrix:

	0	1
0	$\alpha_i^e(0), \alpha_j^e(0)$	$\beta_i^e(0), \beta_j^e(1)$
1	$\beta_i^e(1), \beta_j^e(0)$	$\alpha_i^e(1), \alpha_j^e(1)$

where the costs for agreements are smaller or equal to the costs for disagreements (even if these costs may vary depending on which strategy a player adopts), i.e. $\beta_k^e(b) \geq \alpha_k^e(b') \geq 0$ for all $b, b' \in \{0, 1\}$ and $k = i, j$.

We say that a strategy profile $\mathbf{x}^\star$ is *optimal* (or the optimum) for a network game if it minimizes $\sum_i c_i(\mathbf{x})$ over all profiles $\mathbf{x}$. Next theorem shows that the optimal profile can be easily computed in two-strategy NCGs. Indeed, it turns out to be equivalent to compute the minimum cut of a suitably built graph.

Theorem 2. *The optimum of a two-strategy NCG is polynomially computable.*

A Best-Response Mechanism. Note that, given a two-strategy network game $\mathcal{G}$ and a profile $\mathbf{x}$, the game $\mathcal{G}_{\mathbf{x}, \boldsymbol{\gamma}}$ defined by $\mathcal{G}, \mathbf{x}$ and a vector of fees $\boldsymbol{\gamma}$ is NBR-solvable with clear outcomes if and only if, in $\mathcal{G}_{\mathbf{x}, \boldsymbol{\gamma}}$ it is possible to schedule players so that, for each player i, choosing x_i is the unique best-response (under opportune tie-breaking) given that players scheduled before i are playing according to $\mathbf{x}$. Formally, given an order π, we say that a vector of fees $\boldsymbol{\gamma}$ is *inducing* a profile $\mathbf{x}$ *according to* π if $\gamma(i) \leq \Gamma_\pi(i) := \sum_{\substack{e=(i,j) \\ j \in N_\pi(i)}} c_i^e(\overline{x}_i, x_j) +$ $\sum_{\substack{e=(i,j) \\ j \notin N_\pi(i)}} \min_{b \in \{0,1\}} c_i^e(\overline{x}_i, b) + p_i(\overline{x}_i) - p_i(x_i)$ for each i. Indeed if i plays $\overline{x}_i$, in addition to her preference cost $p_i(\overline{x}_i)$, for an edge $e = (i, j)$ she will pay for sure $c_i^e(\overline{x}_i, x_j)$ if $j \in N_\pi(i)$ and at least $\min_{b \in \{0,1\}} c_i^e(\overline{x}_i, b)$ if $j \notin N_\pi(i)$. We say that a vector of fees $\boldsymbol{\gamma}$ is *inducing* the profile $\mathbf{x}$ if there exists at least one order π such that $\boldsymbol{\gamma}$ is inducing $\mathbf{x}$ according to π. Then, the game $\mathcal{G}_{\mathbf{x}, \boldsymbol{\gamma}}$ is NBR-solvable with clear outcomes if and only if $\boldsymbol{\gamma}$ is inducing $\mathbf{x}$. We also say an order π is *feasible* for $\mathbf{x}$ if there exists a vector of fees of null cost that is both inducing $\mathbf{x}$ and deficit-free according to π. Similarly, a vector of fees $\boldsymbol{\gamma}$ is *valid* for $\mathbf{x}$ *according to* π if it makes π feasible for $\mathbf{x}$. Finally, a vector of fees $\boldsymbol{\gamma}$ is *valid* for $\mathbf{x}$ if there is an order π such that $\boldsymbol{\gamma}$ is valid for π according to $\mathbf{x}$.

Given these preliminary definitions, we show that a budget-balanced incentive compatible best-response mechanism for the optimal profile of a two-strategy NCG can be easily designed. Given such a game and its optimal profile $\mathbf{x}^\star$, we define the *base value* $B(i)$ of a player i as $B(i) = p_i(\overline{x}_i^\star) - p_i(x_i^\star) + \sum_{e=(i,j)} \alpha_i^e(\overline{x}_i^\star)$. That is, $B(i)$ is the maximum fee that may be assigned to i by a vector of fees inducing $\mathbf{x}^\star$ according to π, where π schedules i as first. Note that the maximum fee assignable to i does not decrease if i is scheduled later, and thus $\Gamma_\pi(i) \geq B(i)$ for any order π. Then, since $\beta_j^e(x_j^\star) - \alpha_j^e(x_j^\star) \geq 0$ for any j, we obtain this lemma.

Lemma 1. *For a two-strategy NCG with optimal profile* $\mathbf{x}^\star$, $\sum_i B(i) \geq C(\mathbf{x}^\star)$.

Consider now the following vector of fees $\boldsymbol{\gamma}$: for each i such that $B(i) \geq 0$, we set $\gamma(i) \leq B(i)$ such that $\sum_{i: B(i) \geq 0} \gamma(i) = C(\mathbf{x}^\star) - \sum_{i: B(i) < 0} B(i)$ (this is possible by Lemma 1). For each remaining player i, we set $\gamma(i) = B(i)$. It follows that $\boldsymbol{\gamma}$ defines a budget-balanced incentive compatible best-response mechanism.

Theorem 3. *For a two-strategy NCG, a vector of fees valid for $\mathbf{x}^\star$ always exists and can be computed in polynomial time.*

Order-Freeness. Suppose two different orders, π and π', are both feasible for $\mathbf{x}^\star$ and let γ and γ' be the corresponding valid vectors of fees. Then, γ and γ' can charge very different costs to the same player. It may be then questionable that a player accepts the proposed fee if there exists another feasible order for which she improves her utility for sure. Thus, we would like to have a vector of fees γ for which no player "envies" another schedule or feels disadvantaged by the mechanism. Formally, let Π denote the set of all orders feasible for $\mathbf{x}^\star$. Since player i can be charged up to $\Gamma_\pi(i)$ in the ordering π (for having γ inducing $\mathbf{x}^\star$ according to π), choosing a fee $\gamma(i)$ which is at most $\Gamma_\pi(i)$ for *any* $\pi \in \Pi$ would ensure "envy-freeness" for player i. Then by defining $\boldsymbol{\Gamma}^\star = (\Gamma^\star(i))_i$ the vector that sets $\Gamma^\star(i) = \min_{\pi \in \Pi} \Gamma_\pi(i)$, we say that a vector of fees γ is *order-free valid* for $\mathbf{x}^\star$ if it is valid for $\mathbf{x}^\star$ and $\gamma(i) \leq \Gamma^\star(i)$ for each player i.

Order-free validity can always be achieved, as stated in the following theorem.

Theorem 4. *For a two-strategy NCG $\mathcal{G}$, a vector of fees that is order-free valid for $\mathbf{x}^\star$ always exists and it can be computed in polynomial time.*

Beyond the existence of one order-free valid vector of fees, we might be interested in determining whether a given vector of fees is order-free valid or not (for instance if we are focused on *fair* vectors of fees). Unfortunately, computing the values $(\Gamma^\star(i))_i$ is NP-hard, as stated by the following theorem. The hardness follows from a reduction of the feedback arc set problem.

Theorem 5. *Given a two-strategy NCG and a player i, establishing whether $\Gamma^\star(i) = B(i)$ is (strongly) NP-complete, even if $\alpha_j^e(0) = \alpha_j^e(1) = \alpha_j^e$ and $\beta_j^e(0) = \beta_j^e(1) = \beta_j^e$ for each player j and each edge e.*

Collusion-Resistance. A best-response mechanism does not prevent some players to collude and jointly move away from $\mathbf{x}^\star$. We wonder about the possibility of inducing a profile $\mathbf{x}^\star$ such that no coalition of players deviates from $\mathbf{x}^\star$, even if side payments are allowed. Specifically, we say that a vector of fees γ is *collusion-resistant* if for every subset $L \subset [n]$ of players and any joint strategy $\mathbf{y}_L \neq \mathbf{x}_L^\star$, we have $\sum_{i \in L} (p_i(x_i^\star) + \gamma(i)) \leq \sum_{i \in L} (p_i(y_i) + h_i(y_i))$, where $h_i(y_i) = \gamma(i)$ if $y_i = x_i^\star$, and $h_i(y_i) = \sum_{\substack{e=(i,j) \\ j \in L}} c_i^e(y_i, y_j) + \sum_{\substack{e=(i,j) \\ j \notin L}} c_i^e(y_i, x_j^\star)$, otherwise. Roughly speaking, we would like to choose a vector of fees γ such that the cumulative cost of a coalition L is minimized by playing according to $\mathbf{x}^\star$ (and thus each of the members gets the corresponding fee from γ), given that the other players are playing according to this optimal profile. We then have the following theorem.

Theorem 6. *For a two-strategy NCG a collusion-resistant vector of fees order-free valid for $\mathbf{x}^\star$ always exists and can be computed in polynomial time.*

We note that for any two-strategy game (not necessarily NCGs) we can give an alternative definition of collusion-resistance based on cooperative cost games.

We consider the cost game v that sets $v([n]) = C(\mathbf{x}^\star)$ and for each $L \subset [n]$,

$$v(L) = \sum_{i \in L} \left(p_i(\overline{x}_i^\star) - p_i(x_i^\star) + \sum_{\substack{e=(i,j) \\ j \in L}} c_i^e(\overline{x}_i^\star, \overline{x}_j^\star) + \sum_{\substack{e=(i,j) \\ j \notin L}} c_i^e(\overline{x}_i^\star, x_j^\star) \right). \quad (1)$$

Then, we have the following characterization.

Lemma 2. *Given a two-strategy game and its optimal profile* $\mathbf{x}^\star$, *a vector of fees* $\boldsymbol{\gamma}$ *is collusion-resistant and has null cost if and only if* $\boldsymbol{\gamma}$ *is in the core of the cooperative cost game* v *defined by* (1).

Shapley-Fairness. In cooperative game theory the *Shapley value* [15] is acknowledged as a standard for fairness in settings where transfer of utility is allowed.

Thus, above characterization motivates us to adopt such a fairness standard also in our setting. Specifically, given a two-strategy NCG and its optimal profile $\mathbf{x}^\star$, we say a vector of fees $\boldsymbol{\gamma}$ is *Shapley-fair* if $\boldsymbol{\gamma}$ is the Shapley value for the cooperative game defined by (1). Next lemma shows that a Shapley-fair vector of fees can be easily computed by suitably decomposing the characteristic function.

Lemma 3. *The Shapley value of* (1) *is computable in polynomial time.*

Using the same decomposition, we can show that the Shapley value corresponds to a collusion-resistant vector of fees of null cost.

Lemma 4. *For a two-strategy NCG, the Shapley value of* (1) *is in the core.*

Unfortunately, it is easy to show that the Shapley value may not be a valid vector of fees for the optimum of an NCG, even if for each edge $e = (i, j)$ we have $\alpha_i^e(0) = \alpha_i^e(1) = \alpha_j^e(0) = \alpha_j^e(1) = \alpha^e$ and $\beta_i^e(0) = \beta_i^e(1) = \beta_j^e(0) = \beta_j^e(1) = \beta^e$. However, for a subclass of NCGs, we can achieve also Shapley-fairness.

Theorem 7. *Consider a two-strategy NCG such that for each edge* $e = (i, j)$ *we have* $\alpha_i^e(0) = \alpha_i^e(1) = \alpha_j^e(0) = \alpha_j^e(1) = \alpha^e$ *and* $\beta_i^e(0) = \beta_i^e(1) = \beta_j^e(0) = \beta_j^e(1) = \beta^e$. *Suppose, moreover, that in the optimal profile* $x^\star$, *all players adopt the same strategy. Then, a collusion-resistant and Shapley-fair vector of fees order-free valid for* $\mathbf{x}^\star$ *always exists and can be easily computed.*

Equal-Fairness, Cost-Fairness and Profile-Fairness. In this section we suggest other fairness measures that may be adopted. Let us start by describing some *ideal* vectors of fees. A first example of fair vector of fees $\gamma_\star^f$ is the one in which each player receives the same fee, i.e. $\gamma_\star^f(i) = \frac{1}{n} C(\mathbf{x}^\star)$ for each player i. As an alternative example, we would like to have that fees are such that each player has the same total cost in the game defined by these fees. That is, we would like to consider the fair vector of fees $\gamma_\star^c$ such that $\gamma_\star^c(i) = \frac{1}{n} \left(C(\mathbf{x}^\star) + \sum_j p_j(x_j^\star) \right) - p_i(x_i^\star)$ for each player i. Yet another example is given by the vector of fees $\gamma_\star^p$ in which each player pays exactly her contribution to $C(\mathbf{x}^\star)$, i.e. $\gamma_\star^p(i) = \sum_{e=(i,j)} c_i^e(x_i^\star, x_j^\star)$. Obviously, there are instances in which these ideal vectors of fees are not valid for $\mathbf{x}^\star$. Then, it is a good trade-off to ask for a vector of fees valid for $\mathbf{x}^\star$ closest to the ideal one. Formally, we say that the vector of fees $\boldsymbol{\gamma}$ is *equal-fair* if it minimizes over all vectors of fees valid for $\mathbf{x}^\star$ the distance $d(\boldsymbol{\gamma}, \boldsymbol{\gamma}_\star^f) = \sum_i \left| \gamma(i) - \gamma_\star^f(i) \right|$. Similarly, $\boldsymbol{\gamma}$ is *cost-fair* if it minimizes $d(\boldsymbol{\gamma}, \boldsymbol{\gamma}_\star^c) = \sum_i |\gamma(i) - \gamma_\star^c(i)|$ and *profile-fair* if it minimizes $d(\boldsymbol{\gamma}, \boldsymbol{\gamma}_\star^p) = \sum_i |\gamma(i) - \gamma_\star^p(i)|$.

Finding these fair vectors of fees is hard, as stated by the next theorem. The hardness follows from reducing the perfectly balanced ordering problem [16].

Theorem 8. *Given a two-strategy NCG and a constant $K > 0$, it is (strongly) NP-complete to decide whether there exists a vector of fees $\boldsymbol{\gamma}$ of null cost inducing the optimal profile $\mathbf{x}^\star$ whose distance $d(\boldsymbol{\gamma}, \boldsymbol{\gamma}_\star^f) \leq K$, even if $\alpha_i^e(0) = \alpha_i^e(1) = \alpha_i^e$ and $\beta_i^e(0) = \beta_i^e(1) = \beta_i^e$ for each player i and each edge e. The claim holds also with $\boldsymbol{\gamma}_\star^c$ or $\boldsymbol{\gamma}_\star^p$ in place of $\boldsymbol{\gamma}_\star^f$.*

Given this hardness result, it makes sense to ask for a fair vector of fees in a subset of the valid ones. For example, it would be interesting to compute the fairest vector of fees among the ones such that $\gamma(i) \leq B(i)$ for each player i. Indeed, Theorem 6 shows that these vectors of fees $\boldsymbol{\gamma}$ can be at the same time collusion-resistant and order-free valid for the optimal profile of an NCG. The next theorem shows that the fairest vector of fees is easy to compute in this case.

Theorem 9. *The equal-fairest, the cost-fairest and the profile-fairest vectors of fees, among any vector $\boldsymbol{\gamma}$ such that $\gamma(i) \leq B(i)$ for each player i, can be computed in polynomial time.*

4   Conclusions and Open Problems

The focus of this work is the design of mechanisms through which an authority may induce optimal states by influencing the bargaining among components of a network. Our mechanisms adopt a "dynamical approach", that is, tends to modify the game so that natural dynamics can converge to the target state. We believe this approach may help in designing mechanisms also for other settings.

The main results of this work refer to the special case of inducing the optimal profile of two-strategy NCGs. In light of our motivating example, it turns out that on-going bargaining with universities can be exploited for inducing optimal pools. Moreover, the administration can avoid envy among academic institutions, coalitions' deviations and, in some cases, the assignment of unfair fees.

As a future direction, we are interested to extend the approach introduced in this paper for NCGs to other classes of games. However, it is easy to see that the optimal profile may be difficult to compute if we allow anti-coordination between players or if we allow more than two strategies for players. These observations motivates to investigate the possibility of inducing non-optimal profiles (for example, approximatively optimal profiles): again, it is not hard to see that this possibility is computationally ruled out (we refer the reader to the full version of this paper [14]). However, it would be still interesting to characterize when a mechanism can be designed for non-coordination games or non-optimal profiles.

In this work we assumed that mechanism knows both preference and communication costs of agents. Even if the amount of data available nowadays, make this assumption less restrictive, it would be, in our opinion, interesting to extend our approach to settings in which the mechanism has only limited informations about agents' costs.

Acknowledgements. This work is supported by French National Agency (ANR), project COCA ANR-09-JCJC-0066-01. The authors would like to thank Jérôme Monnot for the useful discussions about the model and the anonymous referees for their interesting suggestions.

References

1. Jackson, M.O.: Social and Economic Networks. Princeton University Press, Princeton (2008)
2. Nowak, M.A.: Evolutionary Dynamics: Exploring the Equations of Life. Harvard University Press (2006)
3. Galam, S., Walliser, B.: Ising model versus normal form game. Physica A: Statistical Mechanics and its Applications 389(3), 481–489 (2010)
4. Kempe, D., Kleinberg, J.M., Tardos, É.: Influential nodes in a diffusion model for social networks. In: Caires, L., Italiano, G.F., Monteiro, L., Palamidessi, C., Yung, M. (eds.) ICALP 2005. LNCS, vol. 3580, pp. 1127–1138. Springer, Heidelberg (2005)
5. Young, H.P.: The diffusion of innovations in social networks. Economics Working Paper Archive number 437, Johns Hopkins University, Department of Economics (2000)
6. Ferraioli, D., Goldberg, P.W., Ventre, C.: Decentralized dynamics for finite opinion games. In: Serna, M. (ed.) SAGT 2012. LNCS, vol. 7615, pp. 144–155. Springer, Heidelberg (2012)
7. Koutsoupias, E., Papadimitriou, C.: Worst-case equilibria. Computer Science Review 3(2), 65–69 (2009)
8. Beckmann, M., McGuire, C., Winsten, C.: Studies in the economics of transportation. Yale University Press (1959)
9. Cole, R., Dodis, Y., Roughgarden, T.: Pricing network edges for heterogeneous selfish users. In: Proceedings of the Thirty-Fifth Annual ACM Symposium on Theory of Computing, STOC 2003, pp. 521–530. ACM (2003)
10. Korilis, Y., Lazar, A., Orda, A.: Achieving network optima using stackelberg routing strategies. IEEE/ACM Transactions on Networking 5(1), 161–173 (1997)
11. Monderer, D., Tennenholtz, M.: k-implementation. In: ACM Conference on Electronic Commerce, pp. 19–28 (2003)
12. Christodoulou, G., Koutsoupias, E., Nanavati, A.: Coordination mechanisms. Theor. Comput. Sci. 410(36), 3327–3336 (2009)
13. Nisan, N., Schapira, M., Valiant, G., Zohar, A.: Best-response mechanisms. In: Innovations in Computer Science (ICS), pp. 155–165 (2011)
14. Escoffier, B., Ferraioli, D., Gourvès, L., Moretti, S.: Designing frugal best-response mechanisms for social network coordination games,
http://www.lamsade.dauphine.fr/~ferraioli/papers/paris_paper.pdf
15. Shapley, L.S.: A value for n-person games. Contributions to the Theory of Games 2, 307–317 (1953)
16. Biedl, T.C., Chan, T.M., Ganjali, Y., Hajiaghayi, M.T., Wood, D.R.: Balanced vertex-orderings of graphs. Discrete Applied Mathematics 148(1), 27–48 (2005)

Inefficiency of Games with Social Context[*]

Aris Anagnostopoulos[1], Luca Becchetti[1], Bart de Keijzer[2], and Guido Schäfer[2]

[1] Sapienza University of Rome, Italy
{aris,becchetti}@dis.uniroma1.it
[2] CWI and VU University Amsterdam, The Netherlands
{b.de.keijzer,g.schaefer}@cwi.nl

Abstract. The study of other-regarding player behavior such as altruism and spite in games has recently received quite some attention in the algorithmic game theory literature. Already for very simple models, it has been shown that altruistic behavior can actually be harmful for society in the sense that the price of anarchy may *increase* as the players become more altruistic. In this paper, we study the severity of this phenomenon for more realistic settings in which there is a complex underlying social structure, causing the players to direct their altruistic and spiteful behavior in a refined player-specific sense (depending, for example, on friendships that exist among the players). Our findings show that the increase in the price of anarchy is modest for congestion games and minsum scheduling games, whereas it is drastic for generalized second price auctions.

1 Introduction

Many practical situations involve a group of strategic decision makers who attempt to achieve their own self-interested goals. It is well known that strategic decision making may result in outcomes that are suboptimal for the society as a whole. The need to gain an accurate understanding of the extent of suboptimality caused by selfish behavior has led to the study of the *inefficiency of equilibria* in algorithmic game theory. In this context, a common inefficiency measure is the *price of anarchy* [21], which relates the worst-case cost of a Nash equilibrium to the one of an optimal outcome.

More recently, quite some attention has been given to more general settings in which the players do not necessarily behave entirely selfishly, but may alternatively exhibit *spiteful* or *altruistic* behavior; see, for instance, [2,4,5,7–9,14,17–19]. Studying such alternative behaviors in games is motivated by the observation that altruism and spite are phenomena that frequently occur in real life (see, for example, [15]). Consequently, it is desirable to incorporate such alternative behavior in game-theoretical analyses.

Previous work on the price of anarchy for spiteful and altruistic games has focused on simple models of spite and altruism, where a spite/altruism level α_i is associated to each player i denoting the extent to which his perceived cost is

---

[*] This research was partially supported by the EU FET project MULTIPLEX 317532.

B. Vöcking (Ed.): SAGT 2013, LNCS 8146, pp. 219–230, 2013.
© Springer-Verlag Berlin Heidelberg 2013

influenced by any nonspecific other player. Already for these simple models it has been observed in a series of papers [5,7,8] that altruistic behavior can actually be harmful in the sense that the price of anarchy may *increase* as players become more altruistic. This observation served as a starting point for the investigations conducted in this paper. The main question that we address here is: How severe can this effect be if one considers more refined models of altruism that capture complex social relationships between the players?

Our Contributions. In the present paper, we study a more general player-specific model of spite and altruism. Our model can be viewed as extending a given strategic game by imposing a *social-network structure* on top of the players, which specifies for each pair of players (i, j) an altruism/spite level α_{ij} signifying how much player i cares about player j; these relations are not necessarily symmetric. This allows us to model more realistically settings in which the behavior of the players depends on a complex underlying social structure, expressing friendships and animosities among the players. Our altruistic games fall into the framework of *social context games* proposed in [1].

For this general model of games with altruism and spite, we are interested in studying the price of anarchy. The *smoothness framework*, originally introduced by Roughgarden [22], has become a standard method for proving upper bounds on the price of anarchy. Basically, this framework shows that such bounds can be derived by establishing a certain smoothness condition. An additional strength of this approach is that the smoothness condition allows to derive upper bounds on the price of anarchy for various solution concepts, ranging from pure Nash equilibria to coarse correlated equilibria; the latter being naturally related to outcomes resulting from natural learning algorithms (see, for example, Young [26]). Here, we extend the smoothness framework such that it can be used conveniently it our setting.

Using this extension, we prove upper bounds on the price of anarchy for altruistic versions of three classes of well-studied games: congestion games, minsum scheduling games, and generalized second price auctions. We show that for unrestricted altruism levels the price of anarchy is unbounded. In particular, this happens if there is a player i who does not care about himself or he cares more about some friend than about himself, that is, $\alpha_{ij} > \alpha_{ii}$. We therefore derive our upper bounds under the mild assumption that each player cares at least a little about himself and he cares about any other player at most as much as he cares about himself; we refer to this as *restricted altruistic social context*. Under this assumption, we derive the following upper bounds on the coarse price of anarchy:

- A bound of 7 for altruistic linear congestion games, and a bound of $\varphi^3 \approx 4.236$ for the special case of singleton linear congestion games, where $\varphi = (1 + \sqrt{5})/2$ denotes the golden ratio.
- A bound of $4 + 2\sqrt{3} \approx 7.4641$ and $12 + 8\sqrt{2} \approx 23.3137$ for altruistic minsum machine scheduling games for related and unrelated machines, respectively.
- A bound of $2(n + 1)$ for altruistic generalized second price auctions, where n is the number of players.

Our results therefore show that for congestion games and minsum scheduling games the price of anarchy cannot drastically increase. Specifically, it remains constant, independently of how complex the underlying altruistic social structure is. On the other hand, for generalized second price auctions the price of anarchy may degrade quite drastically: we prove an upper bound of $O(n)$, as opposed to a small constant which is known for the purely selfish setting [6].

We derive most upper bounds using a simple proof template: we decompose the altruistic game into a selfish part and an altruistic part and prove smoothness for each part separately. We can import known smoothness results for the selfish part and only need to focus on the altruistic part.

Our upper bound proof for singleton congestion games uses a novel proof approach: We use a refined *amortized* argument by distributing some additional "budget" unevenly among the facilities. We believe that this approach might be of independent interest.

Related Work. There are several papers that propose models of altruism and spite [4, 5, 7–9, 14, 17–19]. All these models are special cases of the one studied here. Among these articles, the inefficiency of equilibria in the presence of altruistic/spiteful behavior was studied for various games in [5, 7–9, 14]. After its introduction in [22], the smoothness framework has been adapted in various directions [23–25], including an extension to a particular model of altruism in [8], which constitutes a special case of the altruistic games considered here.

Biló et al. [2] also studied social context congestion games, in the case where the perceived cost of a player is the minimum, maximum, or sum of the immediate cost of his neighbors. [2] establishes, among other results, an upper bound of $17/3$ on the pure price of anarchy of linear congestion games for a special case of the setting we study here.

Related but different from our setting, is the concept of *graphical congestion games* [3, 16]. Here the cost and the strategy set of a player depends only on a subset of the players.

2   Preliminaries

Altruistic Extensions of Games. We study the effect of altruistic behavior in strategic games. To model the complex altruistic relationships between the players, we equip the underlying game with an *altruistic social context*. More precisely, let $\Gamma = (N, \{\Sigma_i\}_{i \in N}, \{c_i\}_{i \in N})$ be a strategic game (termed *base game*), where $N = \{1, \ldots, n\}$ is the set of players, Σ_i is the strategy set of player i, and $c_i : \Sigma \to \mathbb{R}$ is the direct cost function of player i that maps every strategy profile $s \in \Sigma = \Sigma_1 \times \cdots \times \Sigma_n$ to a real value. Unless stated otherwise, we assume that Γ is a cost minimization game, that is, every player i wants to minimize his individual cost function c_i. Further, we assume that an *altruistic social context* is given by an $n \times n$ matrix $\alpha \in \mathbb{R}^{n \times n}$.

Given a base game Γ and an altruistic social context α, the *α-altruistic extension* of Γ is defined as the strategic game $\Gamma^\alpha = (N, \{\Sigma_i\}_{i \in N}, \{c_i^\alpha\}_{i \in N})$, where for all $i \in N$ and $s \in \Sigma$, the *perceived cost* $c_i^\alpha(s)$ of player i is given by

$$c_i^\alpha(s) = \sum_{j=1}^{n} \alpha_{ij} c_j(s). \tag{1}$$

Thus, the perceived cost of player i in the α-altruistic extension is the α_{ij}-weighted sum of the individual direct costs of all players in the base game. A positive (negative) α_{ij} value signifies that player i cares positively (negatively) about the direct cost of player j, which can be interpreted as an altruistic (spiteful) attitude of i towards j. Note that α_{ii} specifies how player i cares about himself; we also call α_{ii} the *self-perception level*. For simplicity, we will often refer to the resulting game Γ^α as the α-*altruistic game*, without explicitly mentioning the base game Γ and the altruistic social context α.

The above viewpoint has a natural interpretation in terms of *social networks*: Suppose the players in N are identified with the nodes of a complete directed graph $G = (N, A)$. The weight of an edge $(i, j) \in A$ is equal to α_{ij}, specifying the extent to which player i cares about the cost of player j.

The main focus of this paper is on altruistic behavior. We distinguish between *unrestricted* and *restricted* altruistic social contexts α. In the *unrestricted* case we assume that $\alpha_{ij} \geq 0$ for every $i, j \in N$; in particular, the self-perception level of a player can be zero. In this case, one can prove trivial lower bounds for the price of anarchy, just by setting $\alpha_{ij} = 0$, for all i, j. For this reason we consider also the more interesting restricted case. In the *restricted* case, every player has a positive self-perception level and cares about himself at least as much as about any other player, namely, $\alpha_{ii} > 0$ and $\alpha_{ii} \geq \alpha_{ij} \geq 0$ for every $i, j \in N$, $i \neq j$. In the latter case, we can normalize α without loss of generality such that $\alpha_{ii} = 1$ for every player i.[1]

Coarse Equilibria and the Price of Anarchy. We are interested in the efficiency loss caused by altruistic behavior. Let $C : \Sigma \to \mathbb{R}$ be a *social cost* function that maps strategy profiles to real numbers. Most of the time in this paper, the social cost will refer to the sum of the direct costs of all players, namely, $C(s) = \sum_{i=1}^{n} c_i(s)$. The motivation therefore is that we are interested in the efficiency of the outcome resulting from altruistic behavior, which is modeled through the altered perceived cost functions.

We focus on the inefficiency of *coarse equilibria*, which are defined as follows: Let σ be a probability distribution over Σ. Let σ_{-i} denote the projection of σ onto $\Sigma_{-i} = \Sigma_1 \times \cdots \times \Sigma_{i-1} \times \Sigma_{i+1} \times \cdots \times \Sigma_n$. Then σ is a *coarse equilibrium* of the altruistic game Γ^α if, for every player i and every strategy $s_i^* \in \Sigma_i$, it holds that $\mathbf{E}_{s \sim \sigma}[c_i^\alpha(s)] \leq \mathbf{E}_{s_{-i} \sim \sigma_{-i}}[c_i^\alpha(s_i^*, s_{-i})]$. We use $CE(\Gamma^\alpha)$ to denote the set of coarse equilibria of Γ^α. Coarse equilibria include several other solution concepts, such as correlated equilibria, mixed Nash equilibria, and pure Nash equilibria.

We study the *price of anarchy* [21] of coarse equilibria of altruistic games. For an altruistic game Γ^α, define $POA(\Gamma^\alpha) = \sup_{s \in CE(\Gamma^\alpha)} C(s)/C(s^*)$, where s^* is a strategy profile that minimizes C. The *coarse price of anarchy* of a class of altruistic games $\mathcal{G}$ is defined as $POA(\mathcal{G}) = \sup_{\Gamma^\alpha \in \mathcal{G}} POA(\Gamma^\alpha)$.

---

[1] To see this, note that, by dividing all α_{ij} by $\alpha_{ii} > 0$, the set of equilibria and the social cost of any outcome remain the same.

Because of space restrictions, some material has been omitted from this extended abstract and will appear in a full version of this work.

3   Smoothness and a Proof Template

Smoothness. Roughgarden [22] introduced a general smoothness framework to derive bounds on the coarse price of anarchy. Next we extend this framework to α-altruistic games with arbitrary social cost functions.

Definition 1. *Let Γ^α be an α-altruistic extension of a cost minimization game with $\alpha \in \mathbb{R}^{n \times n}$ and social cost function C. Further, let s^* be a strategy profile that minimizes C. Γ^α is (λ, μ)-smooth if there exists a strategy profile $\bar{s} \in \Sigma$ such that for every strategy profile $s \in \Sigma$ it holds that*

$$\sum_{i=1}^{n} \sum_{j=1}^{n} \alpha_{ij}(c_j(\bar{s}_i, s_{-i}) - c_j(s)) \leq \lambda C(s^*) + (\mu - 1)C(s). \tag{2}$$

The following theorem shows that (λ, μ)-smoothness implies a bound on the coarse price of anarchy.

Theorem 1. *Let Γ^α be an α-altruistic extension of a cost minimization game with $\alpha \in \mathbb{R}^{n \times n}$ and social cost function C. If Γ^α is (λ, μ)-smooth with $\mu < 1$, then the coarse price of anarchy of Γ^α is at most $\lambda/(1 - \mu)$.*

The above smoothness definition allows us to import some additional results from [22] (e.g., on the efficiency of natural learning algorithms). The proof of Theorem 1 and further discussion wll appear in a full version of this work.[2]

Proof Template. Most of our smoothness results are based on the following decomposition idea. Recall that for restricted altruistic social contexts we have $\alpha_{ii} = 1$. Suppose that the underlying base game is known to be (λ_1, μ_1)-smooth (in the purely selfish setting), that is, there is some $\bar{s} \in \Sigma$ such that

$$\sum_{i=1}^{n} c_i(\bar{s}_i, s_{-i}) \leq \lambda_1 C(s^*) + \mu_1 C(s), \tag{3}$$

and that $C(s) \leq \sum_i c_i(s)$. Then, to establish $(\lambda, \mu) = (\lambda_1 + \lambda_2, \mu_1 + \mu_2)$-smoothness for the altruistic game Γ^α, it suffices to prove that for $\bar{s}$ it holds

$$\sum_{i=1}^{n} \sum_{j \neq i} \alpha_{ij}(c_j(\bar{s}_i, s_{-i}) - c_j(s)) \leq \lambda_2 C(s^*) + \mu_2 C(s). \tag{4}$$

---

[2] In the purely selfish setting (i.e., when $\alpha_{ii} = 1$ and $\alpha_{ij} = 0$ for every $i, j \in N$, $i \neq j$) our smoothness definition is slightly more general than the one in [22] where (2) is required to hold for any arbitrary strategy profile s^* and with $\bar{s} = s^*$. Also, in [22] the analogue of Theorem 1 is shown under the additional assumption that C is *sum-bounded*, that is, $C(s) \leq \sum_i c_i(s)$. Here, we get rid of this assumption.

4   Congestion Games

In a congestion game $\Gamma = (N, E, \{d_e\}_{e \in E}, \{\Sigma_i\}_{i \in N})$ we are given a set of players $N = \{1, \ldots, n\}$, a set of *facilities* E with a *delay function* $d_e : \mathbb{N} \to \mathbb{R}$ for every facility $e \in E$, and a strategy set $\Sigma_i \subseteq 2^E$ for every player $i \in N$. For a strategy profile $s \in \Sigma = \Sigma_1 \times \cdots \times \Sigma_n$, define $x_e(s)$ as the number of players using facility $e \in E$, that is, $x_e(s) = |\{i \in N : e \in s_i\}|$. The direct cost of player i is defined as $c_i(s) = \sum_{e \in s_i} d_e(x_e(s))$ and the social cost function is given by $C(s) = \sum_{i=1}^{n} c_i(s)$. In a *linear* congestion game, the delay function of every facility $e \in E$ is of the form $d_e(x) = a_e x + b_e$, where $a_e, b_e \in \mathbb{Q}_{\geq 0}$ are nonnegative rational numbers.

4.1   General Linear Congestion Games

Theorem 2. *Every α-altruistic extension of a linear congestion game with restricted altruistic social context α is $(\frac{7}{3}, \frac{2}{3})$-smooth. Therefore, the coarse price of anarchy is at most 7 for these games.*

We need the following simple lemma for the proof of Theorem 2. Its proof will appear in a full version of this paper.

Lemma 1. *For every two integers $x, y \in \mathbb{N}$, $xy \leq \frac{2}{3}y^2 + \frac{1}{3}x^2$.*

Proof (Theorem 2). Let s be an arbitrary strategy profile and let s^* be a strategy profile that minimizes C. We can assume without loss of generality that $d_e(x) = x$ for all $e \in E$.

The base game is known to be $(\lambda_1, \mu_1) = (\frac{5}{3}, \frac{1}{3})$-smooth for $\bar{s} = s^*$ [10, 11, 22]. Using our proof template, it is sufficient to show that (4) holds with $(\lambda_2, \mu_2) = (\frac{2}{3}, \frac{1}{3})$.

Let x_e and x_e^* refer to $x_e(s)$ and $x_e(s^*)$, respectively. Fix some player $i \in N$ and let $x_e' = x_e(s_i^*, s_{-i})$. Note that $x_e' = x_e + 1$ for $e \in s_i^* \setminus s_i$, $x_e' = x_e - 1$ for $e \in s_i \setminus s_i^*$ and $x_e' = x_e$ otherwise. Using these relations, we obtain

$$\sum_{j \neq i} \alpha_{ij}(c_j(s_i^*, s_{-i}) - c_j(s)) = \sum_{j \neq i} \left(\sum_{e \in s_j \cap (s_i^* \setminus s_i)} \alpha_{ij} - \sum_{e \in s_j \cap (s_i \setminus s_i^*)} \alpha_{ij} \right)$$

$$= \sum_{e \in s_i^* \setminus s_i} \sum_{j \neq i : e \in s_j} \alpha_{ij} - \sum_{e \in s_i \setminus s_i^*} \sum_{j \neq i : e \in s_j} \alpha_{ij}.$$

Summing over all players and exploiting that in the restricted case $0 \leq \alpha_{ij} \leq 1$ for every $i, j \in N$, $i \neq j$, we can bound

$$\sum_{i=1}^{n} \left(\sum_{e \in s_i^* \setminus s_i} \sum_{j \neq i : e \in s_j} \alpha_{ij} - \sum_{e \in s_i \setminus s_i^*} \sum_{j \neq i : e \in s_j} \alpha_{ij} \right) \leq \sum_{i=1}^{n} \sum_{e \in s_i^*} \sum_{j : e \in s_j} 1 = \sum_{e \in E} x_e x_e^*.$$

Using Lemma 1, we conclude that $\sum_{e \in E} x_e x_e^* \leq \frac{2}{3}C(s^*) + \frac{1}{3}C(s)$ as desired. $\square$

4.2   Singleton Congestion Games

We derive a better smoothness result for *singleton congestion games with identical delay functions*, that is, when $\Sigma_i \subseteq E$ for every $i \in N$, so that for each strategy $s \in \Sigma_i$ we have that $|s| = 1$.

Theorem 3. *Every α-altruistic extension of a singleton linear congestion game with identical delay functions on all facilities under restricted altruistic social context α is $(1 + \varphi, 1/\varphi^2$-smooth, where $\varphi = (1 + \sqrt{5})/2$ is the golden ratio. Therefore, the coarse price of anarchy is at most $\varphi^3 \approx 4.236$ for these games.*

To prove this theorem, we use a novel proof approach. In most existing proofs one first massages the smoothness condition to derive an equivalent condition summing over all facilities (instead of players), and then establishes smoothness by reasoning for each facility separately. If we follow this approach here, we again obtain an upper bound of 7. Instead, we use an *amortized* argument here to derive our improved bound.

A careful analysis (details omitted because of paucity of space) can show that the smoothness definition (2) for singleton linear congestion games with $\bar{s} = s^*$ is equivalent to

$$\sum_{i=1}^{n} \sum_{j \neq i} \left(\lambda \left| s_i^* \cap s_j^* \right| + (\mu + \alpha_{ij}) \left| s_i \cap s_j \right| - (1 + \alpha_{ij}) \left| s_i^* \cap s_j \right| \right) + (\lambda + \mu - 1)n \geq 0.$$

(5)

We translate the proof of this inequality to a coloring problem on a suitably defined graph. We construct an *extended social network* as follows: For every player $i \in N$ we introduce two nodes i and i^* representing player i under s and s^*, respectively. We call the former type of nodes *s-nodes* and the latter type of nodes *s^*-nodes*. For every two players $i, j \in N$ with $i \neq j$ we introduce four edges: (i, j) with weight $2\mu + \alpha_{ij} + \alpha_{ji}$, (i^*, j^*) with weight 2λ, (i^*, j) with weight $-(1 + \alpha_{ij})$, and (i, j^*) with weight $-(1 + \alpha_{ji})$. We identify the set of facilities E with a set of m colors, such that $E = [m]$. The colors assigned to i and i^* are s_i and s_i^*, respectively. Call an edge $e = (u, v)$ in the extended network *c-monochromatic* if both u and v have color c. In addition, we distribute a total budget of $(\lambda + \mu - 1)n$ among the $2n$ nodes of the extended network.

With the viewpoint of the previous paragraph, the left-hand side of (5) is equal to the total weight of all *c*-monochromatic edges (summed over all colors c) plus the total budget of all nodes. The idea now is to argue that we can fix λ and μ such that for each color $c \in [m]$ the total weight of all *c*-monochromatic edges plus the respective node budget is at least 0. The crucial insight to derive our improved bound is that the budget is split unevenly among the nodes: we assign a budget of $(\lambda + \mu - 1)$ to every *s*-node and 0 to every *s^*-node.

Fix some color $c \in [m]$ and consider the subgraph of the extended network induced by the nodes having color c. Partition the nodes into the set S_c of *s*-nodes and the set S_c^* of *s^*-nodes. Imagine we draw this subgraph with all nodes in S_c put on the left-hand side and all nodes in S_c^* put on the right-hand side. The edges from S_c to S_c^* are called *crossing edges*. The edges that stay within

S_c or S_c^* are called *internal* edges. Let $x = x_c = |S_c|$ and $y = y_c = |S_c^*|$. Note that the internal edges in S_c constitute a complete graph on x nodes. Similarly, the internal edges in S_c^* constitute a complete graph on y nodes. Note that the crossing edges constitute a $K_{x,y}$ with a few edges missing, namely the pairs (i, i^*) representing the same player i (which are nonexistent by construction). Let $z = z_c$ be the number of such pairs.

In the worst case, $\alpha_{ij} = 0$ for all internal edges and $\alpha_{ij} = 1$ for all crossing edges. The total contribution to the left-hand side of (5) that we can account for color c is then

$$2\mu \cdot \tfrac{1}{2}x(x-1) + 2\lambda \cdot \tfrac{1}{2}y(y-1) - 2 \cdot (xy - z) + (\lambda + \mu - 1) \cdot x$$
$$= \mu x^2 + \lambda y^2 - 2xy + (\lambda - 1)x - \lambda y + 2z. \tag{6}$$

We need the following lemma, whose proof will appear in a full version of this work. It is actually tight, implying that under the smoothness framework we cannot show a better bound. It is a small variation of Lemma 1 in [12].

Lemma 2. *Let $\varphi = \frac{1+\sqrt{5}}{2}$ be the golden ratio. For every two integers $x, y \in \mathbb{N}$, $2xy - \varphi x + \varphi^2 y \le \frac{1}{\varphi^2}x^2 + \varphi^2 y^2$.*

Fix $\lambda = 1 + \varphi$ and $\mu = 1/\varphi^2$. Then (6) is nonnegative by Lemma 2. Summing over all colors $c \in [m]$ proves (5). Given our choices of $\lambda = 1 + \varphi$ and $\mu = 1/\varphi^2$ we obtain a bound on the coarse price of anarchy of $\varphi^3 \approx 4.236$. □

5   Minsum Machine Scheduling

In a scheduling game, we deal with a set of machines $[m]$, and a set of jobs $[n]$ that are to be scheduled on the machines. For each job $i \in [n]$ and machine $k \in [m]$, we are given a *processing time* $p_{i,k} \in \mathbb{R}_{\ge 0}$, which is the time it takes to run job i on machine k.

There are many ways in which a machine may execute the set of jobs it gets assigned. We restrict ourselves here to a popular policy where the jobs on a machine are executed one-by-one, in order of increasing processing time. Ties are broken deterministically, and we write $i \prec_k j$ if $p_{i,k} < p_{j,k}$ or $p_{i,k} = p_{j,k}$ and the tie breaking rule schedules job i before job j on machine k. A *schedule* is a vector $s = (s_1, \ldots, s_n)$, where for $i \in [n]$, s_i is the machine on which job i is to be ran. We define the value $N(i, k, s)$ to be the number of jobs j on machine k under strategy profile s for which it holds that $i \prec_k j$. Given s, the *completion time* of a job i under s is $p_{i,s_i} + \sum_{j:j \prec_{s_i} i, s_j = s_i} p_{j,s_j}$. The jobs take the role of the players: the strategy set of a player is $[m]$, so the strategy profiles are schedules. The cost $c_j(s)$ of a job $j \in [n]$ under strategy profile s is the completion time of j under s.

We define the social cost function for this game to be the sum of the completion times of the jobs. The social cost can be written as
$C(s) = \sum_{k=1}^{m} \sum_{i:s_i=k} (N(i, k, s) + 1)p_{i,k}.$

If the processing times are not restricted, we speak of *unrelated machine scheduling games*. We speak of *related machine scheduling games* if the processing times are defined as follows: For each machine $k \in [m]$, there is a *speed* $t_k \in \mathbb{R}_{>0}$ and for each job $j \in [n]$ there is a *length* $p_j \in \mathbb{R}_{\geq 0}$ such that $p_{i,k} = p_j/t_k$ for all $i \in [n]$, $k \in [m]$.

Next, we prove constant upper bounds on the price of anarchy for restricted altruistic social contexts.

Theorem 4. *Every α-altruistic extension of a machine scheduling game with restricted altruistic social context α is $(2+x, 1/x)$-smooth for related machines and $(2+x, 1/2+1/x)$-smooth for unrelated machines for every $x \in \mathbb{R}_{>0}$. Therefore, the coarse price of anarchy is at most $4 + 2\sqrt{3} \approx 7.4641$ (choosing $x = 1 + \sqrt{3}$) and $12 + 8\sqrt{2} \approx 23.3137$ (choosing $x = 2 + 2\sqrt{2}$) for these games, respectively.*

Proof. We only give the main steps of the proof here. All missing details will appear in a full version of this extended abstract. In [20] it is proved that the base game for the case of related machines is $(2,0)$-smooth, and from [13], it follows that the base game for the case of unrelated machines is $(2, 1/2)$-smooth. Let s^* be an optimal schedule and let s be any schedule. We show that for all $x > 0$

$$\sum_{i=1}^{n} \sum_{j \neq i} \alpha_{ij}(c_j(s_i^*, s_{-i}) - c_j(s)) \leq xC(s^*) + \frac{C(s)}{x}.$$

Let $P_1 = \{(i,j) : s_i^* = s_j, s_i^* \neq s_i, i \prec_{s_i^*} j\}$. Informally, P_1 is the set of pairs of jobs (i,j) such that i's strategy change from s_i to s_i^* makes j become scheduled later. After some derivations, we obtain

$$\sum_{i=1}^{n} \sum_{j \neq i} \alpha_{ij}(c_j(s_i^*, s_{-i}) - c_j(s)) \leq \sum_{(i,j)\in P_1} p_{i,s_i^*}.$$

Turning the last expression into a summation over the machines, and again after a series of calculations, we obtain

$$\sum_{(i,j)\in P_1} p_{i,s_i^*} \leq \sum_{k=1}^{m} \sum_{i:s_i^*=k} (x(N(i,k,s^*)+1)-1)p_{i,k}$$

$$+ \sum_{k=1}^{m} \sum_{\substack{i:s_i^*=k, s_i \neq k, \\ N(i,k,s) > xN(i,k,s^*)+x-1}} \lceil N(i,k,s) - xN(i,k,s^*) - x + 1 \rceil p_{i,k}.$$

Consider a job i and machine k such that it holds that $s_i^* = k$, $s_i \neq k$, and $N(i,k,s) > xN(i,k,s^*)+x-1$. Let $S(i,k)$ be the set of $\lceil N(i,k,s)-xN(i,k,s^*)-x \rceil$ smallest jobs $j \succ_k i$ such that $s_j = k$. Note that $S(i,k)$ is well defined in the sense that this number of jobs exists because $N(i,k,s) > xN(i,k,s^*) + x - 1$ implies $\lceil N(i,k,s) - xN(i,k,s^*) - x \rceil \geq 0$, and because there exist $N(i,k,s) \geq |S(i,k)|$ jobs $j \succ_k i$ with $s_j = k$. Note that for every job $j \in S(i,k)$ it holds that

$N(j, k, s) \geq N(i, k, s) - |S(i, k)| > xN(i, k, s^*) + x - 1$. We use this to upper bound the above and eventually obtain:

$$\sum_{(i,j)\in P_1} p_{i,s_i^*} \leq xC(s^*) + \sum_{k=1}^{m} \sum_{j:s_j=k} \sum_{\substack{i:s_i^*=k, s_i \neq k, i \prec_k j, \\ N(j,k,s) > xN(i,k,s^*)+x-1}} p_{j,k}. \qquad (7)$$

The next step in the derivation is made by observing that for each job j and each machine k such that $s_j = k$, there are at most $\lceil (N(j, k, s) - x + 1)/x \rceil$ jobs $i \prec_k j$ such that $s_i^* = k$, $s_i \neq k$ and $N(j, k, s) > xN(i, k, s^*) + x - 1$. To see this, assume for contradiction that there are *more* than $\lceil (N(j, k, s) - x + 1)/x \rceil$ jobs $i \prec_k j$ such that $s_i^* = k$, $s_i \neq k$ and $N(j, k, s) > xN(i, k, s^*) + x - 1$. Let i be the $(\lceil (N(j, k, s) - x + 1)/x \rceil + 1)$-th largest job for which these three properties hold. Then, there are at least $(\lceil (N(j, k, s) - x + 1)/x \rceil + 1)$ jobs scheduled on machine k that have these properties and that are scheduled after i on machine k under strategy s^*. Therefore, we have that $xN(i, k, s^*) + x - 1 \geq x\lceil (N(j, k, s) - x + 1)/x \rceil + x - 1 \geq N(j, k, s)$, which is a contradiction. Exploiting this observation, we derive that the right-hand side of (7) is at most $xC(s^*) + \frac{C(s)}{x}$, which concludes the proof. $\square$

6 Generalized Second Price Auctions

We study auctions where a set $N = [n]$ of n bidders compete for k slots. Each bidder $i \in N$ has a valuation $v_i \in \mathbb{R}_{\geq 0}$ and specifies a bid $b_i \in \mathbb{R}_{\geq 0}$. Each slot $j \in [k]$ has a *click-through rate* $\gamma_j \in \mathbb{R}_{\geq 0}$. Without loss of generality, we assume that the slots are sorted according to their click-through rates such that $\gamma_1 \geq \cdots \geq \gamma_k$ and that $k = n$.[3]

We consider the *generalized second price auction (GSP)* as the underlying mechanism. Given a bidding profile $b = (b_1, \ldots, b_n)$, GSP orders the bidders by nonincreasing bids and assigns them in this order to the slots. Each bidder pays the next highest bid for his slot. More precisely, let $b_1 \geq \cdots \geq b_n$ be the ordered list of bids. We assume without loss of generality that if $b_i = b_j$ for two bidders $i > j$ then i precedes j in the order. Then bidder i is assigned to slot i and has to pay b_{i+1}, where we define $b_{n+1} = 0$. The utility of player i for bidding profile b is defined as $u_i(b) = \gamma_i(v_i - b_{i+1})$. The *social welfare* for a bidding profile b is defined as $\Pi(b) = \sum_{i=1}^{n} \gamma_i v_i$.

A standard assumption we make in this setting is that bidders do not *overbid* their valuations, that is, $\Sigma_i = [0, v_i]$. This assumption is made for reasons related to individual rationality.

We prove that the coarse price of anarchy of α-altruistic GSP auctions is $O(n)$ if the altruistic social context is restricted. Note that we consider a profit maximization game here. Definition 1, Theorem 1, and our proof template extend naturally to profit maximization games. The details will appear in a full version of this paper. We are able to prove the following theorem.

---

[3] If $k < n$ we can add $n - k$ dummy slots with click-through rate 0; if $k > n$ we can remove the $k - n$ last slots.

Theorem 5. *Every α-altruistic extension of a generalized second price auction with restricted altruistic social context α is $(\frac{1}{2}, n)$-smooth. Therefore, the coarse price of anarchy is at most $2n + 1$ for these games.*

Proof. Let b^* and b be two bidding profiles. By renaming, we assume that for all j, bidder j gets assigned to slot j under bidding profile b.

The base game is known to be $(\lambda_1, \mu_1) = (\frac{1}{2}, 1)$-smooth [23]. That is, for every two bidding profiles b, b^*, it holds that $\sum_{i \in N} u_i(b_i^*, b_{-i}) \geq \frac{1}{2}\Pi(b^*) - \Pi(b)$.

It remains to bound

$$\sum_{i=1}^{n}\sum_{j \neq i} \alpha_{ij}(u_j(b_i^*, b_{-i}) - u_j(b)) \geq \sum_{i=1}^{n}\sum_{j \neq i} \alpha_{ij}(-u_j(b)) \geq \sum_{i=1}^{n}\sum_{j \neq i} \alpha_{ij}(-\gamma_j v_j)$$

$$\geq \sum_{i=1}^{n}\sum_{j \neq i} -\gamma_j v_j \geq -(n-1)\Pi(b).$$

Combining these inequalities proves $(\lambda, \mu) = (\frac{1}{2}, n)$-smoothness.   □

As in the case of congestion games, the analysis is essentially tight. Details will appear in an extended version of this work.

Concluding Remarks. The main focus of this paper was put on deriving upper bounds on the price of anarchy that are *independent* of the underlying social network structure. An interesting open question is whether one can derive refined bounds by exploiting *structural properties* of the underlying social network.

Our model of altruistic games and the smoothness definition introduced in Section 2 allows us to incorporate spiteful player behavior. We leave it as an interesting open direction for future research to pursue such analyses for spiteful behavior.

References

1. Ashlagi, I., Krysta, P., Tennenholtz, M.: Social context games. In: Papadimitriou, C., Zhang, S. (eds.) WINE 2008. LNCS, vol. 5385, pp. 675–683. Springer, Heidelberg (2008)
2. Bilò, V., Celi, A., Flammini, M., Gallotti, V.: Social context congestion games. In: Kosowski, A., Yamashita, M. (eds.) SIROCCO 2011. LNCS, vol. 6796, pp. 282–293. Springer, Heidelberg (2011)
3. Bilò, V., Fanelli, A., Flammini, M., Moscardelli, L.: Graphical congestion games. Algorithmica 61(2), 274–297 (2011)
4. Brandt, F., Sandholm, T., Shoham, Y.: Spiteful bidding in sealed-bid auctions. In: Proc. 20th Intl. Joint Conf. on Artifical Intelligence, pp. 1207–1214 (2007)
5. Buehler, R., et al.: The price of civil society. In: Chen, N., Elkind, E., Koutsoupias, E. (eds.) WINE 2011. LNCS, vol. 7090, pp. 375–382. Springer, Heidelberg (2011)
6. Caragiannis, I., Kaklamanis, C., Kanellopoulos, P., Kyropoulou, M.: On the efficiency of equilibria in generalized second price auctions. In: Proc. 12th Conf. on Electronic Commerce, pp. 81–90 (2011)

7. Caragiannis, I., Kaklamanis, C., Kanellopoulos, P., Kyropoulou, M., Papaioannou, E.: The impact of altruism on the efficiency of atomic congestion games. In: Wirsing, M., Hofmann, M., Rauschmayer, A. (eds.) TGC 2010, LNCS, vol. 6084, pp. 172–188. Springer, Heidelberg (2010)

8. Chen, P.A., de Keijzer, B., Kempe, D., Schäfer, G.: The robust price of anarchy of altruistic games. In: Chen, N., Elkind, E., Koutsoupias, E. (eds.) WINE 2011. LNCS, vol. 7090, pp. 383–390. Springer, Heidelberg (2011)

9. Chen, P.A., Kempe, D.: Altruism, selfishness, and spite in traffic routing. In: Proc. 9th Conf. on Electronic Commerce, pp. 140–149 (2008)

10. Christodoulou, G., Koutsoupias, E.: On the price of anarchy and stability of correlated equilibria of linear congestion games. In: Brodal, G.S., Leonardi, S. (eds.) ESA 2005. LNCS, vol. 3669, pp. 59–70. Springer, Heidelberg (2005)

11. Christodoulou, G., Koutsoupias, E.: The price of anarchy of finite congestion games. In: Proc. 37th Symp. on the Theory of Computing, pp. 67–73 (2005)

12. Christodoulou, G., Mirrokni, V.S., Sidiropoulos, A.: Convergence and approximation in potential games. Theoretical Computer Science 438, 13–27 (2012)

13. Cole, R., Correa, J.R., Gkatzelis, V., Mirrokni, V., Olver, N.: Inner product spaces for minsum coordination mechanisms. In: Proc. 43rd Symp. on the Theory of Computing, pp. 539–548 (2011)

14. Elias, J., Martignon, F., Avrachenkov, K., Neglia, G.: Socially-aware network design games. In: Proc. 29th Conf. on Computer Communications, pp. 41–45 (2010)

15. Fehr, E., Schmidt, K.M.: The Economics of Fairness, Reciprocity and Altruism: Experimental Evidence and New Theories. Handbook on the Economics of Giving, Reciprocity and Altruism, vol. 1, ch. 8, pp. 615–691 (2006)

16. Fotakis, D., Gkatzelis, V., Kaporis, A.C., Spirakis, P.G.: The impact of social ignorance on weighted congestion games. In: Leonardi, S. (ed.) WINE 2009. LNCS, vol. 5929, pp. 316–327. Springer, Heidelberg (2009)

17. Hoefer, M., Skopalik, A.: Altruism in atomic congestion games. In: Fiat, A., Sanders, P. (eds.) ESA 2009. LNCS, vol. 5757, pp. 179–189. Springer, Heidelberg (2009)

18. Hoefer, M., Skopalik, A.: Stability and convergence in selfish scheduling with altruistic agents. In: Leonardi, S. (ed.) WINE 2009. LNCS, vol. 5929, pp. 616–622. Springer, Heidelberg (2009)

19. Hoefer, M., Skopalik, A.: Social context in potential games. In: Goldberg, P.W. (ed.) WINE 2012. LNCS, vol. 7695, pp. 364–377. Springer, Heidelberg (2012)

20. Hoeksma, R., Uetz, M.: The price of anarchy for minsum related machine scheduling. In: Solis-Oba, R., Persiano, G. (eds.) WAOA 2011. LNCS, vol. 7164, pp. 261–273. Springer, Heidelberg (2012)

21. Koutsoupias, E., Papadimitriou, C.: Worst-case equilibria. In: Meinel, C., Tison, S. (eds.) STACS 1999. LNCS, vol. 1563, p. 404. Springer, Heidelberg (1999)

22. Roughgarden, T.: Intrinsic robustness of the price of anarchy. In: Proc. 41st Symp. on the Theory of Computing, pp. 513–522 (2009)

23. Roughgarden, T.: The price of anarchy in games of incomplete information. In: Proc. 13th Conf. on Electronic Commerce, pp. 862–879 (2012)

24. Syrgkanis, V.: Bayesian games and the smoothness framework. CoRR abs/1203.5155 (2012)

25. Syrgkanis, V., Tardos, É.: Composable and efficient mechanisms. In: Proc. 45th Symp. on the Theory of Computing (2013)

26. Young, H.P.: Strategic Learning and its Limits (1995)

Copula-Based Randomized Mechanisms for Truthful Scheduling on Two Unrelated Machines

Xujin Chen[1], Donglei Du[2], and Luis F. Zuluaga[3]

[1] Institute of Applied Mathematics, AMSS, Chinese Academy of Sciences,
Beijing 100190, China
xchen@amss.ac.cn
[2] Faculty of Business Administration, University of New Brunswick,
Fredericton NB Canada E3B 9Y2
ddu@unb.ca
[3] Department of Industrial and Systems Engineering, Lehigh University,
Bethlehem, PA, USA 18015
luis.zuluaga@lehigh.edu

Abstract. We design a Copula-based generic randomized truthful mechanism for scheduling on two unrelated machines with approximation ratio within [1.5852, 1.58606], offering an improved upper bound for the two-machine case. Moreover, we provide an upper bound 1.5067711 for the two-machine two-task case, which is almost tight in view of the lower bound of 1.506 for the scale-free truthful mechanisms [6]. Of independent interest is the explicit incorporation of the concept of Copula in the design and analysis of the proposed approximation algorithm. We hope that techniques like this one will also prove useful in solving other problems in the future.

Keywords: Algorithmic mechanism design, Random mechanism, Copula, Truthful scheduling.

1 Introduction

The main focus of this work is to offer randomized truthful mechanisms with improved approximation for minimizing makespan on unrelated parallel machines: $R2||C_{\max}$, a central problem extensively investigated in both the classical scheduling theory and the more recent algorithmic mechanism design initiated by the seminal work of Nisan and Roenn [11].

Formally, we are interested in the following scheduling problem: there are n tasks to be processed by m machines. Machine $i \in \{1, 2, \ldots, m\}$ takes t_{ij} time to process task $j \in \{1, 2, \ldots, n\}$. The objective is to schedule these tasks non-preemptively on these machines to minimize the makespan – the latest completion time among all the tasks. An allocation for the scheduling problem is specified by a set of binary variables x_{ij} such that $x_{ij} = 1$ if and only if task j is allocated to machine i.

B. Vöcking (Ed.): SAGT 2013, LNCS 8146, pp. 231–242, 2013.

Different from traditional approximation algorithms for the scheduling problem, we focus on the class of weakly monotonic algorithms [5] defined as follows: an allocation or a scheduling algorithm is *weakly monotonic* if for any two instances of the scheduling problem t_{ij} and $\tilde{t}_{ij}$ ($i = 1, 2, \ldots, m$ and $j = 1, 2, \ldots, n$) differing only on a single machine, the allocation x_{ij} and $\tilde{x}_{ij}$ returned by the algorithm satisfies $\sum_{j=1}^{n} (x_{ij} - \tilde{x}_{ij})(t_{ij} - \tilde{t}_{ij}) \leq 0$ for all $i = 1, 2, \ldots, m$.

The interest in monotonic algorithms stems from its connection to truthful mechanism design, where selfish agents maximize their profit by revealing their true private information. In this particular scheduling problem, a mechanism consists of two algorithms, an *allocation* algorithm which allocates tasks to machines and a *payment* algorithm which specifies the payment every machine receives. Each machine is a selfish agent who knows its own processing time for every task and wants to maximize its own payoff – the payment received minus the total execution time for the tasks allocated to it. A mechanism is *truthful* if it is a dominant strategy for each machine to reveal its true processing time. It is well-known that the weak monotonicity property above characterizes the allocation algorithm in any truthful mechanism for the scheduling problem on-hand (see e.g., [4]). In this paper, we are concerned with the approximation ratio of weakly monotonic allocation algorithms. When the allocation algorithm is randomized, i.e., the binary variables x_{ij} ($i = 1, 2, \ldots, m$, $j = 1, 2, \ldots, n$) output by the algorithm are random variables, we call the allocation algorithm *weakly monotonic* if it is a probability distribution over a family of deterministic weakly monotonic allocation algorithms. Every weakly monotonic randomized allocation algorithm gives rise to a (universally) truthful mechanism [11].

As usual, the *approximation ratio* of an allocation algorithm is the worst-case ratio between the makespan of the allocation output by the algorithm and the optimal makespan. One fundamental open problem on the mechanism design for scheduling is to find the exact approximation ratios R_{DET} and R_{RAN} among all weakly monotonic deterministic and randomized allocation algorithms respectively [11]. The current best bounds are $2.618 \approx 1 + \phi \leq R_{\text{DET}} \leq m$ with the upper and lower bounds established by Nisan and Ronen [11] and Koutsoupias and Vidali [4], respectively, and $2 - 1/m \leq R_{\text{RAN}} \leq 0.83685m$ with the upper and lower bounds proved by Mu'alem and Schapira [10] and Lu and Yu [7], respectively.

In view of the unbounded gap between the lower and upper bounds for the general m machines, a lot of research efforts have been devoted to the special case of $m = 2$ machines (see e.g., [2,6,11]), which is highly nontrivial and suggests more insights for resolving the general problem. In this paper, we will focus on the two-machine case. The deterministic approximation is exactly 2 as shown by Nisan and Ronen [11]. The currently best randomized approximation ratio is shown to lie between 1.5 and 1.6737. The upper bound due to Lu and Yu was proved by introducing a unified framework for designing truthful mechanisms [7]. This improved the ratio 1.75 of Nisan and Ronen's mechanism [11] by 0.0763. Later, Lu and Yu [8] provided an improved ratio of 1.5963, whose proof unfortunately is incorrect as shown in this paper later in Section 3.1. Dobzinski and Sundararajan [3] and Christodoulou et al. [2] independently showed that any weakly monotonic allocation algorithm for two machines with a finite approximation ratio is *weakly task-independent*, meaning that, for any task, its allocation does not change as long as none of its own processing time on machines changes. The weak task-independence is strengthened to be a *strong* one if the random variables x_{ij} output by the allocation algorithm are independent between different tasks [6].

In this paper, we use the concept of Copula to address the correlations among random outputs of the allocation algorithm under Lu and Yu's framework [7]. Our main contribution is to offer a Copula-based generic randomized mechanism for two-machine scheduling with approximation ratio within [1.5852, 1.58606], reducing the existing best upper bound [7] by more than 0.0876. Moreover, we provide an upper bound of 1.5067711 for the two-machine two-task case, which improves upon the previous 1.5089 bound given in [6] and is almost tight in view of the lower bound of 1.506 for the so called scale-free weakly monotonic allocation algorithm [6].

To our best knowledge, we are unaware of any extant work on the explicit usage of the concept of Copula in the design and analysis of approximation algorithms. We hope that techniques like this one will also prove useful in solving other problems in the future.

The rest of the paper is organized as follows: We present the Copula-based generic randomized mechanism in Section 2. We then analyze the mechanism for strongly independent tasks and weakly independent tasks in Section 3 and Section 4 respectively. Finally, we conclude the paper with some remarks on our choice of Copula in Section 5. The omitted details in the extended abstract can be found in the full version [1].

2   A Generic Randomized Mechanism Based on Copula

Given any real α, we use α^+ to denote the nonnegative number $\max\{0, \alpha\}$. Let $F : \mathbb{R}_+ \to [0, 1]$ be a non-decreasing function satisfying $F(0) = 0$ and $\lim_{x \to \infty} F(x) = 1$. Write $\bar{F}(x)$ for $1 - F(x)$. Let $X_1, X_2, \ldots, X_n$ be n dependent random variables with joint distribution function $\Pr(X_1 \leq x_1, X_2 \leq x_2, \ldots, X_n \leq x_n)$ given by the Clayton Copula

$$G(x_1, x_2, \ldots, x_n) = \left[\left(\left(\sum_{i=1}^{n} \sqrt[n-1]{F(x_i)} \right) - n + 1 \right)^+ \right]^{n-1} . \qquad (2.1)$$

It is easy to see that for any $1 \leq i < j \leq n$, the joint distribution of X_i and X_j is given by

$$H(x_i, x_j) = G(\infty, \ldots, \infty, x_i, \infty, \ldots, \infty, x_j, \infty, \ldots, \infty)$$

$$= \left[\left(\sqrt[n-1]{F(x_i)} + \sqrt[n-1]{F(x_j)} - 1 \right)^+ \right]^{n-1} . \qquad (2.2)$$

We also study identically independent distributions for which

$$G(x_1, x_2, \ldots, x_n) = \prod_{i=1}^{n} F(x_i) \text{ and } H(x_i, x_j) = F(x_i)F(x_j). \qquad (2.3)$$

Using a joint distribution satisfying Clayton's Copula in (2.1) or the independence condition in (2.3) gives the following specification of the randomized allocation algorithm introduced by Lu and Yu [8].

Algorithm 1. INPUT: A processing time matrix $t \in \mathbb{R}_+^{2 \times n}$.
OUTPUT: A randomized allocation $x \in \{0, 1\}^{2 \times n}$.

1. Choose random variables $X_1, X_2, \ldots, X_n$ according to distribution function G
2. **For** each task $j = 1, 2, \ldots, n$ **do**
3.   if $t_{1j}/t_{2j} < X_j$ then $x_{1j} \leftarrow 1$ else $x_{1j} \leftarrow 0$
4. $x_{2j} \leftarrow 1 - x_{1j}$
5. **End-for**
6. Output x

Let real function $\varphi : \mathbb{R}_+ \times \mathbb{R}_+ \to \mathbb{R}$ be defined by

$$\varphi(x, y) = 1 + y - \min\{1, 1 - \tfrac{1}{x} + y\}F(x) - yF(y) + \min\{1 + \tfrac{1}{x}, 1 + y\}H(x, y) \quad (2.4)$$

Theorem 1. *The approximation ratio of Algorithm 1 is at most* $\max\{\varphi(x, y) : x, y \in \mathbb{R}_+\}$.

Proof. For every $j \in [n]$, let $r_j = t_{1j}/t_{2j}$. It has been shown by Lu and Yu [8] that the approximation ratio of Algorithm 1 is bounded above by $\max\{\rho_{jk} : j, k \in [n]\}$, where for every pair of distinct indices $i, j \in [n]$,

$$\rho_{jk} = \Pr(x_{1j} = 1) + r_k \cdot \Pr(x_{1k} = 1) + (\tfrac{1}{r_j} - r_k)^+ \cdot \Pr(x_{2j} = 1, x_{1k} = 1)$$
$$+ (1 + \tfrac{1}{r_j}) \cdot \Pr(x_{2j} = 1, x_{2k} = 1).$$

Notice that $X_j \leq r_j \Leftrightarrow x_{1j} = 0 \Leftrightarrow x_{2j} = 1$. Hence

$$\begin{aligned}
\rho_{jk} &= \Pr(X_j > r_j) + r_k \cdot \Pr(X_k > r_k) + (\tfrac{1}{r_j} - r_k)^+ \cdot \Pr(X_j \leq r_j, X_k > r_k) \\
&\quad + (1 + \tfrac{1}{r_j}) \cdot \Pr(X_j \leq r_j, X_k \leq r_k) \\
&= \bar{F}(r_j) + r_k \cdot \bar{F}(r_k) + (\tfrac{1}{r_j} - r_k)^+ \cdot (F(r_j) - H(r_j, r_k)) + (1 + \tfrac{1}{r_j}) \cdot H(r_j, r_k) \\
&= 1 + r_k - \left(1 - (\tfrac{1}{r_j} - r_k)^+\right) F(r_j) - r_k F(r_k) + \left(1 + \tfrac{1}{r_j} - (\tfrac{1}{r_j} - r_k)^+\right) H(r_j, r_k) \\
&= 1 + r_k - \min\{1, 1 - \tfrac{1}{r_j} + r_k\}F(r_j) - r_k F(r_k) + \min\{1 + \tfrac{1}{r_j}, 1 + r_k\}H(r_j, r_k),
\end{aligned}$$

showing that $\rho_{jk} = \varphi(x_j, x_k)$. ☐

3   Strongly Independent Tasks

In this section, we consider tasks being allocated strongly independently. Therefore, the joint distribution takes the form $H(x, y) = F(x)F(y)$, giving

$$\varphi(x, y) = 1 + y - \min\left\{1, 1 - \tfrac{1}{x} + y\right\} F(x) - y F(y) + \min\left\{1 + \tfrac{1}{x}, 1 + y\right\} F(x) F(y), \quad (3.1)$$

from which the following symmetry can be proved by elementary mathematics.

Lemma 1. *Let G satisfy (2.3). If $F(x) = 1 - F(1/x)$ for any $x \geq 0$, then $\varphi(x, y) = \varphi(1/y, 1/x)$ for any $x, y \in \mathbb{R}_+$.*   □

In Section 3.1, we point out a mistake of [8] in estimating over a transcendental function, which invalidates the ratio 1.5963 claimed. In Section 3.2, we introduce an algebraic piecewise function to construct a class of joint distributions of independent random variables. Then, we prove that using this class of independent distributions in Algorithm 1 gives an improved ratio 1.58606. In Section 3.3, we show the limitation of Algorithm 1 for strongly independent tasks, from which no ratio better than 1.5852 can be expected.

3.1   Lu and Yu's Transcendental Function

Lu and Yu [8] considered function $F(x) = 1 - \frac{1}{2^{x^{2.3}}}$. For any $\alpha_1, \alpha_2 \in \mathbb{R}_+$, let $\beta_1 = F(\alpha_1)$ and $\beta_2 = F(1/\alpha_2)$. By Theorem 4 and in particular the instance on page 410 of [8], Lu and Yu's mechanism has approximation ratio at least

$$\theta(\alpha_1, \alpha_2)$$
$$= (1 + \alpha_2)\beta_1\beta_2 + \beta_1(1 - \beta_2) + (1 + \alpha_1)(1 - \beta_1)(1 - \beta_2) + \max\{\alpha_1, \alpha_2\}\beta_2(1 - \beta_1)$$
$$= \begin{cases} (1 + \alpha_2)\beta_1\beta_2 + \beta_1(1 - \beta_2) + (1 + \alpha_1)(1 - \beta_1)(1 - \beta_2) + \alpha_2\beta_2(1 - \beta_1), & \text{if } \alpha_2 \geq \alpha_1; \\ (1 + \alpha_2)\beta_1\beta_2 + \beta_1(1 - \beta_2) + (1 + \alpha_1)(1 - \beta_1)(1 - \beta_2) + \alpha_1\beta_2(1 - \beta_1), & \text{if } \alpha_1 \geq \alpha_2. \end{cases}$$

They claimed in Theorem 5 of [8] that under this $F(x)$, $\theta(\alpha_1, \alpha_2) \leq 1.5963$. However, a contradiction is given by $\theta(0.87793459260323, 2.09409917605545) > 1.64$. In view of this, the previously best known approximation ratio for truthful scheduling on two unrelated machines was 1.6737 in Lu and Yu's earlier conference paper [7]. In this paper we reduce the ratio to 1.58606.

3.2   An Algebraic Piecewise Function

Suppose that $a \in [1.7, 3]$ and $b \in [0.7, 1]$ are constants. We study the following *continuous* piecewise algebraic function

$$F(x) = \begin{cases} 1, & x \in I_1 = [a, +\infty), \\ 1 - \frac{2(1-b)(a-x)}{a-1}, & x \in I_2 = [\frac{a+1}{2}, a), \\ \frac{1}{2} + \frac{(2b-1)(x-1)}{a-1}, & x \in I_3 = [1, \frac{a+1}{2}), \\ \frac{1}{2} - \frac{(2b-1)(1/x-1)}{a-1}, & x \in I_4 = [\frac{2}{a+1}, 1), \\ \frac{2(1-b)(a-1/x)}{a-1}, & x \in I_5 = [\frac{1}{a}, \frac{2}{a+1}), \\ 0, & x \in I_6 = [0, \frac{1}{a}), \end{cases} \quad (3.2)$$

where the five *demarcation points* $\frac{1}{a}$, $\frac{2}{a+1}$, 1, $\frac{a+1}{2}$, a divide the domain $[0, +\infty)$ into six *intervals* $I_1, I_2, \ldots, I_6$. The function $F(\cdot)$, when plugged into (2.3), gives an improvement 0.08764 over the previous best ratio of 1.6737 [7]. Notice that $F(\cdot)$ enjoys the property that

$$F(x) + F(1/x) = 1 \text{ for any } x \geq 0. \tag{3.3}$$

An immediate corollary is $F(1) = 0.5$.

Theorem 2. *Let $a = 1.715$ and $b = 0.76$. Using $F(x)$ in (3.2) and $G(x_1, \ldots, x_n)$ in (2.3), Algorithm 1 achieves approximation ratio 1.58606.*

Proof. By Theorem 1, it suffices to show that the maximum of $\varphi(x, y)$ in (3.1) is no more than $\rho^* = 1.58606$. By (3.3) and Lemma 1, we may assume $xy \geq 1$, for which the function $\varphi(x, y)$ to be maximized takes the form of

$$\varphi(x, y) = 1 + y - F(x) - yF(y) + \left(1 + \tfrac{1}{x}\right) F(x)F(y). \tag{3.4}$$

Note that $\varphi(x, y)$ is continuous in $\mathbb{R}_+ \times \mathbb{R}_+$. Suppose $x^*, y^* \in \mathbb{R}_+$ with $x^* y^* \geq 1$ attains the maximum, i.e., $(x^*, y^*) \in \arg\max_{xy \geq 1} \varphi(x, y)$.

We will show that $\varphi(x^*, y^*) < \rho^*$ by considering the different possible domains of the variables x and, y in a case by case basis. When x^* or y^* does not belong to the domain associated with a given case, we say that (x^*, y^*) does not *belong to the case*. We will show that, for any case $x \in I_i$, $y \in I_j$ $(1 \leq i, j \leq 6)$ to which (x^*, y^*) may belong, the function's value $\varphi(x, y)$ is smaller than ρ^* by upper bounding its value at critical points (i.e., when the derivatives of $\varphi(x, y)$ are equal to zero) and that at demarcation points.

CASE 1. $x \geq a$. It follows from (3.2) that $F(x) = 1$ and from (3.4) that $\varphi(x, y) = y + (1 + \tfrac{1}{x} - y)F(y)$. If $y \leq 1 + \tfrac{1}{x}$ or $y \geq a$, then $\varphi(x, y) \leq y + (1 + \tfrac{1}{x} - y) = 1 + \tfrac{1}{x} \leq 1 + \tfrac{1}{a} < 1.584$. If $y > 1 + \tfrac{1}{x}$ and $y < a$, then $y \in (1, a)$.

In case of $y \in [\frac{a+1}{2}, a)$, since $\frac{\partial \varphi}{\partial x}(x, y) = -\frac{2400y - 541}{3575x^2} < 0$, from KKT condition, we deduce that (x^*, y^*) does not belong to this case unless $x^* = a$. When $x = a$, note that $\varphi\left(a, \frac{a+1}{2}\right) < 1.53$, and that $\varphi(a, y)$ has a unique critical point $y = \frac{a^2 + a + 1}{2a}$ in $(\frac{a+1}{2}, a)$ with corresponding critical value less than 1.58602.

In case of $y \in (1, \frac{a+1}{2})$, it suffices to consider the case where $x = a$ as $\frac{\partial \varphi}{\partial x}(x, y) = -\frac{16y - 5}{22x^2} < 0$. Note that $\frac{\partial \varphi}{\partial y}(a, y) = \frac{17949}{7546} - \frac{16}{11}y > 2.3 - \frac{16}{11}(\frac{a+1}{2}) > 0$, which excludes the possibility of (x^*, y^*) belonging to this case.

CASE 2. $y \geq a > x \geq 0$. Note that $\varphi(x, y) = 1 + y - F(x) - y + (1 + \tfrac{1}{x})F(x) = 1 + \frac{F(x)}{x}$ is a function of single variable x. It is easy to check that the derivative of $1 + \frac{F(x)}{x}$ is positive for all $x \in (\frac{1}{a}, a) - \{\frac{2}{a+1}, 1, \frac{a+1}{2}\}$. The continuity of φ implies $\varphi(x, y) \leq \varphi(a, y) = 1 + \tfrac{1}{a} < 1.584$ for all $x \in (\frac{1}{a}, a)$. When $x \in (0, \frac{1}{a}]$, it is clear that $\varphi(x, y) = 1$.

Cases 1 and 2 above show that $\varphi(x, y) < \rho^*$ when x or y belongs to I_1. For the remaining cases, we have $x, y < a$. As $xy \geq 1$, we have $x, y > \frac{1}{a}$ both contained in $(\cup_{i=2}^4 I_i) \cup (I_5 - \{\frac{1}{a}\})$. We distinguish among Cases 3 – 6, where Case $i + 1$ deals with for $x \in I_i$, $i = 2, 3, 4$ and Case 6 deals with $x \in I_5 - \{\frac{1}{a}\}$.

CASE 3. $x \in I_2 = [\frac{a+1}{2}, a)$. We distinguish among four subcases for $y \in [\frac{a+1}{2}, a)$, $y \in [1, \frac{a+1}{2})$, $y \in [\frac{2}{a+1}, 1)$, and $y \in (\frac{1}{a}, \frac{2}{a+1})$, respectively.

CASE 3.1. $y \in [\frac{a+1}{2}, a)$. In case of $x, y \in (\frac{a+1}{2}, a)$, solving $\frac{\partial \varphi}{\partial x}(x, y) = 0 = \frac{\partial \varphi}{\partial y}(x, y)$ gives $\frac{987.84x^2 + 29.2681}{576x^2 - 129.84} = y = \frac{2400x^2 + 7990.125x - 541}{7150x}$, which implies $2764800x^4 - 4921488x^3 + 1656341.83x - 140486.88 = 0$. Among the four real roots of the biquadratic equation, only one root $x_0 \doteq 1.5419$ belongs $I_2 = [\frac{a+1}{2}, a)$. So function $\varphi(x, y)$ has a unique critical point (x_0, y_0), where $y_0 = \frac{987.84x_0^2 + 29.2681}{576x_0^2 - 129.84} \doteq 1.586$, giving critical value $\varphi(x_0, y_0) < 1.585$.

In case of $x = \frac{a+2}{2}$, function $\varphi(\frac{a+1}{2}, y)$ has a unique critical point $y_0 = \frac{a^2 + a + ab + 3b}{2a + 2} \doteq 1.5174$ in $(\frac{a+1}{2}, a)$, giving critical value smaller than 1.586059. Note that $\varphi(\frac{a+1}{2}, \frac{a+1}{2}) < 1.57$.

In case of $y = \frac{a+1}{2}$ and $x \in (\frac{a+1}{2}, a)$, the derivative of $\varphi(x, \frac{a+1}{2})$ is $\frac{10279}{89375x^2} - \frac{576}{3575} < \frac{10279}{89375} - \frac{576}{3575} < 0$, saying that (x^*, y^*) does not belong to this case.

CASE 3.2. $y \in [1, \frac{a+1}{2})$. Similar arguments to that in Case 3.1 show the following: In case of $x \in (\frac{a+1}{2}, a)$ and $y \in (1, \frac{a+1}{2})$, function φ attains its critical value $\varphi(x_0, y_0) < 1.583$ at $x_0 \doteq 1.5249$, $y_0 = \frac{1053x_0^2 + 43.95625}{624x_0^2 + 140.66} \doteq 1.566$. In case of $x = \frac{a+1}{2}$, function φ attains its critical value $\varphi(\frac{a+1}{2}, y_0) < 1.585$ at $y_0 = \frac{a^2 - 3 + 2ab - 2b - 2a + 4ab^2 + 12b^2}{4(a+1)(2b-1)} \doteq 1.5037$; at the boundary, $\varphi(\frac{a+1}{2}, 1) < 1.4$. In case of $y = 1$ and $x \in (\frac{a+1}{2}, a)$, (x^*, y^*) does not belong to this case.

CASE 3.3. $y \in [\frac{2}{a+1}, 1)$. If $\frac{\partial \varphi}{\partial x}(x, y) = 0$, then $x^2 = \frac{5.41}{24}\left(\frac{70.4}{16 - 5y} - 5.4\right) < \frac{5.41}{24}\left(\frac{70.4}{16-5} - 5.4\right) < 0.3$, contradicting the hypothesis $x \in [\frac{a+1}{2}, a)$ of Case 3. Thus $\frac{\partial \varphi}{\partial x}(x, y) \neq 0$, and it suffices to consider the case where $x = \frac{a+1}{2}$. Note that the derivative of $\varphi(\frac{a+1}{2}, y)$ is $\frac{143336}{149325y^2} - \frac{5}{22} > \frac{143336}{149325} - \frac{5}{22} > 0$. We deduce that (x^*, y^*) does not belong to Case 3.3.

CASE 3.4. $y \in (\frac{1}{a}, \frac{2}{a+1})$. It can be shown that (x^*, y^*) does not belong to this case by arguments similar to that in Case 3.3.

CASE 4. $x \in I_3 = [1, \frac{a+1}{2})$. It follows from $xy \geq 1$ that $y > \frac{2}{a+1}$ for which we distinguish among three subcases for $y \in [\frac{a+1}{2}, a)$, $y \in [1, \frac{a+1}{2})$ and $y \in [\frac{2}{a+1}, 1)$, respectively.

In case of $y \in [\frac{a+1}{2}, a)$, for $x \in (1, \frac{a+1}{2})$ and $y \in (\frac{a+1}{2}, a)$, function φ attains critical value $\varphi(x_0, y_0) < 1.5854$ at the unique critical point (x_0, y_0), where $x_0 \doteq 1.2027$ and $y_0 = \frac{26754x_0^2 + 35165/32}{15600x_0^2 + 4875} \doteq 1.4504$. Note that $\varphi(1, \frac{a+1}{2}) = 1.5858$. For $x = 1$ and $y \in (\frac{a+1}{2}, a)$, the derivative of $\varphi(1, y)$ is negative. For $y = \frac{a+1}{2}$ and $x \in (1, \frac{a+1}{2})$, the derivative of $\varphi(x, \frac{a+1}{2})$ is negative. It follows that (x^*, y^*) belongs to neither of the two cases.

In case of $y \in [1, \frac{a+1}{2})$, if $\frac{\partial \varphi}{\partial x}(x, y) = 0$, then $x^2 = \frac{6.875}{27 - 16y} - \frac{5}{16} < \frac{6.875}{27 - 8(a+1)} - \frac{5}{16} < 1$, contradicting the hypothesis $x \geq 1$ of Case 4. So it suffices to consider $x = 1$. Within $y \in (1, \frac{a+1}{2})$, function $\varphi(1, y)$ attains its unique critical value $\varphi(1, \frac{43}{32}) < 1.586$ at $y = \frac{43}{32}$. At the boundary, we have $\varphi(1, 1) = 1.5$.

In case of $y \in (\frac{2}{a+1}, 1)$, if $\frac{\partial \varphi}{\partial x}(x, y) = 0$, then $x^2 = \frac{80 - 135y}{80y - 256}$, which along with $x \geq 1$ enforces $y \geq \frac{336}{215}$, a contradiction to $y < 1$. Therefore we may assume $x = 1$. Since the derivative of $\varphi(1, y)$ is $\frac{16 - 5y^2}{22y^2} > 0$, we deduce that (x^*, y^*) does not belong to this case.

CASE 5. $x \in I_4 = [\frac{2}{a+1}, 1)$. It follows from $xy \geq 1$ that $y > 1$. We distinguish between two subcases depending on whether y is at least $\frac{a+1}{2}$ or not.

In case of $y \in [\frac{a+1}{2}, a)$, if $\frac{\partial \varphi}{\partial y}(x, y) = 0$, then $y = \frac{2524.47x^2 + 429x - 624}{1716x^2}$, which along with $y \geq \frac{a+1}{2}$ enforces $5x^2 + 11x - 16 \geq 0$ implying $x \leq -3.2$ or $x \geq 1$, a contradiction to the hypothesis $x \in I_4$ of Case 5. So we may assume $y = \frac{a+1}{2}$. Within $x \in (\frac{2}{a+1}, 1)$, the unique critical point of $\varphi(x, \frac{a+1}{2})$ is $x = \frac{608}{609}$, giving critical value less than 1.586. At the boundary, we have $\varphi(\frac{2}{a+1}, \frac{a+1}{2}) < 1.52$.

In case of $y \in (1, \frac{a+1}{2})$, when $x \in (\frac{2}{a+1}, 1)$, function φ attains its unique critical value $\varphi(x_0, y_0) < 1.58603$ at $x_0 \doteq 0.985$, $y_0 = \frac{50193x_0/16 + 1690}{5408 - 1859x_0} \doteq 1.3364$. When $x = \frac{2}{a+1}$, function $\varphi(\frac{2}{a+1}, y)$ has a unique critical value $\varphi(\frac{2}{a+1}, 1.12665) < 1.56$.

CASE 6. $x \in I_5 - \{\frac{1}{a}\} = (\frac{1}{a}, \frac{2}{a+1})$. It follows from $xy \geq 1$ that $y \in (\frac{a+1}{2}, a)$. If $\frac{\partial \varphi}{\partial x}(x, y) = 0$, then $x = \frac{2400y - 541}{637637/400 + 858y}$, which along with $x \leq \frac{2}{a+1}$ enforces $y \leq \frac{4657}{4800} < 1$, a contradiction. Since $\frac{\partial \varphi}{\partial x}(x, y)$ is a continuous function, we deduce that $\frac{\partial \varphi}{\partial x}(x, y)$ is always positive or always negative, implying that (x^*, y^*) does not belong to Case 6.

Among all cases analyzed above, the bottleneck 1.586058... ($< \rho^*$) is attained by Case 3.1 with $\varphi\left(\frac{a+1}{2}, \frac{a^2 + a + ab + 3b}{2a + 2}\right)$.   □

3.3   The Limitation of Algorithm 1

It was announced in [8] and proved in its full paper that, for strongly independent tasks, the performance ratio of Algorithm 1 cannot be better than 1.5788. We improve the lower bound by 0.0074, which nearly closes the gap between the lower and upper bounds for Algorithm 1.

Theorem 3. *Let G in (2.3) be defined by any non-decreasing function F : $\mathbb{R}_+ \to [0, 1]$ with $F(0) = 0$ and $\lim_{x \to \infty} F(x) = 1$. The approximation ratio of Algorithm 1 is at least 1.5852.*

Proof. Suppose that there exists function F such that Algorithm 1 achieves an approximation ratio less than 1.5852. It follows from (3.1) that for any $x, y \in \mathbb{R}_+$,

$$1.5852 > \varphi(x, y)$$
$$= \begin{cases} 1 + y - F(x) - yF(y) + \left(1 + \frac{1}{x}\right) F(x)F(y), & xy \geq 1; \\ 1 + y - \left(1 - \frac{1}{x} + y\right) F(x) - yF(y) + (1 + y)F(x)F(y), & xy \leq 1. \end{cases} \quad (3.5)$$

Let $\alpha = 1.352$ and $\beta = 1.532$. We examine $\varphi(x, y)$ for some values of x, y in $\{\alpha, \beta, 1, 1/\alpha, 1/\beta\}$, and derive a contradiction from $\varphi(x, y) < 1.5852$.   □

4   Weakly Independent Tasks

We assume function $F(\cdot)$ takes the form of (3.2). The weak independence is specified by the joint distribution $H(x_i, x_j) = [(\sqrt[n-1]{F(x_i)} + \sqrt[n-1]{F(x_j)} - 1)^+]^{n-1}$ as in (2.2).

Using the Copula based distribution, Algorithm 1 can guarantee approximation 1.5067711 for $n = 2$ tasks, as proved in Section 4.1. We study the case of $n \geq 3$ tasks in Section 4.2, where MATLAB's global solver is used to solve the optimization problems involved in the computer conducted search/proof of the approximation ratio. Our results show that the Clayton Copula based algorithm outperforms the strong independent-task allocation, and the former converges to the later as n approaches to infinity.

4.1   The Case of $n = 2$

In this subsection, we reduce Lu's upper bound $\frac{1}{6}(\sqrt{25 - 12\sqrt{3}} + 7) \doteq 1.5089$ [6] for two tasks, and narrow the gap from the lower bound 1.506 [6] to be 0.0007711. For the case of $n = 2$, we have $H(x_1, x_2) = (F(x_1) + F(x_2) - 1)^+$ and $\varphi(x, y) = 1 + y - \min\{1, 1 - \frac{1}{x} + y\} F(x) - yF(y) + \min\{1 + \frac{1}{x}, 1 + y\} (F(x) + F(y) - 1)^+$.

Lemma 2. *Let G satisfy (2.1). When $n = 2$, $\varphi(x, y) = \varphi(1/y, 1/x)$ for any $x, y \in \mathbb{R}_+$.*

Proof. Without loss of generality we may assume $xy \geq 1$. Since $F(\cdot)$ is non-decreasing and satisfies (3.3), we have $F(x) \geq F(1/y) = 1 - F(y)$, and $\varphi(x, y) = 1 + y - F(x) - yF(y) + (1 + 1/x)(F(x) + F(y) - 1)$

On the other hand, $F(1/y) + F(1/x) \leq F(1/y) + F(y) = 1$ implies $\varphi(1/y, 1/x) = 1 + 1/x - (1 - y + 1/x)F(1/y) - F(1/x)/x = 1 + 1/x - (1 - y + 1/x)(1 - F(y)) - (1 - F(x))/x$. Now it is easy to check that $\varphi(x, y) = \varphi(1/y, 1/x)$.   □

Using Lemma 2 and a similar proof to that of Theorem 2 (see [1]), we establish a 1.5068 performance ratio for Algorithm 1.

Theorem 4. *Let $F(\cdot)$ be defined as in (3.2) with $a = 2.2468$ and $b = 0.7607$. For $n = 2$, using $G(x_1, x_2)$ in (2.1), Algorithm 1 achieves an approximation ratio of 1.5067711.*   □

4.2   The Case of General n

In this subsection, we mainly discuss the multiple task case $n \geq 3$. We look for a distribution function $F(\cdot)$ of form (3.2) which minimizes the maximum of the binary function

$$\varphi(x, y) = 1 + y - \min\{1, 1 - \frac{1}{x} + y\} F(x) - yF(y)$$
$$+ \min\{1 + \frac{1}{x}, 1 + y\} \left[(\sqrt[n-1]{F(x)} + \sqrt[n-1]{F(y)} - 1)^+ \right]^{n-1}. \quad (4.1)$$

To accomplish the task, we need determine the maximum of φ for any given constants a and b. Theoretically, this can be done in a way similar to the proofs of Theorems 2 and 4. In practice, computer-assisted arguments turn out more suitable, as explained below.

- The previous case by case analyses are simplified by the property $\varphi(x,y) = \varphi(1/y, 1/x)$ (see Lemmas 1 and 2), which allows us to only focus on the case of $xy \geq 1$. For $n \geq 3$, this property is generally lost due to the *complicated term* $[(\sqrt[n-1]{F(x)} + \sqrt[n-1]{F(y)} - 1)^+]^{n-1}$ in (4.1). As a result, it might be much more tedious to discuss all possible combinations for x, y from six intervals $[0, \frac{1}{a})$, $[\frac{1}{a}, \frac{2}{a+1})$, ..., $[a, +\infty)$ where $F(\cdot)$ is described by different algebraic expressions.
- Finding the critical points of $\varphi(x,y)$ becomes more and more challenging as n increases. One has to resort to software for solving equations of high degrees resulting from the complicated term.

We conduct a case analyses using MATLAB's global optimization tool GLOB-ALSEARCH (cf., [12]) to help us to solve the nonlinear program $\max_{x,y} \varphi(x,y)$ subject to four constrains $xy \leq$ (or $\geq$) 1, $\sqrt[n-1]{F(x)} + \sqrt[n-1]{F(y)} \leq$ (or $\geq$) 1, $l_1 \leq x \leq u_1$, $l_2 \leq y \leq u_2$ for different choices of n, a, and b, where l_1, u_1, l_2, u_2 specify the intervals containing x and y. The computational results are summarized in Table 1. For each input triplet of n, a, b, Table 1 provides the values of (x^*, y^*) which attain the largest value of $\varphi(x,y)$ after GLOBALSEARCH is employed to solve the nonlinear program 10 times. The difference δ between the largest value of $\varphi(x,y)$ and the smallest one among the 10 computations is also recorded. From the last column of Table 1 we observe that δ does not exceed 1.4066×10^{-7}, showing the stability of the computational results.

As the second line (when $n = 2$) in Table 1 illustrates, GLOBALSEARCH finds the optimal solution established in Theorem 4 within numerical tolerance. The second to last column of Table 1 shows that $\varphi(x^*, y^*)$ increases as n grows, interpreting the common sense that achieving truthfulness with respect to more tasks costs more. The increasing property of approximation ratios with respect to n can be visualized when we connect points $(n, \varphi(x^*, y^*))$ with a curve (see [1]), where we have the following observations.

- The curve makes a "large" jump at $n = 3$, from 1.5068 to 1.5413;
- The increasing speed is tiny after $n = 30$, which attains $\varphi(x^*, y^*) \doteq 1.5828$;
- The curve looks flat after $n = 100$; in particular the average slope is less than 5×10^{-6} for $n \in [100, 200]$.

More interesting phenomena are observed from the first three columns of Table 1: the optimal value of a decreases with n, and approaches a limit approximately equal to 1.7149; while starting from $n = 3$ the optimal value of b increases with n, and approaches a limit approximately equal to 0.7599. Note the limit values of a and b in Table 1 "coincide" with the values of $a = 1.715$ and $b = 0.76$ used in Theorem 2 for strongly independent tasks. The reason is that for any distribution function $F(\cdot)$, function $\varphi(x,y)$ in (4.1) is always upper bounded by function $\varphi(x,y)$ in (3.1), and the former approaches the latter as n tends to infinity. This fact is proved in [1].

Table 1. Computational results on minimizing the maximum of φ (choosing a and b to minimize the maximum $\varphi(x^*, y^*)$), where the data in the last row for $n = \infty$ are taken from the proof of Theorem 2.

n	a	b	x^*	y^*	$\varphi(x^*, y^*)$	δ
2	2.2468	0.7607	$\frac{a+1}{2} = 1.6234$	1.9313955486	1.5067710964	1.5499×10^{-13}
3	1.9328	0.7418	1.9105670668	1.7231009560	1.5412707361	5.4073×10^{-9}
4	1.8442	0.7453	$a = 1.8442$	1.6932202823	1.5559952305	8.8818×10^{-16}
5	1.8070	0.7487	1.1418758036	1.5193285944	1.5634859375	1.8911×10^{-9}
6	1.7863	0.7510	1.1468400067	1.4989121029	1.5679473463	3.4101×10^{-9}
7	1.7734	0.7526	1.1447309125	1.4845715829	1.5709131851	2.7397×10^{-8}
8	1.7646	0.7536	1.1192295299	1.4661575387	1.5730320737	4.9022×10^{-9}
9	1.7581	0.7543	$\frac{a+1}{2} = 1.37905$	1.5499380481	1.5746303803	2.1302×10^{-9}
10	1.7530	0.7548	1.0673757071	1.4334673997	1.5758769995	4.5725×10^{-8}
15	1.7410	0.7570	1.0190835924	1.3975512392	1.5795353027	3.2335×10^{-8}
20	1.7326	0.7573	0.9997077878	1.3798783532	1.5811826690	8.8565×10^{-10}
30	1.7267	0.7582	$\frac{a+1}{2} = 1.36335$	1.5259350403	1.5828322598	2.3226×10^{-13}
45	1.7225	0.7587	0.9879452462	1.3491108561	1.5839252561	1.4493×10^{-9}
70	1.7199	0.7592	0.9868820343	1.3445069231	1.5846893837	3.2863×10^{-9}
100	1.7183	0.7594	$\frac{a+1}{2} = 1.35915$	1.5197905945	1.5850948285	3.1020×10^{-13}
200	1.7167	0.7597	$\frac{a+1}{2} = 1.35835$	1.5186228330	1.5855735653	7.5118×10^{-13}
500	1.7156	0.7598	0.9851752572	1.3375636313	1.5858603200	1.8349×10^{-9}
1000	1.7153	0.7599	$\frac{a+1}{2} = 1.35765$	1.5176140596	1.5859488980	3.2567×10^{-12}
5000	1.7150	0.7599	0.9849521898	1.3365770913	1.5860275919	2.9110×10^{-9}
10^4	1.7149	0.7599	$a = 1.7149$	1.6490128248	1.5860403769	1.4479×10^{-11}
10^5	1.7149	0.7599	0.9849621198	1.3364590898	1.5860442151	5.2509×10^{-9}
10^6	1.7149	0.7599	0.9849513401	1.3364514130	1.5860456086	1.0466×10^{-7}
$\vdots$	$\vdots$	$\vdots$	$\vdots$	$\vdots$	$\vdots$	$\vdots$
∞	1.715	0.76	$\frac{a+1}{2} = 1.3575$	$\frac{a^2+a+ab+3b}{2a+2} \doteq 1.5174$	1.5860582220	0

5 Concluding Remark

We note that the choice of the Clayton Copula in (2.1) is not accidental. We wish to choose the Copula that leads to the best approximation ratio for our mechanism. However, the Clayton Copula is the best lower bound among all Archimedean copulas [9]. Therefore, any hope to improve the bounds presented in this work will have to resort to non-Archimedean copulas, which usually lack the nice closed-form property of Archimedean copulas.

Acknowledgements. This work was done while Xujin Chen was visiting Faculty of Business Administration, University of New Brunswick. Xujin Chen was supported in part by NNSF of China (11222109), NSERC grants (283103, 290377) and CAS Program for Cross & Cooperative Team of Science & Technology Innovation. Donglei Du was supported in part by NSERC grant 283106.

References

1. Chen, X., Du, D., Zuluaga, L.F.: Copula-based randomized mechanisms for truthful scheduling on two unrelated machines. CoRR abs/1306.3909 (2013)
2. Christodoulou, G., Koutsoupias, E., Vidali, A.: A characterization of 2-player mechanisms for scheduling. In: Halperin, D., Mehlhorn, K. (eds.) ESA 2008. LNCS, vol. 5193, pp. 297–307. Springer, Heidelberg (2008)
3. Dobzinski, S., Sundararajan, M.: On characterizations of truthful mechanisms for combinatorial auctions and scheduling. In: Proceedings of the 9th ACM Conference on Electronic Commerce, EC 2008, pp. 38–47 (2008)
4. Koutsoupias, E., Vidali, A.: A lower bound of $1 + \phi$ for truthful scheduling mechanisms. In: Kučera, L., Kučera, A. (eds.) MFCS 2007. LNCS, vol. 4708, pp. 454–464. Springer, Heidelberg (2007)
5. Lavi, R., Mu'Alem, A., Nisan, N.: Towards a characterization of truthful combinatorial auctions. In: Proceedings of the 44th Annual IEEE Symposium on Foundations of Computer Science, FOCS 2003, pp. 574–583 (2003)
6. Lu, P.: On 2-player randomized mechanisms for scheduling. In: Leonardi, S. (ed.) WINE 2009. LNCS, vol. 5929, pp. 30–41. Springer, Heidelberg (2009)
7. Lu, P., Yu, C.: An improved randomized truthful mechanism for scheduling unrelated machines. In: Proceedings of the 25th International Symposium on Theoretical Aspects of Computer Science, STACS 2008, pp. 527–538 (2008)
8. Lu, P., Yu, C.: Randomized truthful mechanisms for scheduling unrelated machines. In: Papadimitriou, C., Zhang, S. (eds.) WINE 2008. LNCS, vol. 5385, pp. 402–413. Springer, Heidelberg (2008)
9. McNeil, A.J., Nešlehová, J.: Multivariate archimedean copulas, d-monotone functions and l_1-norm symmetric distributions. The Annals of Statistics 37(5), 3059–3097 (2009)
10. Mu'alem, A., Schapira, M.: Setting lower bounds on truthfulness: extended abstract. In: Proceedings of the Eighteenth Annual ACM-SIAM Symposium on Discrete Algorithms, SODA 2007, pp. 1143–1152 (2007)
11. Nisan, N., Ronen, A.: Algorithmic mechanism design. Games and Economic Behavior 35, 166–196 (2001)
12. Ugray, Z., Lasdon, L., Plummer, J.C., Glover, F., Kelly, J., Martí, R.: Scatter search and local nlp solvers: A multistart framework for global optimization. INFORMS Journal on Computing 19(3), 328–340 (2007)

Imperfect Best-Response Mechanisms

Diodato Ferraioli[1] and Paolo Penna[2]

[1] PSL*, Université Paris Dauphine, Lamsade and CNRS UMR 7243
diodato.ferraioli@dauphine.fr
[2] Institute of Theoretical Computer Science, ETH Zurich
paolo.penna@inf.ethz.ch

Abstract. Best-response mechanisms (Nisan, Schapira, Valiant, Zohar, 2011) provide a unifying framework for studying various distributed protocols in which the participants are instructed to repeatedly best respond to each others' strategies. Two fundamental features of these mechanisms are convergence and incentive compatibility.

This work investigates convergence and incentive compatibility conditions of such mechanisms when players are not guaranteed to always best respond but they rather play an *imperfect* best-response strategy. That is, at every time step every player deviates from the prescribed best-response strategy according to some probability parameter. The results explain to what extent convergence and incentive compatibility depend on the assumption that players never make mistakes, and how robust such protocols are to "noise" or "mistakes".

1 Introduction

One of the key issues in designing a distributed protocol (algorithm) is its convergence to a stable state, also known as self-stabilization. Intuitively, starting from any initial (arbitrarily corrupted) state, the protocol should eventually converge to the "correct state" as intended by the designer. Incentive compatibility considerations have been also become important in the study of distributed protocols since the participants cannot be assumed to altruistically implement the protocol if that is not beneficial for themselves.

A unifying game-theoretic approach for proving both convergence and incentive compatibility has been recently proposed by Nisan et al. [14]. They consider so-called *best-response mechanisms* or *dynamics* in which the protocol prescribes that each participant (or player) should simply best-respond to the strategy currently played by the other players. Essentially the same base *game* is played over and over (or until some equilibrium is reached), with players updating their strategies in *some* (unspecified) order. Nisan et al. [14] proved that for a suitable class of games the following happens:

- *Convergence.* The dynamics eventually reaches a unique equilibrium point (a unique pure Nash equilibrium) of the base game regardless of the order in which players respond (including concurrent responses).

B. Vöcking (Ed.): SAGT 2013, LNCS 8146, pp. 243–254, 2013.
© Springer-Verlag Berlin Heidelberg 2013

- *Incentive compatibility.* A player who deviates from the prescribed best-response strategy can only worsen his/her final utility, that is, the dynamics will reach a different state that yields weakly smaller payoff.

These two conditions say that the protocol will eventually "stabilize" if implemented correctly, and that the participants are actually willing to do so. Convergence itself is a rather strong condition because no assumption is made on how players are scheduled for updating their strategies, a typical situation in *asynchronous* settings. Incentive compatibility is also non-trivial because a best-response is a *myopic* strategy which does not take into account the future updates of the other players. In fact, neither of these conditions can be guaranteed on general games.

Nisan et al. [14] showed that several protocols and mechanisms arising in computerized and economics settings are in fact best-response mechanisms over the *restricted* class of games for which convergence and incentive compatibility are always guaranteed. Their applications include: (1) the Border Gateway Protocol (BGP) currently used in the Internet, (2) a game-theoretic version of the TCP protocol, and (3) mechanisms for the classical cost-sharing and stable roommates problems studied in micro-economics.

In this work we address the following question:

What happens to these protocols/mechanisms if players do not always best respond?

Is it possible that when players sometimes deviate from the prescribed protocol (e.g., by making mistakes in computing their best-response or by scarce knowledge about other players' actions) then the protocol does not converge anymore? Can such mistakes induce some other player to adopt a non-best-response strategy that results in a better payoff? Such questions arise naturally from fault tolerant considerations in protocol design, and have several connections to equilibria computation and bounded rationality issues in game theory.

Our Contribution. We investigate convergence and incentive compatibility conditions of mechanisms (dynamics) in [14] when players are not guaranteed to always best respond but they rather play an *imperfect* best-response strategy. That is, at every time step every player deviates from the prescribed best-response strategy according to some probability parameter $p \geq 0$. The parameter p can be regarded as the probability of making a mistake every time the player updates his/her strategy.

Our results indicate to what extent convergence and incentive compatibility depend on the assumption that players never make mistakes, and provide necessary and sufficient conditions for the robustness of these mechanisms/dynamics:

- *Convergence.* Because of mistakes convergence can be achieved only in a probabilistic sense. We give bounds on the parameter p in order to guarantee convergence with sufficiently good probability.
  One might think that for small values of p our dynamics behaves (approximately) as the dynamics without mistakes, i.e. it converges to an equilibrium

point regardless of the order in which players respond. However, it turns out this is not the case. Indeed, our first negative result (Theorem 2) shows that even when p is exponentially small in the number n of players the dynamics does not converge, i.e., the probability of being in the equilibrium is always small (interestingly, such negative result applies also to certain instances of BGP in the realistic model of Gao and Rexford [6]).

The proof of this result shows the existence of a particularly "bad" schedule that amplifies the probability that the imperfect dynamics deviates from the perfect one. This highlights that imperfect dynamics differ from their perfect counterpart in which convergence results *must consider* how players are scheduled. Indeed, we complement the negative result above with a general positive result (Theorem 3) saying that convergence can be guarantee whenever p is polynomially small in some parameters defining both the game and the schedule of the players. For such values of p, the upper bound on the convergence time of dynamics without mistakes is (nearly) an upper bound for the *imperfect* best-response dynamics.

- *Incentive compatibility.* We first observe that games that are incentive compatible for dynamics without mistakes, may no longer be incentive compatible for imperfect best-response dynamics (Theorem 4). In other words, a player who deviates incidentally from the mechanism induces another player to *deliberately* deviate. A sufficient condition for incentive compatibility of imperfect best-response mechanisms (Theorem 5) turns out to be a quantitative version of the one given in [14]. Roughly speaking, if the payoffs of the Nash equilibrium are *sufficiently* larger than the other possible payoffs, then incentive compatibility holds. As the probability p of making mistakes vanishes, the class of games for which convergence and incentive compatibility holds tends to the class of games in [14].

Our focus is on the same class of (base) games of [14] since this is the only known general class for which best-response dynamics converge (regardless of the schedule of the players) and are incentive compatible. In our view this class is important as it describes accurately certain protocols that are implemented in practice and it unifies several results in game theory. In particular, the mathematical model of how the commercial relationships between Autonomous Systems (the Gao-Rexford model [6]) leads to games in this class and, ultimately, to the fact that BGP converges and is incentive compatible [12,14]. Considering more general games for the analysis of BGP would in fact produce "wrong" results (constructing unrealistic examples for which the protocol does not converge or is not incentive compatible).

We nevertheless take one step further and apply the tools from [14] (and this work) to a natural generalization of their games. Intuitively speaking, these games guarantee only that best-response converge to a *sub-game*. In this case, the dynamics of the original game can be approximated by the dynamics of the sub-game (Theorem 6). Unfortunately, this "reduction" cannot be pushed further simply because the sub-game can be an arbitrary game and different p-imperfect best-response dynamics lead to different equilibria (even for the same p). However, when the dynamics on the sub-game are well-understood, then we can infer their behavior also on the original game. One applications of this approach is on PageRank games [9] which turn out to be reducible to a sub-game for which the well-studied logit equilibrium [2] has a closed formula.

Due to space limitations, we only give proof sketches. We refer the reader to the full version of the paper [5] for complete proofs and more detailed descriptions of the results.

Related Work. Convergence of best-response dynamics is a main topic in game theory. It relates to the so-called problem of equilibrium selection (how can the players converge to an equilibrium?). Noisy versions of such dynamics have been studied in order to consider the effects of bounded rationality and limited knowledge of the players (which limits their ability to compute their best responses).

Our imperfect best response dynamics are similar to the *mutation model* by Kandori et al. [10], and to the *mistakes model* by Young [15], and Kandori and Rob [11]. A related model is the *logit dynamics* of Blume [4] in which the probability of a mistake depends on the payoffs of the game. All of these works assume a particular schedule of the players (the order in which they play in the dynamics). Whether such an assumption effects the selected equilibrium is the main focus of a recent work by Alos-Ferrer and Netzer [1]. They studied convergence of these dynamics on general games when the parameter p vanishes, and provide a characterization of the resulting equilibria in terms of a kind of potential function of the game. Convergence results that take into account non-vanishing p are only known for fixed dynamics on specific class of games.

Incentive compatibility of best-response dynamics provide a theoretical justification for several protocols and auctions widely adopted in practice. Levin et al. [12] proved convergence and incentive compatibility of the intricate BGP protocol in the current Internet (based on the mathematical model by Gao and Rexford [6] that captures the commercial structure that underlies the Internet and explains convergence of BGP). The theoretical analysis of TCP-inspired games by Godfrey et al. [7] shows that certain variants of the current TCP protocol converge (the flow rate stabilizes) and are incentive compatible on arbitrary networks (this property assumes routers adopt specific queuing policy). The so-called Generalized Second-Price auctions used in most of ad-auctions is another example of incentive compatible best-response mechanism as proved by Nisan et al. [13]. All of these problems (and others) and results have been unified by Nisan et al. [14] in their framework.

2 Definitions

We consider an n-player base game G in which each player i has a finite set of strategies S_i, and a utility function u_i. Each player can select a strategy $s_i \in S_i$ and the vector $s = (s_1, \ldots, s_n)$ is the corresponding strategy profile, with $u_i(s)$ being the payoff of player i. To stress the dependency of the utility u_i on the strategy z_i of player i we adopt the standard notation (z_i, s_{-i}) to denote the vector $(s_1, \ldots, s_{i-1}, z_i, s_{i+1}, \ldots, s_n)$.

(Imperfect) Best-Response Dynamics. A *game dynamics* consists of a (possibly infinite) sequence of strategy profiles $s^0, s^1, \ldots$, where s^0 is an arbitrarily chosen profile and the profile s^t is obtained from s^{t-1} by letting some of the players updating their strategies. Therefore a game dynamics is determined by a *schedule* of the players specifying, for each time step, the subset of players that are selected for updating their strategies and a *response rule*, which specifies how a player updates her strategy (possibly depending on the past history and on the current strategy profile). In this work we focus on dynamics based on the following kind of schedules and response rules.

As for the response rule, we consider a scenario in which a selected player can deviates from the (prescribed) best-response.

Definition 1 (*p*-imperfect response rule). *A response rule is p-imperfect if a player update her strategy to the best-response with probability at least $1 - p$.*

Examples of these rules are given in the mutation [10] or mistakes models [15,11]. The best-response rule is obviously 0-imperfect, which we also denote as *perfect*. The response rule in logit dynamics [4] is *p*-imperfect with

$$p \le \frac{m-1}{m-1+e^\beta}$$

for all games in which the payoff between a non-best and a best-response differs by at least one[1] and each player has at most m strategies.

In order to avoid trivial impossibility results on convergence we need to consider a non-adaptive adversarial schedule that satisfies some reasonable fairness condition. We allow both deterministic and randomized schedules satisfying the following definition.

Definition 2 ((R, ε)-fair schedule). *A schedule is (R, ε)-fair if there exists a nonnegative integer R such that, for any interval of R time steps, all players are selected at least once in this interval with probability at least $1 - \varepsilon$, i.e. for every player i and any time step a we have*

$$\Pr(SEL_{i,a,R}) \ge 1 - \varepsilon,$$

where $SEL_{i,a,R}$ is the event that player i is selected at least once in the interval $[a+1, a+R]$.

Scheduling players in round-robin fashion or concurrently corresponds to $(n, 0)$-fair and $(1, 0)$-fair schedules, respectively. Selecting a player at random at each time step is (R, ε)-fair with $R = \mathcal{O}(n \log n)$. Observe that if a schedule is (R, ε)-fair, then, for every $0 < \delta < \varepsilon$, all players are selected at least once with probability at least $1 - \delta$ in an interval of $R \cdot \left\lceil \frac{\log(1/\delta)}{\log(1/\varepsilon)} \right\rceil$ time steps (this holds because the probability $1 - \varepsilon$ is guaranteed for any interval of R time steps). We also denote with η the maximum number of players selected for update in one step by the schedule. Note that $\eta \le n$.

---

[1] When the minimum difference is δ this extends by taking $\beta_\delta = \beta \cdot \delta$ in place of β.

Henceforth, we always refer as *imperfect best-response dynamics* to any dynamics whose schedule is (R, ε)-fair and whose response rule is p-imperfect, and as *imperfect best-response mechanisms* to the class of all imperfect best-response dynamics. We highlight that we do not put any other constraint on the way the dynamics run. In particular we allow both the schedule and the response rule to depend on the *status* of the game, that is on a set of information other than the current strategy profile.

Convergence and Incentive Compatibility. We say that a game dynamics for a game G *converges* if it eventually converges to a Nash equilibrium of the game, i.e. there exists $t > 0$ such that the strategy profile of players at time step t coincides with a Nash equilibrium of G.

Let us denote with X^t the random variable that represent the strategy profile induced by a dynamics on a game G after t time steps. If there is a finite time step T after which the dynamics is terminated, then the *total utility* of a player i is defined as $\Gamma_i = E\left[u_i\left(X^T\right)\right]$; otherwise, that is if the dynamics does not terminate after finite time, the total utility is defined as $\Gamma_i = \limsup_{t \to \infty} E\left[u_i\left(X^t\right)\right]$. Then, a dynamics for a game G is *incentive compatible* if playing this dynamics is a pure Nash equilibrium in a new game $G^\star$ in which players' strategies are all possible response rule that may be used in G and players' utilities are given by their total utilities. That is, a dynamics for a game G is incentive compatible if every player does not improves her total utilities by playing according to a response rule different from the one prescribed, given that each other player does not deviate from the prescribed response rule.

Never Best-Response and the Main Result in [14]. Nisan et al. [14] analyzed the convergence and incentive compatibility of the (perfect) best-response dynamics. Before stating their result, let us now recall some definitions.

Definition 3 (never best-response). *A strategy s_i is a never best-response (NBR) for player i if, for every s_{-i}, there exists s_i' such that $u_i(s_i, s_{-i}) < u_i(s_i', s_{-i})$.*[2]

Note that according to a p-imperfect response rule, a player updates her strategy to a NBR with probability at most p.

Definition 4 (elimination sequence). *An elimination sequence for a game G consists of a sequence of sub-games*

$$G = G_0 \supset G_1 \supset \cdots \supset G_r = \hat{G},$$

where any game G_{k+1} is obtained from the previous one by letting a player $i^{(k)}$ eliminate strategies which are NBR in G_k.

---

[2] Nisan et al. [14] assume that each player has also a *tie breaking rule* $\prec_i$, i.e., a total order on S_i, that depends solely on the player's private information. In the case that a tie breaking rule $\prec_i$ has been defined for player i, then s_i is a NBR for i also if $u_i(s_i, s_{-i}) = u_i(s_i', s_{-i})$ and $s_i \prec_i s_i'$. However, such tie-breaking rule can be implemented in a game by means of suitable perturbations of the utility function: with such an implementation our definition is equivalent to the one given in [14].

The length of the shortest elimination sequence for a game G is denoted with ℓ_G (we omit the subscript when it is clear from the context). It is easy to see that for each game $\ell_G \leq n(m-1)$, where m is the maximum number of strategies of a player.

Our results will focus on the following classes of games.

Definition 5 (NBR-reducible and NBR-solvable games). *The game G is NBR-reducible to $\hat{G}$ if there exists an elimination sequence for G that ends in $\hat{G}$. The game G is NBR-solvable if it is NBR-reducible to $\hat{G}$ and $\hat{G}$ consists of a unique profile.*

For example, consider a 2-player game with strategies $\{0, 1, 2\}$ and the following utilities:

$$
\begin{array}{c|c|c|c|}
 & 0 & 1 & 2 \\
\hline
0 & 0,0 & 0,0 & 0,-2 \\
\hline
1 & 0,0 & -1,-1 & -1,-2 \\
\hline
2 & -2,0 & -2,-1 & -2,-2 \\
\hline
\end{array}
\tag{1}
$$

Notice that strategy 2 is a NBR for both players. Hence, there exists an elimination sequence of length 2 that reduces above game in its upper-left 2×2 sub-game with strategy set $\{0, 1\}$ for each player. Therefore, this game is NBR-reducible. If we modify the utilities in this upper-left 2×2 sub-game as follows

$$
\begin{array}{c|c|c|}
 & 0 & 1 \\
\hline
0 & 0,0 & 0,-\delta \\
\hline
1 & -\delta,0 & -1,-1 \\
\hline
\end{array}
$$

then the game reduces further to the profile $(0, 0)$ and hence it is NBR-solvable. Observe that the unique profile at which the game G is reduced in an NBR-solvable game is also the unique Nash equilibrium of the original game.

While the convergence result of [14] holds for the class of NBR-solvable games, in order to guarantee incentive compatibility they introduce the following condition on the payoffs:

Definition 6 (NBR-solvable with clear outcome). *A NBR-solvable game is said to have a clear outcome if, for every player i, there is a player-specific elimination sequence such that the following holds. If i appears the first time in this sequence at position k, then in the sub-game G_k the profile that maximizes the utility of player $i = i^{(k)}$ is the Nash equilibrium.*

Theorem 1 (main result of [14]). *Best-response dynamics of every NBR-solvable game G converge to a pure Nash equilibrium of the game and, if G has clear outcome, are incentive compatible. Moreover, convergence is guaranteed in ℓ_G rounds for any schedule, where a round is a sequence of consecutive time steps in which each player is selected for update at least once.*

Note that convergence and incentive compatibility holds regardless of the schedule of players. Moreover, the theorem implies that for a specific (R, ε)-fair schedule the dynamics converges in $O(R \cdot \ell_G)$ time steps. Note also that convergence does *not* require a clear outcome and this condition is only needed for incentive compatibility.

3   Convergence Properties

In this section we will show that the result about convergence of the best-response dynamics in NBR-solvable games given in [14] is not resistant to the introduction of "noise", i.e., there is a NBR-solvable games and an imperfect best-response dynamics that never converges to the Nash equilibrium even for values of p very small. Specifically we will prove the following theorem.

Theorem 2. *For every $0 < \delta < 1$, there exists a n-player NBR-solvable game G and an imperfect best-response dynamics with parameter p exponentially small in n such that for every integer $t > 0$ the dynamics converges after t steps with probability at most δ.*

Proof (Sketch). Consider a game with n players with strategies 0 and 1. Each player i has better payoff for strategy 1 if all previous players, $1, 2, \ldots, i-1$, also play strategy 1; in all other cases, player i has better payoff for strategy 0. It is easy to check that the shortest elimination sequence has length $\ell = n$ and it leads to all players playing strategy 1.

It is possible to define a deterministic *non-adaptive* schedule which consists of a suitable sequence of length $R = 2^{n-1}$ in which all players are selected at least once. The schedule repeats this sequence over and over. For a particular p-imperfect best-response rule, if a player i does a "mistake", that is, chooses strategy 0 within this sequence, then this mistake will "propagate" to all subsequent players, $i+1, \ldots, n$, and at the end of the sequence player n will play strategy 0. Therefore, convergence requires p to be smaller than $1/R$ and thus exponentially small in n.   □

We remark that the above impossibility result can be easily instantiated with certain instances of BGP games or with specific dynamics such as the logit dynamics (see [5] for detailes).

We next complement above negative result with an upper bound on the value of p necessary for having the game being in the equilibrium with high probability within finite time T. This in particular means that, by terminating the dynamics after T steps, it is very likely that the system ends in the Nash equilibrium.

Theorem 3. *For any NBR-solvable game G and any small $\delta > 0$, there is a time $T = \mathcal{O}(R \cdot \ell \log \ell)$ such that every p-imperfect best-response dynamics is in the Nash equilibrium of G with probability at least $1 - \delta$, whenever $p \leq \frac{c}{\eta R \cdot \ell \log \ell}$, for a suitably chosen constant $c = c(\delta)$.*

Proof (Sketch). It is possible to show that, if the dynamics is in a profile in G_k at time t, then the probability that it is in the next sub-game, G_{k+1}, after further R steps is at least $1 - \eta p R - \varepsilon$. This implies that for any starting configuration the dynamics is in G_k after $k \cdot R$ time steps with probability at least $1 - k \cdot (\eta p R + \varepsilon)$. For $k = \ell$ this probability can be made larger than $1 - \delta$ by observing that every (R, ε)-fair adversary is also $(T, \delta/2\ell)$-fair with $T = \left\lceil \frac{\log(2\ell/\delta)}{\log(1/\varepsilon)} \right\rceil$, and by setting $p \leq \frac{\delta}{2} \cdot \frac{1}{\eta T \ell}$.   □

4   Incentive Compatibility Property

In this section we ask if the incentive compatibility property holds also in presence of noise, that is, if deviating from a p-imperfect best-response rule is not beneficial for the player. Note that adopting a p'-imperfect response rule, with $p' < p$, should be not considered a deviation, since this rule is also a p-imperfect response rule.

The following theorem shows that the incentive compatibility property is not resistant to the introduction of noise.

Theorem 4. *There is a NBR-solvable game with clear outcome and an imperfect best-response dynamics whose response rule is not incentive compatible.*

Proof (Sketch). Consider the following game G:

	left	right
top	$c + 2, 1$	$1, 0$
bottom	$0, 0$	$0, c$

Consider the logit dynamics for G (we already noted that the logit dynamics is an example of imperfect best-response dynamics). Since this game is a potential game, a closed formula for the stationary distribution reveals that the column player has a better expected payoff by playing always strategy `right`.   □

As done for convergence, we now investigate for sufficient conditions for incentive compatibility. We will assume that utilities are non-negative: note that there are a lot of response rules that are invariant with respect to the actual value of the utility function and thus, in these cases, this assumption is without loss of generality. Recall that we denote as $i^{(k)}$ and G_k the first occurrence of the player and the corresponding sub-game in the elimination sequence given by the definition of game with clear outcome (Definition 6).

It turns out that we need a "quantitative" version of the definition of clear outcome, i.e., that whenever the player i has to eliminate a NBR her utility in the Nash equilibrium is sufficiently larger than the utility of any other profile in the sub-game she is actually playing. Specifically, we have the following theorem.

Theorem 5. *For any NBR-solvable game G with clear outcome and any small $\delta > 0$, playing according to a p-imperfect rule is incentive compatible for player $i = i^{(k)}$ as long as $p \leq \frac{c}{\eta R \cdot \ell \log \ell}$, for a suitable constant $c = c(\delta)$, the dynamics terminates after $\Theta\left(R \cdot \ell \log \ell\right)$ steps and*

$$u_i(NE) \geq \frac{1}{1 - 2\delta}\left(2\delta \cdot u_i^* + u_i^k\right),$$

where $u_i(NE)$ is the utility of i in the Nash equilibrium, $u_i^k = \max_{\mathbf{x} \in G^{(k)}} u_i(\mathbf{x})$ and $u_i^* = \max_{\mathbf{x} \in G} u_i(\mathbf{x})$.

We can summarize the intuition behind the proof of Theorem 5 as follows:

- If player i always updates according to the p-imperfect response rule, then when the dynamics terminates the game will be in the Nash equilibrium with high probability and hence her expected utility almost coincides with the Nash equilibrium utility;
- Suppose, instead, player i does not update according to a p-imperfect response rule. Notice that elimination of strategies up to G_k is not affected by what player i does. Therefore profiles of $G \setminus G_k$ will be played only with small probability (but i can gain the highest possible utility from these profiles), whereas the game will be in a profile of G_k with the remaining probability.

5   NBR-Reducible Games

For general games it is not possible to prove convergence to a pure Nash equilibrium without making additional assumptions on the schedule and on the response rule. Such negative result applies also to NBR-reducible games, the natural extension of NBR-solvable ones.

We shall see below that, for NBR-reducible games, several questions on the dynamics of a game G can be answered by studying the dynamics of the reduced game $\hat{G}$. Before formally stating this fact, let us introduce some useful concepts.

The Dynamics as a Markov Chain. We say that the game is in a *status–profile pair* $(h, \mathbf{x})$ if h is the set of information currently available and $\mathbf{x}$ is the profile currently played. We denote with H the set of all status–profile pairs $(h, \mathbf{x})$ and with $\hat{H}$ only the ones with $\mathbf{x} \in \hat{G}$. Let X_t be the random variable that represents the status–profile pair $(h, \mathbf{x})$ of the game after t steps of the imperfect best-response dynamics. Then, for every $(h, \mathbf{x}), (z, \mathbf{y}) \in H$ we set

$$P\big((h, \mathbf{x}), (z, \mathbf{y})\big) = \Pr\Big(X_1 = (z, \mathbf{y}) \mid X_0 = (h, \mathbf{x})\Big).$$

That is, P is the transition matrix of a Markov chain on state space H and it describes exactly the evolution of the dynamics. Note that we are not restricting the dynamics to be memoryless, since in the status we can save the history of all previous iterations. For a set $A \subseteq H$ we also denote $P\big((h, \mathbf{x}), A\big) = \sum_{(z, \mathbf{y}) \in A} P\big((h, \mathbf{x}), (z, \mathbf{y})\big)$.

The Restricted Dynamics. As mentioned above, we will compare the original dynamics with a specific restriction on the subset $\hat{H}$ of status–profile pairs. Now we describe how this restriction is obtained. Henceforth, whenever we refer to the restricted dynamics, we use $\hat{X}_t$ and $\hat{P}$ in place of X_t and P. Then, the restricted dynamics is described by a Markov chain on state space H with transition matrix $\hat{P}$ such that for every $(h, \mathbf{x}), (z, \mathbf{y}) \in H$

$$\hat{P}\big((h, \mathbf{x}), (z, \mathbf{y})\big) = \begin{cases} \frac{P((h,\mathbf{x}),(z,\mathbf{y}))}{P((h,\mathbf{x}),\hat{H})}, & \text{if } (h, \mathbf{x}), (z, \mathbf{y}) \in \hat{H}; \\ 0, & \text{otherwise.} \end{cases}$$

Thus, the restricted dynamics is exactly the same as the original one except that the first never leaves the sub-game $\hat{G}$, whereas in the latter, at each time step, there is probability at most p to leave this sub-game.

Status–Profile Events. We now describe the kind of questions about imperfect best-response mechanisms and NBR-reducible games for which a reduction can be beneficial. Roughly speaking, these are all questions about the occurrence (and the time needed for it) of events that can be described only by looking at status–profile pairs.

Specifically, a *status–profile set event* for an imperfect best-response dynamics is a set of status–profile pairs. A *status–profile distro event* for an imperfect best-response dynamics is a distribution on the status–profile pairs. More generally, we refer to *status–profile event* if we do not care whether it is a set or a distro event. Note that many equilibrium concepts can be described as status–profile events, like Nash equilibria, sink equilibria [8], correlated equilibria [3] or logit equilibria [2]: in any case we should simply list the set of states or the distribution over states at which we are interested in. Properties like "a profile that is visited for k times" or "a cycle of length k visited" are other examples of status–profile events. We remark that in these examples it is crucial that the equilibrium is defined on the status–profile pairs and not just on the profiles: indeed, the status can remember the history of the game and identify such events, whereas they are impossible to recognize if we only know the current profile.

For an NBR-reducible game G, a status–profile set event is *reducible* if the set of status–profile pairs that represent the event contains some profile from $\hat{G}$. A status–profile distro event is *reducible* if status–profile pairs on which is defined the distribution that represent the event contains only profiles of $\hat{G}$. It turns out that each one of the equilibria concepts described above is a reducible status-profile event: indeed, since all profiles not in $\hat{G}$ contain NBR strategies, they are not in the support of any Nash, any sink and any correlated equilibrium; as for the logit equilibrium (that assigns non-zero probability to profiles not in $\hat{G}$) it is not difficult to show that the logit equilibrium of G is close to the logit equilibrium of $\hat{G}$.

A status–profile set event *occurs* if the imperfect best-response dynamics reaches a status–profile pair in the set of pairs that represent the event. Similarly, a status–profile distro event *occurs* if the distribution on the set of profiles generated by the dynamics is close to the one that represent the event. The *occurrence time* of a status–profile event is the first time step in which it occurs.

We are now in a position to state the main result on NBR-reducible games:

Theorem 6. *For any NBR-reducible game G and any small $\delta > 0$, if a reducible status–profile event for an imperfect best-response dynamics occurs in the restricted dynamics, then it occurs with probability at least $1 - \delta$. Moreover, let us denote with τ the occurrence time of the event E in the restricted dynamics. Then, E occurs in the original dynamics in $\mathcal{O}(R \cdot \ell \log \ell + \tau)$ steps with probability at least $1 - \delta$, whenever $p \leq \min\left\{\frac{c_1}{\eta R \cdot \ell \log \ell}, \frac{c_2}{\eta \tau}\right\}$, for suitable constants $c_1 = c_1(\delta)$ and $c_2 = c_2(\delta)$.*

The above theorem can be applied to analyze the logit dynamics for a slightly variation of the *PageRank game* by Hopcroft and Sheldon [9] in which the players are the web pages (nodes) and their strategies is to create links to other pages (nodes). The payoffs are determined by the link structure (according to a random walk of the graph) and the resulting game is NBR-reducible. The sub-game turns out to be a potential game, and thus we obtain an accurate description of the dynamics for the reduced game. This and the theorem above gives an accurate description of the original game.

References

1. Alós-Ferrer, C., Netzer, N.: The logit-response dynamics. Games and Economic Behavior 68(2), 413–427 (2010)
2. Auletta, V., Ferraioli, D., Pasquale, F., Persiano, G.: Mixing time and stationary expected social welfare of logit dynamics. In: Kontogiannis, S., Koutsoupias, E., Spirakis, P.G. (eds.) SAGT 2010. LNCS, vol. 6386, pp. 54–65. Springer, Heidelberg (2010)
3. Aumann, R.J.: Subjectivity and correlation in randomized strategies. Journal of Mathematical Economics 1(1), 67–96 (1974)
4. Blume, L.E.: The statistical mechanics of strategic interaction. Games and Economic Behavior 5, 387–424 (1993)
5. Ferraioli, D., Penna, P.: Imperfect best-response mechanisms. CoRR, abs/1208.0699 (2012)
6. Gao, L., Rexford, J.: Stable internet routing without global coordination. IEEE/ACM Trans. Netw. 9(6), 681–692 (2001)
7. Godfrey, B., Schapira, M., Zohar, A., Shenker, S.: Incentive compatibility and dynamics of congestion control. In: SIGMETRICS, pp. 95–106 (2010)
8. Goemans, M., Mirrokni, V., Vetta, A.: Sink equilibria and convergence. In: Proc. of the 46th Annual IEEE Symposium on Foundations of Computer Science (FOCS), pp. 142–154. IEEE (2005)
9. Hopcroft, J., Sheldon, D.: Network reputation games (2008)
10. Kandori, M., Mailath, G.J., Rob, R.: Learning, mutation, and long run equilibria in games. Econometrica 61(1), 29–56 (1993)
11. Kandori, M., Rob, R.: Evolution of equilibria in the long run: A general theory and applications. Journal of Economic Theory 65(2), 383–414 (1995)
12. Levin, H., Schapira, M., Zohar, A.: Interdomain routing and games. SIAM J. Comput. 40(6), 1892–1912 (2011)
13. Nisan, N., Schapira, M., Valiant, G., Zohar, A.: Best-response auctions. In: Proc. of the 12th ACM Conference on Electronic Commerce (EC), pp. 351–360 (2011)
14. Nisan, N., Schapira, M., Valiant, G., Zohar, A.: Best-response mechanisms. In: Proc. of the 2nd Symp. on Innovations in Computer Science (ICS), pp. 155–165 (2011)
15. Peyton Young, H.: The evolution of conventions. Econometrica 61(1), 57–84 (1993)

Pricing Ad Slots with Consecutive Multi-unit Demand*

Xiaotie Deng[1,2,**], Paul Goldberg[3,***], Yang Sun[4],
Bo Tang[2], and Jinshan Zhang[2,***]

[1] Department of Computer Science, Shanghai Jiaotong University, China
dengxiaotie@gmail.com
[2] Department of Computer Science, University of Liverpool, UK
{Bo.Tang,Jinshan.Zhang}@liverpool.ac.uk
[3] Department of Computer Science, Oxford University, UK
Paul.Goldberg@cs.ox.ac.uk
[4] Department of Computer Science, City University of Hong Kong, Hong Kong
sunyang@live.hk

Abstract. We consider the optimal pricing problem for a model of the rich media advertisement market, as well as other related applications. In this market, there are multiple buyers (advertisers), and items (slots) that are arranged in a line such as a banner on a website. Each buyer desires a particular number of *consecutive* slots and has a per-unit-quality value v_i (dependent on the ad only) while each slot j has a quality q_j (dependent on the position only such as click-through rate in position auctions). Hence, the valuation of the buyer i for item j is $v_i q_j$. We want to decide the allocations and the prices in order to maximize the total revenue of the market maker.

A key difference from the traditional position auction is the advertiser's requirement of a fixed number of consecutive slots. Consecutive slots may be needed for a large size rich media ad. We study three major pricing mechanisms, the Bayesian pricing model, the maximum revenue market equilibrium model and an envy-free solution model. Under the Bayesian model, we design a polynomial time computable truthful mechanism which is optimum in revenue. For the market equilibrium paradigm, we find a polynomial time algorithm to obtain the maximum revenue market equilibrium solution. In envy-free settings, an optimal solution is presented when the buyers have the same demand for the number of consecutive slots. We conduct a simulation that compares the revenues from the above schemes and gives convincing results.

Keywords: mechanism design, revenue, advertisement auction.

---

* The full version of the paper is available at: http://arxiv.org/abs/1308.1382
** Supported by the National Science Foundation of China (Grant No. 61173011) and a Project 985 grant of Shanghai Jiaotong University.
*** Supported by EPSRC project EP/K01000X/1 "Efficient Algorithms for Mechanism Design without Monetary Transfer".

B. Vöcking (Ed.): SAGT 2013, LNCS 8146, pp. 255–266, 2013.

1   Introduction

Ever since the pioneering studies on pricing protocols for sponsored search advertisement, especially with the generalized second price auction (GSP), by Edelman, Ostrovsky, and Schwarz [9], as well as Varian [16], market making mechanisms have attracted much attention from the research community in understanding their effectiveness for the revenue maximization task facing platforms providing Internet advertisement services. In the traditional advertisement setting, advertisers negotiate ad presentations and prices with website publishers directly. An automated pricing mechanism simplifies this process by creating a bidding game for the buyers of advertisement space over an IT platform. It creates a complete competition environment for the price discovery process. Accompanying the explosion of the online advertisement business, there is a need to have a complete picture on what pricing methods to use in practical terms for both advertisers and Ad space providers.

In addition to search advertisements, display advertisements have been widely used in webpage advertisements. They have a rich format of displays such as text ads and rich media ads. Unlike sponsored search, there is a lack of systematic studies on its working mechanisms for decision makings. The market maker faces a combinatorial problem of whether to assign a large space to one large rich media ad or multiple small text ads, as well as how to decide on the prices charged to them. We present a study of the allocation and pricing mechanisms for displaying slots in this environment where some buyers would like to have one slot and others may want several consecutive slots in a display panel. In addition to webpage ads, another motivation of our study is TV advertising where inventories of a commercial break are usually divided into slots of a few seconds each, and slots have various qualities measuring their expected number of viewers and the corresponding attractiveness.

We discuss three types of mechanisms and consider the revenue maximization problem under these mechanisms, and compare their effectiveness in revenue maximization under a dynamic setting where buyers may change their bids to improve their utilities. Our results make an important step toward the understanding of the advantages and disadvantages of their uses in practice. Assume the ad supplier divides the ad space into small enough slots (pieces) such that each advertiser is interested in a position with a fixed number of *consecutive* pieces. In modelling values to the advertisers, we modify the position auction model from the sponsored search market [9,16] where each ad slot is measured by the Click Through Rates (CTR), with users' interest expressed by a click on an ad. Since display advertising is usually sold on a per impression (CPM) basis instead of a per click basis (CTR), the quality factor of an ad slot stands for the expected impression it will brings in unit of time. Unlike in the traditional position auctions, people may have varying demands (need different spaces to display their ads) in a rich media ad auction for the market maker to decide on slot allocations and their prices.

We will lay out the the specific system parameters and present our results in the following subsections.

1.1   Our Modeling Approach

We have a set of *buyers* (advertisers) and a set of *items* to be sold (the ad slots on a web page). We address the challenge of computing prices that satisfy certain desirable properties. Next we describe the elements of the model in more detail.

- **Items.** Our model considers the geometric organization of ad slots, which commonly has the slots arranged in some sequence (typically, from top to bottom in the right-hand side of a web page). The slots are of variable quality. In the study of sponsored search auctions, a standard assumption is that the quality (corresponding to click-through rate) is highest at the beginning of the sequence and then monotonically decreases. Here we consider a generalization where the quality may go down and up, subject to a limit on the total number of local maxima (which we call *peaks*), corresponding to focal points on the web page. As we will show later, without this limit the revenue maximization problem is NP-hard.
- **Buyers.** A buyer (advertiser) may want to purchase multiple slots, so as to display a larger ad. Note that such slots should be *consecutive* in the sequence. Thus, each buyer i has a fixed *demand* d_i, which is the number of slots she needs for her ad. Two important aspects of this are
  - ◇ *sharp* multi-unit demand, referring to the fact that buyer i should be allocated d_i items, or none at all; there is no point in allocating any fewer
  - ◇ *consecutiveness* of the allocated items, in the pre-existing sequence of items.

  These constraints give rise to a new and interesting combinatorial pricing problem.
- **Valuations.** We assume that each buyer i has a parameter v_i representing the value she assigns to a slot of unit quality. Valuations for multiple slots are additive, so that a buyer with demand d_i would value a block of d_i slots to be their total quality, multiplied by v_i. This valuation model has been considered by Edelman et al. [9] and Varian [16] in their seminal work for keywords advertising.

Pricing Mechanisms. Given the valuations and demands from the buyers, the market maker decides on a price vector for all slots and an allocation of slots to buyers, as an output of the market. The question is one of which output the market maker should choose to achieve certain objectives. We consider two approaches:

- **Truthful Mechanism** whereby the buyers report their demands (publicly known) and values (private) to the market maker; then prices are set in such a way as to ensure that the buyers have the incentive to report their true valuations. We give a revenue-maximizing approach (i.e., maximizing the total price paid), within this framework.
- **Competitive Equilibrium** whereby we prescribe certain constraints on the prices so as to guarantee certain well-known notions of fairness and envy-freeness.

- **Envy-Free Solution** whereby we prescribe certain constraints on the prices and allocations so as to achieve envy-freeness, as explained below.

The mechanisms we exhibit are computationally efficient. We also performed experiments to compare the revenues obtained from these three mechanisms.

1.2   Related Works

The theoretical study of position auctions (of a single slot) under the generalized second price auction was initiated in [9,16]. There has been a series of studies of position auctions in deterministic settings [12]. Our consideration of position auctions in the Bayesian setting fits in the general one dimensional auction design framework. Our study considers continuous distributions on buyers' values. For discrete distributions, [4] presents an optimal mechanism for budget constrained buyers without demand constraints in multi-parameter settings and very recently they also give a general reduction from revenue to welfare maximization in [5]; for buyers with both budget constraints and demand constraints, 2-approximate mechanisms [1] and 4-approximate mechanisms [3] exist in the literature.

There are extensive studies on multi-unit demand in economics, see for example [2,6,10]. In an earlier paper [7] we considered sharp multi-unit demand, where a buyer with demand d should be allocated d items or none at all, but with no further combinatorial constraint, such as the consecutiveness constraint that we consider here. The sharp demand setting is in contrast with a "relaxed" multi-unit demand (i.e., one can buy a subset of at most d items), where it is well known that the set of competitive equilibrium prices is non-empty and forms a distributive lattice [11,15]. This immediately implies the existence of an equilibrium with maximum possible prices; hence, revenue is maximized. Demange, Gale, and Sotomayor [8] proposed a combinatorial dynamics which always converges to a revenue maximizing (or minimizing) equilibrium for unit demand; their algorithm can be easily generalized to relaxed multi-unit demand. A strongly related work to our consecutive settings is the work of Rothkopf et al. [14], where the authors presented a dynamic programming approach to compute the maximum social welfare of consecutive settings when all the qualities are the same. Hence, our dynamic programming approach for general qualities in Bayesian settings is a non-trivial generalization of their settings.

1.3   Organization

This paper is organized as follows. In Section 2 we describe the details of our rich media ads model and the related solution concepts. In Section 3, we study the problem under the Bayesian model and provide a Bayesian Incentive Compatible auction with optimal expected revenue for the special case of the single peak in quality values of advertisement positions. Then in Section 4, we extend the optimal auction to the case with limited peaks/valleys and show that it is NP-hard to maximize revenue without this limit. Next, in Section 5, we turn to the full information setting and propose an algorithm to compute the competitive equilibrium with maximum revenue. In Section 6, NP-hardness of envy-freeness for consecutive multi-unit demand buyers is shown. We also design a polynomial time solution for the special case where all advertisers demand the same number of ad slots. For simulations, we refer readers to read the full version of the paper.

2 Preliminaries

In our model, a rich media advertisement instance consists of n advertisers and m advertising slots. Each slot $j \in \{1, \ldots, m\}$ is associated with a number q_j which can be viewed as the quality or the desirability of the slot. Each advertiser (or buyer) i wants to display her own ad that occupies d_i consecutive slots on the webpage. In addition, each buyer has a private number v_i representing her valuation and thus, the i-th buyer's value for item j is $v_{ij} = v_i q_j$.

Throughout this thesis, we will often say that slot j is assigned to a buyer set B to denote that j is assigned to some buyer in B. We will call the set of all slots assigned to B the allocation to B. In addition, a buyer will be called a winner if he succeeds in displaying his ad and a loser otherwise. We use the standard notation $[s]$ to denote the set of integers from 1 to s, i.e. $[s] = \{1, 2, \ldots, s\}$. We sometimes use $\sum_i$ instead of $\sum_{i \in [n]}$ to denote the summation over all buyers and $\sum_j$ instead of $\sum_{j \in [m]}$ for items, and the terms $\mathrm{E_v}$ and $\mathrm{E}_{v_{-i}}$ are short for $\mathrm{E}_{v \in V}$ and $\mathrm{E}_{v_{-i} \in V_{-i}}$.

The vector of all the buyers' values is denoted by v or sometimes $(v_i; v_{-i})$ where v_{-i} is the joint bids of all bidders other than i. We represent a feasible assignment by a vector $x = (x_{ij})_{i,j}$, where $x_{ij} \in \{0, 1\}$ and $x_{ij} = 1$ denotes item j is assigned to buyer i. Thus we have $\sum_i x_{ij} \leq 1$ for every item j. Given a fixed assignment x, we use t_i to denote the quality of items that buyer i is assigned, precisely, $t_i = \sum_j q_j x_{ij}$. In general, when x is a function of buyers' bids v, we define t_i to be a function of v such that $t_i(v) = \sum_j q_j x_{ij}(v)$.

When we say that slot qualities have a single peak, we mean that there exists a peak slot k such that for any slot $j < k$ on the left side of k, $q_j \geq q_{j-1}$ and for any slot $j > k$ on the right side of k, $q_j \geq q_{j+1}$.

2.1 Bayesian Mechanism Design

Following the work of [13], we assume that all buyers' values are distributed independently according to publicly known bounded distributions. The distribution of each buyer i is represented by a Cumulative Distribution Function (CDF) F_i and a Probability Density Function (PDF) f_i. In addition, we assume that the concave closure or convex closure or integration of those functions can be computed efficiently.

An auction $M = (x, p)$ consists of an allocation function x and a payment function p. x specifies the allocation of items to buyers and $p = (p_i)_i$ specifies the buyers' payments, where both x and p are functions of the reported valuations v. Our objective is to maximize the expected revenue of the mechanism is $Rev(M) = \mathrm{E}_v \left[\sum_i p_i(v) \right]$ under Bayesian incentive compatible mechanisms.

Definition 1. *A mechanism* M *is called* Bayesian Incentive Compatible *(BIC)* *iff the following inequalities hold for all* i, v_i, v_i'.

$$\mathrm{E}_{v_{-i}}[v_i t_i(\boldsymbol{v}) - p_i(\boldsymbol{v})] \geq \mathrm{E}_{v_{-i}}[v_i t_i(v_i'; v_{-i}) - p_i(v_i'; v_{-i})] \tag{1}$$

Besides, we say M *is* Incentive Compatible *if* M *satisfies a stronger condition that* $v_i t_i(\boldsymbol{v}) - p_i(\boldsymbol{v}) \geq v_i t_i(v_i'; v_{-i}) - p_i(v_i'; v_{-i})$, *for all* $\boldsymbol{v}, i, v_i'$,

To put it in words, in a BIC mechanism, no player can improve her *expected* utility (expectation taken over other players' bids) by misreporting her value. An IC mechanism satisfies the stronger requirement that no matter what the other players declare, no player has incentives to deviate.

2.2 Competitive Equilibrium and Envy-free Solution

In Section 5, we study the revenue maximizing competitive equilibrium and envy-free solution in the full information setting instead of the Bayesian setting. An outcome of the market is a pair $(\boldsymbol{X}, \boldsymbol{p})$, where $\boldsymbol{X}$ specifies an allocation of items to buyers and $\boldsymbol{p}$ specifies prices paid. Given an outcome $(\boldsymbol{X}, \boldsymbol{p})$, recall $v_{ij} = v_i q_j$, let $u_i(\boldsymbol{X}, \boldsymbol{p})$ denote the *utility* of i.

Definition 2. *A tuple* $(\boldsymbol{X}, \boldsymbol{p})$ *is a* consecutive envy-free pricing *solution if every buyer is consecutive envy-free, where a buyer* i *is consecutive envy-free if the following conditions are satisfied:*

- *if* $X_i \neq \emptyset$, *then (i)* X_i *is* d_i *consecutive items.* $u_i(\boldsymbol{X}, \boldsymbol{p}) = \sum_{j \in X_i} (v_{ij} - p_j) \geq 0$, *and (ii) for any other subset of consecutive items* T *with* $|T| = d_i$, $u_i(\boldsymbol{X}, \boldsymbol{p}) = \sum_{j \in X_i} (v_{ij} - p_j) \geq \sum_{j \in T} (v_{ij} - p_j)$;
- *if* $X_i = \emptyset$ *(i.e.,* i *wins nothing), then, for any subset of consecutive items* T *with* $|T| = d_i$, $\sum_{j \in T} (v_{ij} - p_j) \leq 0$.

Definition 3. *(Competitive Equilibrium) We say an outcome of the market* $(\boldsymbol{X}, \boldsymbol{p})$ *is a* competitive equilibrium *if it satisfies two conditions.*

- $(\boldsymbol{X}, \boldsymbol{p})$ *must be consecutive envy-free.*
- *The unsold items must be priced at zero.*

We are interested in the revenue maximizing competitive equilibrium and envy-free solutions.

3   Optimal Auction for the Single Peak Case

The goal of this section is to present our optimal auction for the single peak case that serves as an elementary component in the general case later. En route, several principal techniques are examined exhaustively to the extent that they can be applied directly in the next section. By employing these techniques, we show that the optimal Bayesian Incentive Compatible auction can be represented by a simple Incentive Compatible one. Furthermore, this optimal auction can be implemented efficiently. Let $T_i(v_i) = E_{v_{-i}}[t_i(v)]$, $P_i(v_i) = E_{v_{-i}}[p_i(v)]$ and $\phi_i(v_i) = v_i - \frac{1-F_i(v_i)}{f_i(v_i)}$. From Myerson' work [13], we obtain the following three lemmas.

Lemma 1 (From [13]). *A mechanism $M = (x, p)$ is Bayesian Incentive Compatible if and only if:*
a) $T_i(x)$ is monotone non-decreasing for any agent i.
b) $P_i(v_i) = v_i T_i(v_i) - \int_{\underline{v_i}}^{v_i} T_i(z)dz$

Lemma 2 (From [13]). *For any BIC mechanism $M = (x, p)$, the expected revenue $E_v[\sum_i P_i(v_i)]$ is equal to the virtual surplus $E_v[\sum_i \phi_i(v_i)t_i(v)]$.*

We assume $\phi_i(t)$ is monotone increasing, i.e. the distribution is regular. Otherwise, Myerson's ironing technique can be utilized to make $\phi_i(t)$ monotone — it is here that we invoke our assumption that we can efficiently compute the convex closure of a continuous function and integration. The following lemma is the direct result of Lemma 1 and 2.

Lemma 3. *Suppose that x is the allocation function that maximizes $E_v[\phi_i(v_i)t_i(v)]$ subject to the constraints that $T_i(v_i)$ is monotone non-decreasing for any bidders' profile v, any agent i is assigned either d_i consecutive slots or nothing. Suppose also that*

$$p_i(v) = v_i t_i(v) - \int_{\underline{v_i}}^{v_i} t_i(v_{-i}, s_i)ds_i \tag{2}$$

Then (x, p) represents an optimal mechanism for the rich media advertisement problem in single-peak case.

We will use dynamic programming to maximize the virtual surplus in Lemma 2. Suppose all the buyers are sorted in a no-increasing order according to their virtual values. We will need the following two useful lemmas. Lemma 4 states that all the allocated slots are consecutive.

Lemma 4. *There exists an optimal allocation x that maximizes $\sum_i \phi_i(v_i)t_i(v)$ in the single peak case, and satisfies the following condition. For any unassigned slot j, it must be that either $\forall j' > j$, slot j' is unassigned or $\forall j' < j$, slot j' is unassigned.*

Next, we prove that this consecutiveness even holds for all set $[s] \subseteq [n]$. That is, there exists an optimal allocation that always assigns the first s buyers consecutively for all $s \in [n]$. For convenience, we say that a slot is "out of" a set of buyers if the slot is not assigned to any buyers in that set. Then the consecutiveness can be formalized in the following lemma.

Lemma 5. *There exists an optimal allocation x in the single peak case, that satisfies the following condition. For any slot j out of $[s]$, it must be either $\forall j' > j$, slot j' is out of $[s]$ or $\forall j' < j$, slot j' is out of $[s]$.*

Since the optimal solution always assigns to $[s]$ consecutively (Lemma 5), we can boil the allocations to $[s]$ down to an interval denoted by $[l, r]$. Let $g[s, l, r]$ denote the maximized value of our objective function $\sum_i \phi_i(v_i) t_i(\boldsymbol{v})$ when we only consider first s buyers and the allocation of s is exactly the interval $[l, r]$. Then we have the following transition function.

$$
g[s, l, r] = \max \begin{cases} g[s-1, l, r] \\ \\ g[s-1, l, r-d_s] + \phi_s(v_s) \sum_{j=r-d_s+1}^{r} q_j \\ \\ g[s-1, l+d_s, r] + \phi_s(v_s) \sum_{j=l}^{l+d_s-1} q_j \end{cases} \tag{3}
$$

Our summary statement is as follows.

Theorem 1. *The mechanism that applies the allocation rule according to Dynamic Programming (3) and payment rule according to Equation (2) is an optimal mechanism for the banner advertisement problem with single peak qualities.*

4   Multiple Peaks Case

Suppose now that there are only h peaks (local maxima) in the qualities. Thus, there are at most $h - 1$ valleys (local minima). Since h is a constant, we can enumerate all the buyers occupying the valleys. After this enumeration, we can divide the qualities into at most h consecutive pieces and each of them forms a single-peak. Then using similar properties as those in Lemma 4 and 5, we can obtain a larger size dynamic programming (still runs in polynomial time) similar to dynamic programming (3) to solve the problem.

Theorem 2. *There is a polynomial algorithm to compute revenue maximization problem in Bayesian settings where the qualities of slots have a constant number of peaks.*

Now we consider the case without the constant peak assumption and prove the following hardness result.

Theorem 3. *(NP-Hardness) The revenue maximization problem for rich media ads with arbitrary qualities is NP-hard.*

5   Competitive Equilibrium

In this section, we study the revenue maximizing competitive equilibrium in the full information setting. To simplify the following discussions, we sort all buyers and items in non-increasing order of their values, i.e., $v_1 \geq v_2 \geq \cdots \geq v_n$.

We say an allocation $\boldsymbol{Y} = (Y_1, Y_2, \cdots, Y_n)$ is efficient if $\boldsymbol{Y}$ maximizes the total social welfare e.g. $\sum_i \sum_{j \in Y_i} v_{ij}$ is maximized over all the possible allocations. We call $\boldsymbol{p} = (p_1, p_2, \cdots, p_m)$ an equilibrium price if there exists an allocation $\boldsymbol{X}$ such that $(\boldsymbol{X}, \boldsymbol{p})$ is a competitive equilibrium. The following lemma is implicitly stated in [11], for completeness, we give a proof below.

Lemma 6. *Let allocation* Y *be efficient, then for any equilibrium price* p, (Y, p) *is a competitive equilibrium.*

By Lemma 6, to find a revenue maximizing competitive equilibrium, we can first find an efficient allocation and then use linear programming to settle the prices. We develop the following dynamic programming to find an efficient allocation. We first only consider there is one peak in the quality order of items. The case with constant peaks is similar to the above approaches, for general peak case, as shown in above Theorem 3, finding one competitive equilibrium is NP-hard if the competitive equilibrium exists, and determining existence of competitive equilibrium is also NP-hard. This is because that considering the instance in the proof of Theorem 3, it is not difficult to see the constructed instance has an equilibrium if and only if 3 partition has a solution.

Recall that all the values are sorted in non-increasing order e.g. $v_1 \geq v_2 \geq \cdots \geq v_n$. $g[s, l, r]$ denotes the maximized value of social welfare when we only consider first s buyers and the allocation of s is exactly the interval $[l, r]$. Then we have the following transition function.

$$g[s, l, r] = \max \begin{cases} g[s-1, l, r] \\ g[s-1, l, r-d_s] + v_s \sum_{j=r-d_s+1}^{r} q_j \\ g[s-1, l+d_s, r] + v_s \sum_{j=l}^{l+d_s-1} q_j \end{cases} \tag{4}$$

By tracking procedure 4, an efficient allocation denoted by $X^* = (X_1^*, X_2^*, \cdots, X_n^*)$ can be found. The price p^* such that (X^*, p^*) is a revenue maximization competitive equilibrium can be determined from the following linear programming. Let T_i be any consecutive number of d_i slots, for all $i \in [n]$.

$$\max \quad \sum_{i \in [n]} \sum_{j \in X_i^*} p_j$$

$$\begin{aligned} s.t. \quad & p_j \geq 0 & \forall \, j \in [m] \\ & p_j = 0 & \forall \, j \notin \cup_{i \in [n]} X_i^* \\ & \sum_{j \in X_i^*} (v_i q_j - p_j) \geq \sum_{j' \in T_i} (v_i q_{j'} - p_{j'}) & \forall \, i \in [n] \\ & \sum_{j \in X_i^*} (v_i q_j - p_j) \geq 0 & \forall i \in [n] \end{aligned}$$

Clearly there is only a polynomial number of constraints. The constraints in the first line represent that all the prices are non negative (no positive transfers). The constraint in the second line means unallocated items must be priced at zero (market clearance condition). And the constraint in the third line contains two aspects of information. First for all the losers e.g. loser k with $X_k = \emptyset$, the utility that k gets from any consecutive number of d_k is no more than zero, which makes all the losers envy-free. The second aspect is that the winners e.g. winner i with $X_i \neq \emptyset$ must receive a bundle with d_i consecutive slots maximizing its utility over all d_i consecutive slots, which together with the constraint in the fourth line (winner's utilities are non negative) guarantees that all winners are envy-free.

Theorem 4. *Under the condition of a constant number of peaks in the qualities of slots, there is a polynomial time algorithm to decide whether there exists a competitive equilibrium or not and to compute a revenue maximizing revenue market equilibrium if one does exist. If the number of peaks in the qualities of the slots is unbounded, both the problems are NP-complete.*

Proof. Clearly the above linear programming and procedure (4) run in polynomial time. If the linear programming output a price $\boldsymbol{p}^*$, then by its constraint conditions, $(\boldsymbol{X}^*, \boldsymbol{p}^*)$ must be a competitive equilibrium. On the other hand, if there exist a competitive equilibrium $(\boldsymbol{X}, \boldsymbol{p})$ then by Lemma 6, $(\boldsymbol{X}^*, \boldsymbol{p})$ is a competitive equilibrium, providing a feasible solution of above linear programming. By the objective of the linear programming, we know it must be a revenue maximizing one.

6   Consecutive Envy-freeness

We first prove a negative result on computing the revenue maximization problem in general demand case. We show it is NP-hard even if all the qualities are the same.

Theorem 5. *The revenue maximization problem of consecutive envy-free buyers is NP-hard even if all the qualities are the same.*

Although the hardness in Theorem 5 indicates that finding the optimal revenue for general demand in polynomial time is impossible , however, it doesn't rule out the very important case where the demand is uniform, e.g. $d_i = d$. We assume slots are in a decreasing order from top to bottom, that is, $q_1 \geq q_2 \geq \cdots \geq q_m$. The result is summarized as follows.

Theorem 6. *There is a polynomial time algorithm to compute the consecutive envy-free solution when all the buyers have the same demand and slots are ordered from top to bottom.*

The proof of Theorem 6 is based on bundle envy-free solutions, in fact we will prove the bundle envy-free solution is also a consecutive envy-free solution by defining price of items properly. Thus, we need first give the result on bundle envy-free solutions. Suppose d is the uniform demand for all the buyers. Let T_i be the slot set allocated to buyer i, $i = 1, 2, \cdots, n$. Let P_i be the total payment of buyer i and p_j be the price of slot j. Let t_i denote the total qualities obtained by buyer i, e.g. $t_i = \sum_{j \in T_i} q_j$ and $\alpha_i = iv_i - (i-1)v_{i-1}, \forall i \in [n]$.

Theorem 7. *The revenue maximization problem of bundle envy-freeness is equivalent to solving the following LP.*

$$
\begin{aligned}
\textit{Maximize:} \quad & \sum_{i=1}^{n} \alpha_i t_i \\
\textit{s.t.} \quad & t_1 \geq t_2 \geq \cdots \geq t_n \\
& T_i \subset [m], \quad T_i \cap T_k = \emptyset \quad \forall i, k \in [n]
\end{aligned}
\tag{5}
$$

Through optimal bundle envy-free solution, we will modify such a solution to consecutive envy-free solution and then prove the Theorem 6.

7   Conclusion and Discussion

The rich media pricing models for consecutive demand buyers in the context of Bayesian truthfulness, competitive equilibrium and envy-free solution paradigm are investigated in this paper. As a result, an optimal Bayesian incentive compatible mechanism is proposed for various settings such as single peak and multiple peaks. In addition, to incorporate fairness e.g. envy-freeness, we also present a polynomial-time algorithm to decide whether or not there exists a competitive equilibrium or and to compute a revenue maximized market equilibrium if one does exist. For envy-free settings, though the revenue maximization of general demand case is shown to be NP-hard, we still provide optimal solution of common demand case. Besides, our simulation shows a reasonable relationship of revenues among these schemes plus a generalized GSP for rich media ads.

Even though our main motivation arises from the rich media advert pricing problem, our models have other potential applications. For example TV ads can also be modeled under our consecutive demand adverts where inventories of a commercial break are usually divided into slots of fixed sizes, and slots have various qualities measuring their expected number of viewers and corresponding attractiveness. With an extra effort to explore the periodicity of TV ads, we can extend our multiple peak model to one involved with cyclic multiple peaks. Besides single consecutive demand where each buyer only have one demand choice, the buyer may have more options to display his ads, for example select a large picture or a small one to display them. Our dynamic programming algorithm (3) can also be applied to this case (the transition function in each step selects maximum value from $2k + 1$ possible values, where k is the number of choices of the buyer).

Another reasonable extension of our model would be to add budget constraints for buyers, i.e., each buyer cannot afford the payment more than his budget. By relaxing the requirement of Bayesian incentive compatible (BIC) to one of approximate BIC, this extension can be obtained by the recent milestone work of Cai et al. [5]. It remains an open problem how to do it under the exact BIC requirement. It would also be interesting to handle it under the market equilibrium paradigm for our model.

References

1. Alaei, S.: Bayesian combinatorial auctions: Expanding single buyer mechanisms to many buyers. In: Proceedings of the 52nd IEEE Symposium on Foundations of Computer Science (FOCS), pp. 512–521 (2011)
2. Ausubel, L., Cramton, P.: Demand revelation and inefficiency in multi-unit auctions. In: Mimeograph, University of Maryland (1996)
3. Bhattacharya, S., Goel, G., Gollapudi, S., Munagala, K.: Budget constrained auctions with heterogeneous items. In: Proceedings of the 42nd ACM Symposium on Theory of Computing, STOC 2010, New York, NY, USA, pp. 379–388 (2010)
4. Cai, Y., Daskalakis, C., Matthew Weinberg, S.: An algorithmic characterization of multi-dimensional mechanisms. In: Proceedings of the 43rd Annual ACM Symposium on Theory of Computing (2012)
5. Cai, Y., Daskalakis, C., Matthew Weinberg, S.: Optimal multi-dimensional mechanism design: Reducing revenue to welfare maximization. In: Proceedings of the 2012 IEEE 53rd Annual Symposium on Foundations of Computer Science, FOCS 2012 (2012)
6. Cantillon, E., Pesendorfer, M.: Combination bidding in multi-unit auctions. C.E.P.R. Discussion Papers (February 2007)
7. Chen, N., Deng, X., Goldberg, P.W., Zhang, J.: On revenue maximization with sharp multi-unit demands. CoRR abs/1210.0203 (2012)
8. Demange, G., Gale, D., Sotomayor, M.: Multi-item auctions. The Journal of Political Economy, 863–872 (1986)
9. Edelman, B., Ostrovsky, M., Schwarz, M.: Internet advertising and the generalized second-price auction: Selling billions of dollars worth of keywords. American Economic Review 97(1), 242–259 (2007)
10. Engelbrecht-Wiggans, R., Kahn, C.M.: Multi-unit auctions with uniform prices. Economic Theory 12(2), 227–258 (1998)
11. Gul, F., Stacchetti, E.: Walrasian equilibrium with gross substitutes. Journal of Economic Theory 87(1), 95–124 (1999)
12. Lahaie, S.: An analysis of alternative slot auction designs for sponsored search. In: Proceedings of the 7th ACM Conference on Electronic Commerce, EC 2006, New York, NY, USA, pp. 218–227 (2006)
13. Myerson, R.B.: Optimal auction design. Mathematics of Operations Research 6(1), 58–73 (1981)
14. Rothkopf, M.H., Pekeč, A., Harstad, R.M.: Computationally manageable combinational auctions. Management Science 44(8), 1131–1147 (1998)
15. Shapley, L.S., Shubik, M.: The Assignment Game I: The Core. International Journal of Game Theory 1(1), 111–130 (1971)
16. Varian, H.R.: Position auctions. International Journal of Industrial Organization 25(6), 1163–1178 (2007)

The Price of Anarchy in Bilateral Network Formation in an Adversary Model

Lasse Kliemann

Christian-Albrechts-Universität zu Kiel
Institut für Informatik
Christian-Albrechts-Platz 4
24118 Kiel, Germany
lki@informatik.uni-kiel.de

Abstract. We study network formation with the bilateral link formation rule (Jackson and Wolinsky 1996) with n *players* and *link cost* $\alpha > 0$.
After the network is built, an adversary randomly destroys one link according to a certain probability distribution. Cost for player v incorporates the expected number of players to which v will become discon-
nected. This model was previously studied for unilateral link formation
(K. 2011).

We prove existence of *pairwise Nash equilibria* under moderate assumptions on the adversary and $n \geq 9$. As the main result, we prove
bounds on the *price of anarchy* for two special adversaries: one destroys
a link chosen uniformly at random, while the other destroys a link that
causes a maximum number of player pairs to be separated. We prove
bounds tight up to constants, namely $O(1)$ for one adversary (if $\alpha > \frac{1}{2}$),
and $\Theta(n)$ for the other (if $\alpha > 2$ considered constant and $n \geq 9$). The
latter is the worst that can happen for any adversary in this model (if
$\alpha = \Omega(1)$).

Keywords: network formation, bilateral link formation, pairwise Nash
equilibrium, pairwise stability, price of anarchy, network robustness.

A full version of this extended abstract is available at:
http://arxiv.org/abs/1308.1832

B. Vöcking (Ed.): SAGT 2013, LNCS 8146, p. 267, 2013.
© Springer-Verlag Berlin Heidelberg 2013

The Query Complexity of Correlated Equilibria

Sergiu Hart[1] and Noam Nisan[1,2]

[1] Hebrew University of Jerusalem
[2] Microsoft Research

Abstract. We consider the complexity of finding a Correlated Equilibrium in an n-player game in a model that allows the algorithm to make queries for players' utilities at pure strategy profiles. Many randomized regret-matching dynamics are known to yield an approximate correlated equilibrium quickly: in time that is polynomial in the number of players, n, the number of strategies of each player, m, and the approximation error, ϵ^{-1}. Here we show that both randomization and approximation are necessary: no efficient deterministic algorithm can reach even an approximate equilibrium and no efficient randomized algorithm can reach an exact equilibrium.

The full version appears in `http://arxiv.org/abs/1305.4874`.

B. Vöcking (Ed.): SAGT 2013, LNCS 8146, p. 268, 2013.
© Springer-Verlag Berlin Heidelberg 2013

The Money Pump as a Measure of Revealed Preference Violations

Bart Smeulders[1], Laurens Cherchye[2], Bram De Rock[3], and Frits C.R. Spieksma[1]

[1] ORSTAT, FEB, KU Leuven, Leuven, Belgium
{bart.smeulders,frits.spieksma}@kuleuven.be
[2] CES, FEB, KU Leuven, Leuven, Belgium
laurens.cherchye@kuleuven.be
[3] Ecares, Université Libre de Bruxelles, Brussels, Belgium
bderock@ulb.ac.be

A commonly cited problem of standard revealed preference tests is that for a given consumer, these tests are bound to give a binary result. Either the data satisfies the revealed preference axioms and he is said to be rational, or they are violated an he is said to be irrational. However, this answer gives no information on the severity of violations, which may in some cases be small or negligible. It is therefore of interest to have information on the severity of observed violations. In a recent and insightful contribution, Echenique et al. [1] proposed a new measure, the *money pump index* (MPI). This measure is based on the vulnerability of irrational consumer behaviour to arbitrage.

Our paper is concerned with the practical computation of the MPI. In principle, an MPI can be computed for every violation of the axioms of revealed preference. This calls for an aggregate MPI that summarizes these MPIs into a single metric. Echenique et al. propose the mean and median MPI as such aggregates. These Mean and Median MPI have an intuitive interpretation in terms of the money lost by the consumer due to irrational behaviour. A first contribution of this note is that we show that no polynomial time algorithm exists for computing the Mean and Median MPI, unless P = NP. Our second contribution is that we propose the Maximum and Minimum MPI as easy-to-apply alternatives. The Maximum MPI gives the percentage of money lost in the most severe violation of rationality, while the Minimum MPI does the same for the least severe violation. Importantly, our newly proposed Maximum and Minimum MPI have clear practical usefulness. We show that the Maximum and Minimum MPI can be computed efficiently. Next, we use the dataset of Echenique et al. to demonstrate the application of the Maximum and Minimum MPI. Here, our particular focus is on assessing the performance of these measures relative to the Mean and Median MPI. In addition, we show that comparing the values of the Maximum and Minimum MPI can reveal interesting information to the empirical analyst. Our working paper can be found at: http://www.econ.kuleuven.be/public/N11086/MP-Working.pdf

Reference

1. Echenique, F., Lee, S., Shum, M.: The money pump as a measure of revealed preference violations. Journal of Political Economy 119(6), 1201–1223 (2011)

B. Vöcking (Ed.): SAGT 2013, LNCS 8146, p. 269, 2013.

Author Index